AF614752

Methods in Molecular Biology™

Series Editor
John M. Walker
School of Life Sciences
University of Hertfordshire
Hatfield, Hertfordshire, AL10 9AB, UK

For further volumes:
http://www.springer.com/series/7651

Bacterial Regulatory RNA

Methods and Protocols

Edited by

Kenneth C. Keiler

Department of Biochemistry and Molecular Biology, The Pennsylvania State University, University Park, PA, USA

Editor
Kenneth C. Keiler
Department of Biochemistry and Molecular Biology
The Pennsylvania State University
University Park, PA, USA

ISSN 1064-3745 ISSN 1940-6029 (electronic)
ISBN 978-1-61779-948-8 ISBN 978-1-61779-949-5 (eBook)
DOI 10.1007/978-1-61779-949-5
Springer New York Heidelberg Dordrecht London

Library of Congress Control Number: 2012939854

Printed on acid-free paper

Humana Press is a brand of Springer
Springer is part of Springer Science+Business Media (www.springer.com)

Preface

The discovery of wide-spread RNA-based regulation in bacteria has led to new evaluations of the importance of bacterial regulatory RNA in every aspect of bacterial physiology, including pathogenesis, quorum sensing, biofilm formation, stress responses, developmental programs, and growth regulation. These studies have employed well-established RNA techniques as well as newly developed genomic and bioinformatic tools. This volume collects many of the most important methods for studying bacterial regulatory RNA written by outstanding experts in the field.

The methods are presented in six sections. Part I covers techniques to identify regulatory RNAs using genomic and experimental approaches. Part II follows with methods for characterizing the function and expression of regulatory RNAs in bacterial cells. A comprehensive overview of RNA structure prediction and methods to determine RNA structure and stability is presented in Part III. Finally, Parts IV–VI contain methods for characterizing interactions between regulatory RNAs and proteins, other RNAs, and larger complexes such as RNA polymerase and the ribosome. Each method is written to guide both new and experienced users and includes a Notes section with advice and tips from the authors.

University Park, PA, USA *Kenneth C. Keiler*

Contents

Part IV RNA-Protein Interactions

Part V RNA–RNA Interactions

Part VI Interactions of Regulatory RNA with RNA Polymerase and the Ribosome

Contributors

EDUARDO ABELIUK • *Department of Bioengineering, Stanford University, Stanford, CA, USA*
HIROJI AIBA • *Suzuka University of Medical Sciences, Suzuka, Japan*
PAUL BABITZKE • *Department of Biochemistry and Molecular Biology, The Pennsylvania State University, University Park, PA, USA*
PHILIP C. BEVILACQUA • *Department of Chemistry and Center for RNA Molecular Biology, The Pennsylvania State University, University Park, PA, USA*
COLIN P. CORCORAN • *Institute for Molecular Infection Biology, University of Würzburg, Würzburg, Germany*
GUILLAUME DESNOYERS • *Department of Biochemistry, University of Sherbrooke, Sherbrooke, QC, Canada*
ELIE J. DINER • *Department of Molecular, Cellular and Developmental Biology, University of California, Santa Barbara, Santa Barbara, CA, USA; Biomolecular Science and Engineering Program, University of California, Santa Barbara, Santa Barbara, CA, USA*
CHRISTINE DREVET • *Institut de Génétique et Microbiologie, Université Paris-Sud, Orsay, France*
CHRISTOPHER S. HAYES • *Department of Molecular, Cellular and Developmental Biology, University of California, Santa Barbara, Santa Barbara, CA, USA; Biomolecular Science and Engineering Program, University of California, Santa Barbara, Santa Barbara, CA, USA*
HYOUTA HIMENO • *Department of Biochemistry and Molecular Biology, Hirosaki University, Hirosaki, Japan*
ERRETT C. HOBBS • *Cell Biology and Metabolism Program, Eunice Kennedy Shriver National Institute of Child Health and Human Development, Bethesda, MD, USA*
BENJAMIN HOFMANN • *Institute for Molecular Infection Biology, University of Würzburg, Würzburg, Germany*
BRIAN D. JANSSEN • *Department of Molecular, Cellular and Developmental Biology, University of California, Santa Barbara, Santa Barbara, CA, USA*
A. WALI KARZAI • *Department of Biochemistry and Cell Biology, Stony Brook University, Centers for Molecular Medicine, Stony Brook, NY, USA; Center for Infectious Diseases, Stony Brook University, Life Sciences Building, Stony Brook, NY, USA*
KENNETH C. KEILER • *Department of Biochemistry and Molecular Biology, The Pennsylvania State University, University Park, PA, USA*
ANDREY S. KRASILNIKOV • *Department of Biochemistry and Molecular Biology, The Pennsylvania State University, University Park, PA, USA*
DAISUKE KURITA • *Department of Biochemistry and Molecular Biology, Hirosaki University, Hirosaki, Japan*
STEPHEN G. LANDT • *Department of Genetics, Stanford University, Stanford, CA, USA*

JONATHAN LIVNY • *The Broad Institute of MIT and Harvard, Cambridge, MA, USA; Channing Laboratory, Brigham and Women's Hospital, Harvard Medical School, Boston, MA, USA*
KIMIKA MAKI • *Division of Biological Science, Nagoya University, Chikusa, Japan*
PIERRE MANDIN • *Laboratoire de Chimie Bactérienne, Aix-Marseille Université, Institut de Microbiologie de la Méditéranée, Marseille, Cedex 20, France*
ERIC MASSÉ • *Department of Biochemistry, University of Sherbrooke, Sherbrooke, QC, Canada*
DAVID H. MATHEWS • *Department of Biochemistry & Biophysics, Center for RNA Biology, University of Rochester Medical Center, Rochester, NY, USA; Department of Biostatistics & Computational Biology, University of Rochester Medical Center, Rochester, NY, USA*
PREETI MEHTA • *Department of Biochemistry and Cell Biology, Stony Brook University, Centers for Molecular Medicine, Stony Brook, NY, USA*
TEPPEI MORITA • *Suzuka University of Medical Sciences, Suzuka, Japan*
AKIRA MUTO • *Department of Biochemistry and Molecular Biology, Hirosaki University, Hirosaki, Japan*
YI PENG • *T. C. Jenkins Department of Biophysics and CMDB Program, Johns Hopkins University, Baltimore, MD, USA*
ANNA PEREDERINA • *Department of Biochemistry and Molecular Biology, The Pennsylvania State University, University Park, PA, USA*
DIMITRI PODKAMINSKI • *Institute for Molecular Infection Biology, University of Würzburg, Würzburg, Germany*
KSENIA POUGACH • *Institutes of Gene Biology and Molecular Genetics, Russian Academy of Sciences, Moscow, Russia*
CHRISTINE POURCEL • *Institut de Génétique et Microbiologie, Université Paris-Sud, Orsay, France*
RENATE RIEDER • *Institute for Molecular Infection Biology, University of Würzburg, Würzburg, Germany*
JAY H. RUSSELL • *Department of Chemical and Biomolecular Engineering, Johns Hopkins University, Baltimore, MD, USA*
MATTHEW G. SEETIN • *Department of Biochemistry & Biophysics, University of Rochester Medical Center, Rochester, NY, USA*
KONSTANTIN SEVERINOV • *Waksman Institute for Microbiology, Rutgers, The State University of New Jersey, Piscataway, NJ, USA*
JOSHUA E. SOKOLOSKI • *Department of Biochemistry and Molecular Biophysics, Washington University St Louis School of Medicine, MO, USA*
TOBY J. SOPER • *T. C. Jenkins Department of Biophysics and CMDB Program, Johns Hopkins University, Baltimore, MD, USA*
GISELA STORZ • *Cell Biology and Metabolism Program, Eunice Kennedy Shriver National Institute of Child Health and Human Development, Bethesda, MD, USA*
BRIAN TJADEN • *Computer Science Department, Wellesley College, Wellesley, MA, USA*
KRITHIKA VENKATARAMAN • *Department of Biochemistry and Cell Biology, Stony Brook University, Centers for Molecular Medicine, Stony Brook, NY, USA*
JÖRG VOGEL • *RNA Biology Group, Institute for Molecular Infection Biology, University of Würzburg, Würzburg, Germany*

KAREN M. WASSARMAN • *Department of Bacteriology, University of Wisconsin-Madison, Madison, WI, USA*
PERRY WOO • *Department of Biochemistry and Cell Biology, Stony Brook University, Centers for Molecular Medicine, Stony Brook, NY, USA; Center for Infectious Diseases, Stony Brook University, Life Sciences Building, Stony Brook, NY, USA*
SARAH A. WOODSON • *T. C. Jenkins Department of Biophysics and CMDB Program, Johns Hopkins University, Baltimore, MD, USA*
ALEXANDER V. YAKHNIN • *Department of Biochemistry and Molecular Biology, The Pennsylvania State University, University Park, PA, USA*
HELEN YAKHNIN • *Department of Biochemistry and Molecular Biology, The Pennsylvania State University, University Park, PA, USA*

Part I

Identification of Regulatory RNAs

Chapter 1

Bioinformatic Discovery of Bacterial Regulatory RNAs Using SIPHT

Jonathan Livny

Abstract

Diverse bacteria encode RNAs that are not translated into proteins but are instead involved in regulating a wide variety of cellular functions. Computational approaches have proven successful in identifying numerous regulatory RNAs in myriad bacterial species but the difficultly of implementing most of these approaches has limited their accessibility to many researchers. Moreover, few of these approaches provide annotations of predicted loci to guide downstream experimental validation and characterization. Here I describe the implementation of SIPHT, a web-accessible program that enables screens for putative loci encoding regulatory RNAs to be conducted in any of nearly 2,000 sequenced bacterial replicons. SIPHT identifies candidate loci by searching for regions of intergenic sequence conservation upstream of predicted intrinsic transcription terminators. Each locus is then annotated for numerous features that provide clues about its potential function and/or enable the most reliable candidates to be identified.

Key words: regRNA, sRNA, SIPHT, Bioinformatics, Annotation

1. Introduction

Following the first seminal bioinformatic discoveries of small, noncoding regulatory RNAs in *Escherichia coli* over a decade ago, numerous studies in diverse bacteria have demonstrated the ubiquitous and central role of RNA-mediated regulation in bacterial physiology and evolution. Bacterial regulatory RNAs (regRNAs) fall into one of two main groups. The first is composed of *trans*-acting small transcripts usually 50–300 nucleotides in length known as sRNAs that form duplexes with specific target mRNAs, altering mRNA stability and/or access of mRNAs to the translational machinery. The second group is composed of *cis*-acting regulatory sequences such as riboswitches that are located in 5′ untranslated regions of mRNAs (r5′UTRs). Interaction of these r5′UTRs with cognate ligands or changes in temperature alter the

Kenneth C. Keiler (ed.), *Bacterial Regulatory RNA: Methods and Protocols*, Methods in Molecular Biology, vol. 905, DOI 10.1007/978-1-61779-949-5_1, © Springer Science+Business Media, LLC 2012

secondary structures of these regulatory regions, modulating mRNA transcriptional elongation, translation, and/or stability.

A number of bacterial regRNAs have been discovered in physical screens (1). Several groups have identified novel regRNAs by size fractionating total RNA then cloning and sequencing small transcripts. However, since most small RNAs in bacterial transcriptomes correspond to a few highly abundant structural RNAs or degradation products of highly expressed mRNAs, these screens have yielded only a handful of abundant regRNAs. Recently, high-throughput sequencing (HTS) has been shown to be a very effective approach for indentifying and quantifying regRNAs in diverse bacteria (2–8). However, the technical and analytical challenges and cost of conducting HTS limits its accessibility to most research groups seeking to discover and study regRNAs in their species of interest. Due to the limitations and challenges of conducting sequencing-based screens for regRNAs, the predominant approach for bacterial regRNA discovery remains bioinformatic prediction of regRNAs followed by targeted experimental validation of candidate loci (9).

Several bioinformatic algorithms have been developed for identifying bacterial regRNAs (9). Nearly all of these algorithms limit their search to intergenic regions (IGRs) of the genome and predict putative regRNA-encoding loci based on their association with (a) intrinsic genomic features, including putative transcription promoters and terminators, predicted secondary structures, and/or certain patterns of di-nucleotide frequency, and/or (b) primary sequence and/or secondary structure conservation among different bacterial species. Algorithms that integrate combinations of primary sequence, secondary structure, and/or conservation information have proven to produce the most reliable predictions of regRNAs (10). However, utilizing these integrative approaches for genome-wide screens requires identification and positional analysis of thousands of predictors generated by several different algorithms, necessitating a level of computer expertise beyond that of most biological researchers. Moreover, genome-wide screens often yield dozens or hundreds of putative regRNA-encoding loci, far exceeding the throughput of most downstream approaches for experimental validation and characterization. Thus, even when predictions of regRNA are available, prioritizing a subset of candidate loci for follow-up studies remains a significant obstacle for many researchers.

To address these challenges, I developed SIPHT, a web-accessible, high-throughput computational tool that enables kingdom-wide predictions and annotations of intergenic regRNA-encoding genes (11). A schematic of the SIPHT workflow is shown in Fig. 1. SIPHT identifies candidate regRNAs by searching for putative Rho-independent terminators downstream of conserved intergenic sequences. Each predicted locus is then annotated for

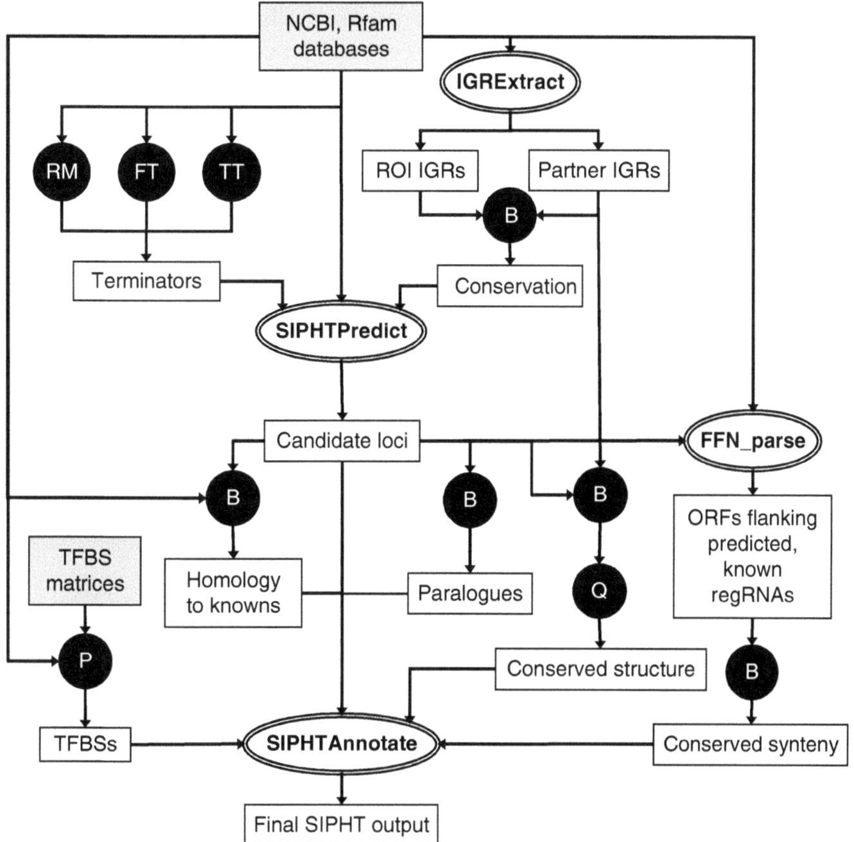

Fig. 1. Schematic of the SIPHT workflow. *Ovals* represent scripts developed specifically for SIPHT; *black circles* correspond to existing programs incorporated into this workflow. *B BLAST; RM* RNAMotif; *FT* FindTerm; *TT* TransTerm; *Q* QRNA; *P* Patser; *TFBS* transcription factor binding sites.

several features, including its conservation in other species, its association with one of several transcription factor binding sites (TFBSs), its position relative to flanking genes, and its homology and/or conserved genomic position with a growing database of previously identified sRNAs. These annotations enable candidate loci that are most likely to correspond to real regRNAs or those of potentially greater biological interest to be identified. SIPHT can be used to conduct screens for regRNAs in hundreds of bacterial strains whose annotated genome sequence is available in the NCBI database. Since the web interface came online, over 100 different users have used it to conduct searches for regRNAs in over 150 different bacterial replicons. Numerous novel regRNAs predicted by SIPHT have been physically confirmed in diverse bacterial species and several studies describing regRNAs identified by SIPHT have been published (8, 12–15).

The objectives of bioinformatic screens for regRNAs can vary considerably. In many instances these screens are conducted to identify a handful of strong and/or biologically relevant candidate loci that are then experimentally validated and characterized. Alternatively, these screens can be undertaken in an effort to design more comprehensive probe sets for microarrays or to help mine putative regRNAs from HTS sequencing datasets. Some of these objectives call for higher specificity (i.e., a lower proportion of false-positive predictions) while for others sensitivity (i.e., a lower proportion of false-negative predictions) is prioritized. With this in mind, SIPHT allows the user to adjust the balance between sensitivity and specificity by changing a variety of search parameters. Moreover, many of the annotations provided by SIPHT can be used to identify the most reliable and/or most biologically relevant candidates from the often large number of predicted loci. This chapter focuses on how each of SIPHT's adjustable parameters can be used to change the stringency of an SIPHT screen and how information in the SIPHT output file can be used to glean information about the reliability of the predicted locus and/or provide insight about its potential biological function.

2. Materials

SIPHT can be accessed at http://newbio.cs.wisc.edu/sRNA/.

3. Methods

The SIPHT introductory page provides two main links. The "Click here for published annotations" directs the user to the results of previous searches in 929 replicons conducted using default search parameters. These published annotations were conducted using the original version of SIPHT that has since undergone several updates and improvements. Thus, to take advantage of the most up-to-date functions and annotations described below one should rerun the SIPHT search in their replicon of interest (ROI), which can be done by following the "Click here to launch a search" link. This link leads to the search launch page containing scroll down menus, radio buttons, and text boxes that enable various SIPHT search parameters to be defined. There are three main stages to defining these parameters: (1) selecting the ROI in which the search is to be conducted, (2) selecting the replicons to be used to identify intergenic conservation in the ROI, and (3) changing any of the 16 adjustable search parameters from their preset default values. In the first stage, a scroll down menu provides identifiers for all available replicons that include its strain name, type (chromosome or plasmid), and corresponding NCBI accession number.

Choices made in the other two stages can have significant impacts on the results of the search and are discussed in more detail below. Once the search is completed, several approaches can be used to mine the SIPHT output file for the strongest candidates or extract clues into the potential functions of predicted loci. These are also described in more detail below.

3.1. Selecting Replicons for BLAST Comparisons

After selection of the ROI in which the SIPHT search is to be conducted, the user comes to the "Select the partner replicons for BLAST", which allows him or her to decide which replicons to include in the query database for BLAST comparison (16) with the ROI (see Note 1). This section has two main options. The first is "Option 1: All replicons", which is the default option and includes all available replicons EXCEPT those in strains of the same species as that of the ROI. A checkbox under the radio button for Option 1 can be used to include replicons in the same species though this usually leads to a high rate of false-positive predictions. Below this radio button is a text box that enables specific replicons to be excluded from the BLAST analysis. This allows the user to exclude BLAST comparison between replicons from strains that, while not sharing the same species names, may be highly related (such as *E. coli* and *Shigella* sp.). The second option in this section is "Option 2: Select individual replicons" that allows conservation analysis to be limited to specific partner replicons. If this option is selected, the user must add replicons to be included in BLAST comparisons to the text box on the right.

3.2. Modifying SIPHT Search Parameters

The SIPHT web interface enables the user to adjust 16 search parameters. The first four influence the BLAST analysis (see Note 1).

- *Maximum BLAST E*: The maximum expected value for BLAST alignments. This affects the stringency of the BLAST analysis. Specifically, the lower the maximum E value, the higher the statistical significance of high-scoring segment pairs (HSPs) must be for them to be included in the SIPHT analysis.
- *Minimum BLAST S*: The minimum score allowed for BLAST alignments. This is similar to maximum BLAST E as it also affects the stringency of the BLAST analysis. Specifically, the higher the minimum S, the higher the statistical significance of HSPs must be for them to be included in the SIPHT analysis. However, unlike E, the value of S is not affected by changes in the query sequences so this threshold will be more consistent among searches using different BLAST partner replicons.
- *Minimum BLAST % identity*: Minimum percent identity for BLAST HSPs. Increasing this parameter lowers the proportion of alignment mismatches in HSPs allowed and thus increases the stringency of the BLAST analysis.

- *Maximum BLAST HSP length*: This allows the user to exclude long HSPs that may correspond to large regions of intergenic sequence homology rather than to the shorter regions of conservation that are more likely to suggest the presence of a regRNA-encoding locus.

The next three adjustable parameters control the stringency of terminator predictions. SIPHT uses three different algorithms to identify putative Rho-independent terminators: RNAMotif v3.0.4 (17), TransTerm v2.05 (18), and FindTerm, a prediction program based on a heuristic algorithm developed by Ruth Hershberg (19). For RNAMotif and FindTerm, the lower the maximum score, the higher the stringency of the predictions while for TransTerm an increased minimum score leads to increased stringency.

The final nine parameters relate to the positional relationships of predicted features to each other or to nearby genomic features (Fig. 2) (see Note 2).

- *Minimum/maximum predicted locus length*: These set the minimum and maximum length of predicted loci. The length is calculated from the 5′-end of the HSPs to the 3′-end of the terminator following the run of Ts.

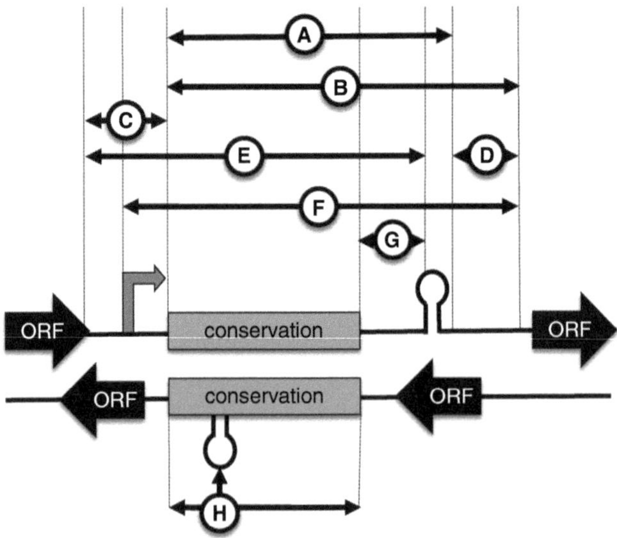

Fig. 2. SIPHT search parameters that can be modified by users and relate to the size and relative genomic location of candidate loci. The *gray arrow* and stem–loop represent a putative promoter and a predicted terminator, respectively. (*A*) Minimum and maximum locus length. (*B*) Minimum distance of locus start to start of ORF on same strand. (*C*) Minimum distance of locus start to end of ORF on same strand. (*D*) Minimum distance of locus end to start of ORF on same strand. (*E*) Minimum distance of locus end to end of ORF on same strand. (*F*) Minimum distance of TFBSs to start of ORF on same strand. (*G*) Maximum gap between end of conservation and start of predicted terminator. (*H*) Minimum locus % of HSP length.

- *Min dist locus start to ORF start*: The minimum distance between the 5′-end of a candidate locus and the annotated start codon of its downstream ORF on the same strand. Increasing this will limit the identification of loci that correspond to conserved 5′UTRs of mRNAs rather than to regRNAs but will also likely lead to the exclusion of *bona fide* r5′UTRs.
- *Min dist locus start to ORF end*: The minimum distance between the 5′-end of a candidate locus and the annotated stop codon of its upstream ORF on the same strand. Increasing this value decreases the identification of candidate loci that likely correspond to conserved 3′UTRs rather than to regRNAs.
- *Min dist locus end to ORF start*: The minimum distance between the 3′-end of a candidate locus (and therefore its associated terminator) and the annotated start codon of the downstream ORF on the same strand. Increasing this value decreases the identification of 5′rUTRs and thus enriches predictions for sRNAs. Moreover, keeping this value positive excludes putative terminators that are within ORFs and are more likely to represent false predictions.
- *Min dist locus end to ORF stop*: The minimum distance between the 3′-end of a candidate locus (and therefore its associated terminator) and the annotated stop codon of the upstream ORF on the same strand. Increasing this value decreases the identification of conserved 3′UTRs. However, it is important to note that some sRNAs are known to share a terminator with an ORF so increasing this value may also limit the detection of real regRNAs.
- *Maximum gap cons→term*: The maximum distance between the 3′-end of conservation and the 5′-end of the predicted terminator. Lowering this value requires regions of conservation and terminators associated with the same locus to be closer together. In some cases, doing this will lead to the exclusion of predicted loci whose conservation in any other replicon does not extend to their associated terminator. In other cases, doing this will not change which loci are predicted, but may exclude annotation of their conservation in more distantly related species in which the conservation of their homologue is limited to the 5′-end.
- *Minimum locus % of HSP length*: This parameter relates to the position of a predicted terminator that is within a region of intergenic conservation. For example, for a locus in which the terminator is near or at the 3′-end of its associated region of conservation, this value would be close to 100. In contrast, for a locus in which the terminator lies in the center of its associated region of conservation, this value would be 50. Presumably,

loci in which the conservation ends at or near the putative 3′-end are more likely to correspond to functional transcripts than those in which the terminator is "floating" in the middle or at the beginning of a large region of intergenic conservation. Thus, increasing this parameter tends to produce more specific but less sensitive predictions.

- *Min dist TFBS from ORF start*: Minimum distance of putative TFBSs from the annotated start codon of the downstream ORF on the same strand. The higher this parameter is set the higher the likelihood that a TFBS associated with a candidate locus controls expression of that locus rather than that of its 3′ORF.

3.3. Extracting Information Regarding Strength and Potential Function of Predicted Loci from SIPHT Output

SIPHT outputs a tab-delimited file that contains numerous annotations for each predicted locus. Many of these annotations provide information about the position of the predicted locus in relation to its flanking genes. Others provide information that can be used to better ascertain the reliability of predicted loci and/or their potential function.

UpGENEdirection, distUpGENE, DnGENEdirection, distDnGENE: One of the main sources of false regRNA predictions by SIPHT are loci that are directly 5′ or 3′ of ORFs and correspond to conserved UTRs. Candidates that are far from and/or on the opposite strand of neighboring ORFs are more likely to represent real sRNAs. These candidates can be identified by sorting predicted loci based on 1) their strand orientation, 2) their distance from flanking ORFs, and 3) the strand orientations of these flanking ORFs.

BLASTscore, BLASTexpect: The BLAST E value and score of intergenic conservation associated with the locus. If conservation was found in multiple replicons, this value corresponds to the lowest E and highest score among these HSPs (see Note 3).

TermType: Another source of inaccuracy in SIPHT searches are false terminator predictions. Each locus is annotated by which terminator prediction program(s) identified its associated putative terminator (see above). In our previous studies we have shown that loci associated with terminators identified by all three programs are enriched for known regRNAs and thus represent stronger candidates for real regRNAs. However, we have also found numerous instances in which terminators linked to experimentally validated regRNAs were predicted by only one or two of these algorithms. Thus, users seeking precision should filter out loci associated with terminators predicted by only one or two programs while those seeking sensitivity should not.

Homology to known sRNAs, conserved synteny with knowns: These annotations enable users to identify putative homologues of known regRNAs based on conserved primary sequence or genomic position, respectively. The latter annotation is often useful as some functionally conserved regRNAs share little sequence homology outside closely related species but may have maintained the same relative genomic position as their functional homologues. Conserved genomic position is particularly helpful in identifying homologues of *cis*-acting regRNAs such as riboswitches whose location and orientation relative to their 3′ORF are necessarily conserved.

NumSearchesPredicted, BLASTpartner: These annotations correspond to the number of replicons in which each locus is conserved and the names of these replicons, respectively, and can provide clues about the potential function of predicted loci. For example, in a SIPHT search of *Vibrio cholerae* using default search parameters, structural and catalytic noncoding RNAs such as tmRNA and RnpB are conserved in over 200 replicons in both Gram-positive and Gram-negative species while most other sRNAs are conserved only among other Gram-negative enteric bacteria or in many cases, only among other members of *Vibrionaceae* (see Note 4).

Potential UTR?, Riboswitch?: The first annotation indicates whether the distance from and strand orientation of the flanking ORFs suggest the predicted locus may be a 5′ and/or 3′UTR. The second is a partially overlapping annotation that denotes whether the locus is likely to be a riboswitch (see Note 5).

Num paralogue, Paralogous candidates, Weaker rev paralogue?: Most sRNAs characterized to date have one or less paralogues in the same replicon. In contrast, many species carry multiple homologues of certain families of riboswitches (20, 21) while others encode numerous copies of conserved intergenic repeat sequences whose biological function is still not well understood (22). Thus, sorting by the number of paralogues can help distinguish sRNAs from riboswitches and/or conserved intergenic repeats. Identifying loci with putative paralogues is particularly important in prioritizing candidate loci for follow up experimental studies, as deletion of a locus with a functionally redundant paralogue is less likely to produce a phenotype. The "Paralogous candidates" annotation provides the names of the predicted paralogues of each locus. Importantly, it also denotes the relative location of each putative paralogue. The annotations "(para)," "(tandem)," and "(anti)" indicate that the paralogue is located in a different intergene, in the same intergene but not overlapping that locus, or in the same intergene antisense to and overlapping the predicted locus, respectively. Paralogues with the "(anti)" designation correspond to loci in which the same region of intergenic conservation is associated with putative terminators on opposite strands. Since the likelihood that both loci correspond to real regRNAs is small, the "Weaker rev paralogue" can be used to quickly identify and discard loci likely to correspond to a false prediction because their antisense paralogue is associated with a stronger putative terminator (based on the number of programs by which it was predicted).

QRNA?: This annotation indicates whether the program QRNA (23) identified predicted secondary structure in the locus that indicates conservation of either secondary structure (RNA) or protein coding sequence (COD). A COD designation suggests the loci may correspond to a small ORF rather than to a regRNA.

Putative TFBSs: Predicted loci are annotated as to their proximity to a variety of putative TFBSs. The specific TFBSs included in each search are dependent on the replicon in which the search is conducted. For each TFBS, its distance from the 3′-end of the locus and the score assigned to it by the Patser program (24) are included. These annotations can be used to identify stronger candidates, as regions of conservation flanked by both a putative promoter and terminator are more likely to correspond to real transcripts. Moreover, the proximity of certain TFBSs, such as those recognized by transcription factors such as Fur and LexA or by alternative sigma factors such as RpoN and RpoE, can also provide insights into the potential biological role of a predicted locus. For example, SIPHT was used to identify putative functional homologues of the iron regulated sRNA RyhB in diverse bacteria based on the location of these loci directly downstream of a putative binding site for the well-conserved iron-regulated transcriptional repressor Fur.

4. Notes

1. The stringency of BLAST analysis and the partner replicons included in BLAST comparisons can have a dramatic effect on the results of SIPHT searches. If BLAST stringency is too high and/or the partner replicons included are too distantly related to the ROI, many regRNAs will be missed, limiting the sensitivity of the search. In contrast, if the BLAST stringency is too low and/or the replicons included are too closely related to the ROI, much of the sequence homology identified may reflect a lack of divergence among closely related species rather than the presence of a regRNA-encoding locus whose primary sequence has been conserved due to functional constraints. This will increase the proportion of false-positive predictions produced. Identifying the most appropriate BLAST parameters

and partner replicons for a particular ROI can often be achieved through iterative SIPHT searches that vary these parameters and include or exclude different subsets of the available partner replicons. By assessing what proportion of the total loci predicted correspond to previously annotated or experimentally validated regRNAs, one can get a sense of the combination of BLAST parameter values and partner replicons that yield the desired balance of sensitivity and specificity.

2. Many of these adjustable parameters described in this section can override and be overridden by others. For example, setting the "Min dist locus start to ORF start" to 400 and the "Maximum predicted locus length" parameter to 200 will limit loci to those ending no closer than 200 bp from the start codon of the downstream ORF even if the "Min dist locus end to ORF start" parameter is set to 100. Similarly, setting "Min dist locus end to ORF start" to 400 and setting "Minimum predicted locus length" to 100 would mean no loci starting closer than 500 bp upstream of an ORF would be identified, regardless of what value is assigned to "Min dist locus start to ORF start." One way to ensure parameters don't override each other is to assign the desired value to one parameter then assign zero to all remaining parameters that may override it.

3. Since only the lowest E and highest S are reported and these are often derived from homology to closely related species, it is difficult to ascertain and sort by the level of conservation of loci among more distantly related species. To do this, one can repeat the search excluding replicons in the same or closely related genera then sort by E or score.

4. These annotations can also be used to identify loci predicted based on spurious BLAST HSPs. For example, IGRs with high AT or GC content have been found to produce high scoring alignments with IGRs in other, evolutionarily distant AT or GC-rich species that are unlikely to correspond to real sequence conservation but rather reflect their shared low sequence complexity. While IGRs with AT or GC content greater than 75% are excluded from SIPHT searches to mitigate this problem, these spurious alignments still occur. Thus, loci that are conserved in evolutionarily distant species but NOT in closely related ones are less likely to correspond to real regRNAs.

5. The annotations of potential 5′UTR and riboswitch are only partially overlapping as the former requires a locus to end no farther than 50 bp from the start of an ORF, whereas the latter requires this maximum distance to be 100 bp.

References

1. Altuvia S (2007) Identification of bacterial small non-coding RNAs: experimental approaches. Curr Opin Microbiol 10:257–261
2. Sorek R, Cossart P (2010) Prokaryotic transcriptomics: a new view on regulation, physiology and pathogenicity. Nat Rev Genet 11:9–16
3. Weissenmayer BA et al (2011) Sequencing illustrates the transcriptional response of *Legionella pneumophila* during infection and identifies seventy novel small non-coding RNAs. PLoS One 6:e17570
4. Liu JM et al (2009) Experimental discovery of sRNAs in *Vibrio cholerae* by direct cloning, 5S/tRNA depletion and parallel sequencing. Nucleic Acids Res 37:e46
5. Livny J, Waldor MK (2010) Mining regulatory 5′UTRs from cDNA deep sequencing datasets. Nucleic Acids Res 38:1504–1514
6. Irnov I et al (2010) Identification of regulatory RNAs in *Bacillus subtilis*. Nucleic Acids Res 38:6637–6651
7. Sharma CM et al (2010) The primary transcriptome of the major human pathogen *Helicobacter pylori*. Nature 464:250–255
8. Mraheil MA et al (2011) The intracellular sRNA transcriptome of *Listeria monocytogenes* during growth in macrophages. Nucleic Acids Res 39:4235–4248
9. Livny J, Waldor MK (2007) Identification of small RNAs in diverse bacterial species. Curr Opin Microbiol 10:96–101
10. Lu X et al (2011) Assessing computational tools for the discovery of small RNA genes in bacteria. RNA 17:1635–1647
11. Livny J et al (2008) High-throughput, kingdom-wide prediction and annotation of bacterial non-coding RNAs. PLoS One 3:e3197
12. Valverde C et al (2008) Prediction of *Sinorhizobium meliloti* sRNA genes and experimental detection in strain 2011. BMC Genomics 9:416
13. Faucher SP et al (2010) *Legionella pneumophila* 6S RNA optimizes intracellular multiplication. Proc Natl Acad Sci USA 107:7533–7538
14. Dichiara JM et al (2010) Multiple small RNAs identified in *Mycobacterium bovis* BCG are also expressed in *Mycobacterium tuberculosis* and *Mycobacterium smegmatis*. Nucleic Acids Res 38:4067–4078
15. Postic G et al (2010) Identification of small RNAs in *Francisella tularensis*. BMC Genomics 11:625
16. Altschul SF et al (1997) Gapped BLAST and PSI-BLAST: a new generation of protein database search programs. Nucleic Acids Res 25: 3389–3402
17. Macke TJ et al (2001) RNAMotif, an RNA secondary structure definition and search algorithm. Nucleic Acids Res 29:4724–4735
18. Kingsford CL et al (2007) Rapid, accurate, computational discovery of Rho-independent transcription terminators illuminates their relationship to DNA uptake. Genome Biol 8:R22
19. Wassarman K et al (2001) Identification of novel small RNAs using comparative genomics and microarrays. Genes Dev 15:1637–1651
20. Barrick JE, Breaker RR (2007) The distributions, mechanisms, and structures of metabolite-binding riboswitches. Genome Biol 8:R239
21. Gardner PP et al (2011) Rfam: Wikipedia, clans and the "decimal" release. Nucleic Acids Res 39:D141–D145
22. Hulton CS et al (1991) ERIC sequences: a novel family of repetitive elements in the genomes of *Escherichia coli*, *Salmonella typhimurium* and other enterobacteria. Mol Microbiol 5:825–834
23. Rivas E et al (2001) Computational identification of noncoding RNAs in *E. coli* by comparative genomics. Curr Biol 11:1369–1373
24. van Helden J (2003) Regulatory sequence analysis tools. Nucleic Acids Res 31:3593–3596

Chapter 2

How to Identify CRISPRs in Sequencing Data

Christine Drevet and Christine Pourcel

Abstract

Clustered regularly interspaced short palindromic repeats (CRISPRs) are DNA sequences composed of a succession of repeats (23–50 bp long) separated by unique sequences called spacers. CRISPRs together with a set of genes called *cas* for CRISPR associated, constitute a defence mechanism against invasion by foreign sequences. We describe protocols and bioinformatics tools that allow the identification of CRISPRs, their comparison and their component determination (the direct repeats and the spacers). A schematic representation of the spacer organization can be produced, allowing an easy comparison between strains.

Key words: CRISPR, Genotyping, Bacteriophage, Database, Spacer, Phylogeny

1. Introduction

Clustered regularly interspaced short palindromic repeats (CRISPRs) loci typically consist of the succession of 23–50 bp direct repeats (DR), separated by variable and non-repetitive sequences called spacers (see Fig. 1). A CRISPR generally possesses at one end a degenerated DR (DR′) and at the other end a complete DR immediately followed by a sequence called the leader, which acts as a promoter (see Fig. 1a). In a single genome several CRISPRs with the same DR can be found, but only one is associated with a group of genes called *cas* (for CRISPR-associated) (see Fig. 1b) (1, 2). The CRISPR/cas system has been identified in a broad range of prokaryotic species, 85% of Archaea and 48% of Bacteria (3).

In the majority of cases, the spacers, when identified, happen to be fragments of bacteriophages or plasmids (4, 5). Different observations suggest that the CRISPR/cas system constitutes a defence system against foreign sequences. The currently sequenced

Kenneth C. Keiler (ed.), *Bacterial Regulatory RNA: Methods and Protocols*, Methods in Molecular Biology, vol. 905, DOI 10.1007/978-1-61779-949-5_2, © Springer Science+Business Media, LLC 2012

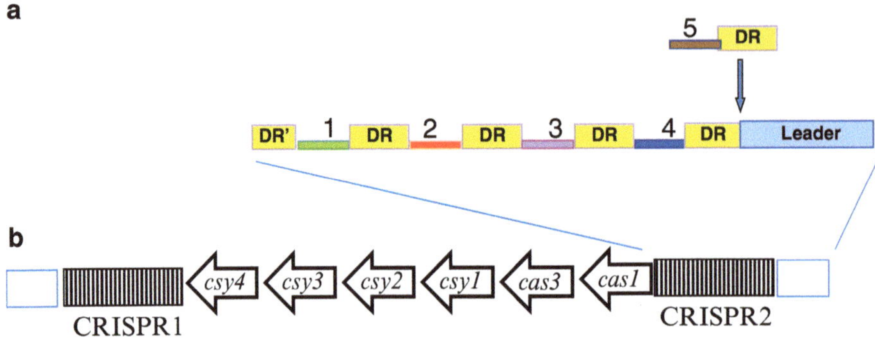

DR CRISPR1: TTTCTTAGCTGCCTATACGGCAGTGAAC
DR CRISPR2: TTTCTTAGCTGCCTACACGGCAGTGAAC

Fig. 1. Schematic structure of a CRISPR and the *Yersinia* subtype CRISPRs in *Pseudomonas aeruginosa*. (**a**) DRs are shown as *yellow boxes* and DR′ represent the degenerate DR on one side of the CRISPR. On the other side is the sequence called leader, which acts as a promoter. Spacers are shown with different colors and are numbered according to the order of their acquisition. A new spacer is added next to the leader after duplication of the last DR element. (**b**) Schematic organization of the CRISPR/cas locus. A set of six *cas* genes are framed by two CRISPR structures. The DR sequences of CRISPR1 and CRISPR2 differ at one nucleotide.

CRISPR structures listed in CRISPRdb (3) show an important variability in the nature of DRs and the number of motifs between and inside species. The largest observed CRISPR to date possesses 588 spacers in *Haliangium ochraceum* DSM 14365. Within a species, strains may or may not possess a given CRISPR (with a particular DR) and the number and nature of spacers may vary considerably. This diversity suggests that the CRISPR structure is continuously evolving, either through the addition of new motifs (a DR and a spacer) or by interstitial deletion of one or several motifs through recombination between two DRs. New motifs are added to the CRISPR in a polarized manner by duplicating the DR next to the leader and adding a new fragment of DNA (4, 6) (see Fig. 1a).

Correctly identifying CRISPRs in a bacterial or archaeal genome is not as straightforward as might be expected. A major challenge is that other types of repeats could be misidentified as CRISPR, so that it is necessary to identify features distinguishing CRISPRs from other repeated sequences. In addition, rules for CRISPR identification deduced from known CRISPRs may be too restrictive to find CRISPR sequences in new datasets. It is difficult to define parameters that will encompass the characteristics of all CRISPRs (including the smaller ones) and, in particular, faithfully define their DRs boundaries. Several programs have been developed to specifically search for CRISPRs, including CRISPRfinder (7), which is used to build the database CRISPRdb (3). Both CRISPRfinder and CRISPRdb are available at the CRISPR web server site http://crispr.u-psud.fr/crispr/. Recent improvements were made to the CRISPR web server by giving access to information on *cas* genes present in the analyzed genomes, and by simplifying the comparison of CRISPR alleles. The database, which is being regularly updated and manually curated, provides files that help in analyzing the diversity of CRISPR elements. The methods below explain how to use these tools to identify CRISPRs in new sequence data, analyze CRISPR components, and compare CRISPRs in different genomes.

2. Materials

The web-based tools necessary for CRISPRs identification and comparison are freely accessible at http://crispr.u-psud.fr.

2.1. CRISPRs Database

The CRISPR web server hosts the CRISPRdb (a MySQL open source database), storing the CRISPRs content from reference genomes of 118 Archaea and 1,582 Bacteria as of April 2012. The database is regularly updated and curated by the authors.

2.2. User Data Analysis

2.2.1. CRISPRfinder

CRISPRfinder is used to identify and characterize CRISPR-like structures. Nucleic acid sequences in FASTA format can be pasted into input text areas or uploaded from a file on the local machine. Multi-sequence files are also allowed by the CRISPRfinder program and each sequence will be treated independently. The output is primarily web-based and viewable on the website. In addition, results can be downloaded as text files.

2.2.2. CRISPRcompar

CRISPRcompar is used to identify allelic CRISPRs from different strains according to their genomic position.

Strains to be compared are selected in the public and private database. Output is viewable on the website. It can be imported into CRISPRtionary for further analysis of spacers.

2.2.3. CRISPRtionary

CRISPRtionary is used to compare spacer arrangement of CRISPRs. The data are a set of allelic sequences of a given CRISPR locus. The output is viewable on the website. In addition, results can be downloaded as text files or as tabulated text files that can be opened with any table viewers, including Excel.

2.3. User Data Storage

Users can create a private account to store (in a MySQL database) and further analyze their own data.

3. Methods

The different tools available on the CRISPR web server allow three main applications: CRISPRs can be identified in users' data directly by submitting sequences to CRISPRfinder (1), CRISPRs can be identified in reference genomes by browsing the CRISPRdb database (2), and CRISPRs in different strains can be compared using the CRISPRcompar and CRISPRtionary tools (3).

3.1. Finding CRISPRs

The smallest CRISPRs detected by CRISPRfinder consist of two DRs (a complete and a degenerated one) separated by a spacer (see Note 1). Large CRISPRs can contain several hundred repeats. The presence of a CRISPR in a strain does not imply its existence in all the members of the species.

1. Submit nucleic acid sequences to the CRISPRfinder program online (http://crispr.u-psud.fr/Server/) by uploading a sequence file or pasting the sequences into the dialog box.
2. Click the FindCRISPR button to launch the CRISPRfinder program with the default parameters (see Note 2). CRISPRfinder advanced version, available from the main CRISPRfinder form, allows modifying the parameters to refine the search (see available options in Note 2). The resulting pages list the identified CRISPRs for each submitted sequence, and separate them into "confirmed" and "questionable" (see Fig. 2; Note 3). Each

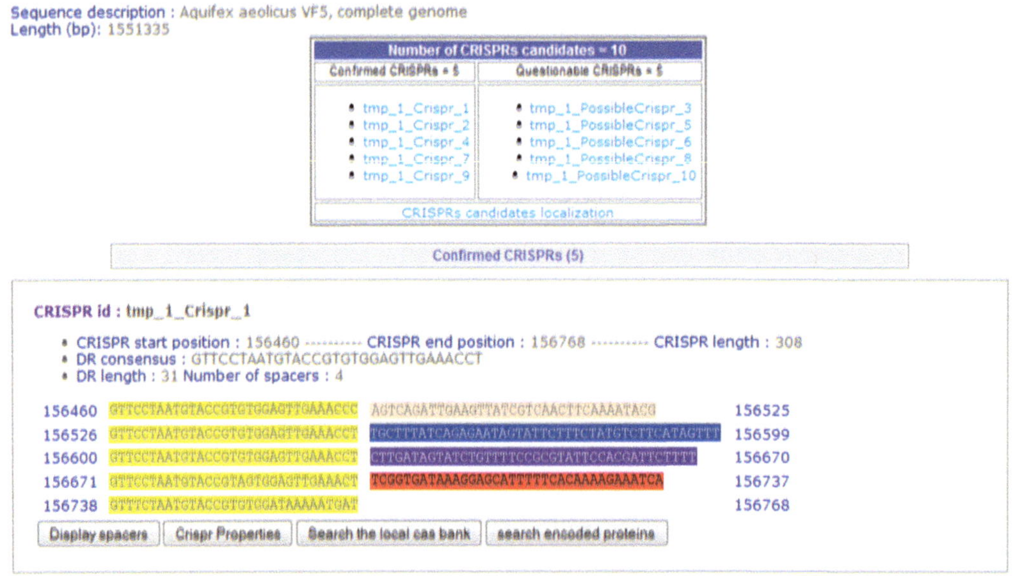

Fig. 2. Output of the CRISPRfinder program. Confirmed and questionable CRISPRs are listed and each of them is depicted in a schematic representation where the DRs are shown in *yellow* and the spacers with different colors.

identified structure is depicted where the DR is shown in yellow and the spacers are shown with different colors.

3. By clicking on the different buttons under the identified CRISPR, one can display the FASTA formatted spacer list (Display spacers), and the CRISPR properties file (CRISPR properties) which are the output of the CRISPRfinder standalone program. The CRISPR properties include DR size, DR consensus sequence, sequences and coordinates of the spacers.
4. Identification of *cas* genes near a questionable CRISPR can provide evidence that the CRISPR is legitimate, as each organisms that contains bona fide CRISPRs should contain a set of *cas* genes. To search for *cas* genes near an identified CRISPR, click the "Search the local *cas* bank" button. Similarities between the submitted data and *cas* genes from a large number of species in the vicinity of the CRISPR (±10,000 bp) will be identified (see Note 4).
5. The "search encoded proteins" function checks whether the CRISPR sequence corresponds to an open reading frame. This tool may help in eliminating structures that are not true CRISPRs, because there is no documented case of CRISPRs including functional proteins (see Note 5).

3.2. Analysis of CRISPR Components

Several features to analyze the CRISPRs DR and spacers are gathered in the pink box on the top right of the page.

1. To identify putative protospacers for CRISPRs, click the "Search protospacers using BLAST" button. The spacers of each CRISPR will be displayed, and BLASTN searches can be performed with each spacer to find similar sequences in the Genbank nr database (see Note 6).
2. Sequences surrounding the CRISPR on each side can be recovered using the "Extract the flanking sequences" button. It is then possible to launch BLASTN searches with these flanking sequences in the Genbank nr database. This feature is helpful when checking whether a CRISPR region is present in other genomes.
3. Clicking on "Search CRISPRs with identical DRs" leads to a table listing strains and CRISPRs identity numbers (Ids) (see Note 7) found in the CRISPR database. These CRISPRs can be visualized by clicking on each Ids. Questionable CRISPRs from Subheading 3.1, step 3, bearing numerous spacers can be validated as true CRISPRs when shared spacers are identified.
4. Sometimes there are small natural variations in DR sequences, even between closely related species. The BLAST CRISPRdb link in the left main menu should be used to identify related DRs from the database.
5. Finally text files of spacers and CRISPR properties can be downloaded.

3.3. CRISPRs Public Database

1. The presence of a CRISPR in a public sequenced genome can be assessed by consulting the CRISPRdb database (http://crispr.u-psud.fr/crispr/). Strains marked in pink color possess confirmed CRISPR, whereas those in gray have only questionable structures and those in yellow have no CRISPR (see Note 8). By default, the strains are presented alphabetically, but choosing the "View the strains taxonomy browser" link shows the strains according to taxonomy (see Note 9).
2. Clicking on a strain leads to a table listing the sequences present in its genome (chromosome and plasmid) and the associated CRISPRs (see Fig. 3a).
3. At this stage it is possible to view annotated *cas* genes in the genomes which harbor a CRISPR (see Note 10). Because the *cas* genes are typically not systematically annotated in genome sequences, a complementary search in the local database of CAS proteins is recommended. Gene members of four CAS protein families, designated Cas1 to Cas4, are expected in the

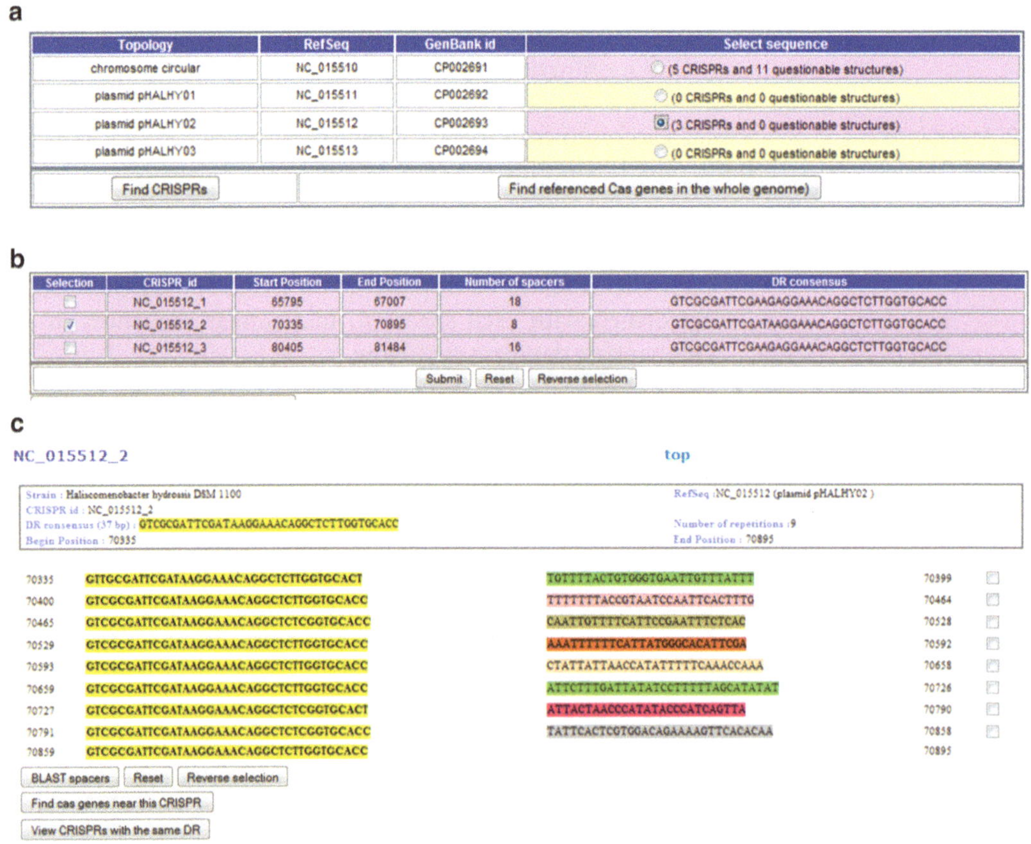

a

Topology	RefSeq	GenBank id	Select sequence
chromosome circular	NC_015510	CP002691	(5 CRISPRs and 11 questionable structures)
plasmid pHALHY01	NC_015511	CP002692	(0 CRISPRs and 0 questionable structures)
plasmid pHALHY02	NC_015512	CP002693	(3 CRISPRs and 0 questionable structures)
plasmid pHALHY03	NC_015513	CP002694	(0 CRISPRs and 0 questionable structures)

Find CRISPRs   Find referenced Cas genes in the whole genome)

b

Selection	CRISPR_id	Start Position	End Position	Number of spacers	DR consensus
	NC_015512_1	65795	67007	18	GTCGCGATTCGAAGAGGAAACAGGCTCTTGGTGCACC
✓	NC_015512_2	70335	70895	8	GTCGCGATTCGATAAGGAAACAGGCTCTTGGTGCACC
	NC_015512_3	80405	81484	16	GTCGCGATTCGAAGAGGAAACAGGCTCTTGGTGCACC

Submit   Reset   Reverse selection

c

NC_015512_2   top

Strain : Haliscomenobacter hydrossis DSM 1100
CRISPR id : NC_015512_2
DR consensus (37 bp) : GTCGCGATTCGATAAGGAAACAGGCTCTTGGTGCACC
Begin Position : 70335
RefSeq :NC_015512 (plasmid pHALHY02)
Number of repetitions :9
End Position : 70895

70335	GTTGCGATTCGATAAGGAAACAGGCTCTTGGTGCACT	TGTTTTACTGTGGGTGAATTGTTTATTT	70399
70400	GTCGCGATTCGATAAGGAAACAGGCTCTTGGTGCACC	TTTTTTTACCGTAATCCAATTCACTTTG	70464
70465	GTCGCGATTCGATAAGGAAACAGGCTCTCGGTGCACC	CAATTGTTTTCATTCCGAATTTCTCAC	70528
70529	GTCGCGATTCGATAAGGAAACAGGCTCTTGGTGCACC	AAATTTTTTCATTATGGGCACATTCGA	70592
70593	GTCGCGATTCGATAAGGAAACAGGCTCTTGGTGCACC	CTATTATTAACCATATTTTTCAAACCAAA	70658
70659	GTCGCGATTCGATAAGGAAACAGGCTCTTGGTGCACC	ATTCTTTGATTATATCCTTTTTAGCATATAT	70726
70727	GTCGCGATTCGATAAGGAAACAGGCTCTCGGTGCACT	ATTACTAACCCATATACCCATCAGTTA	70790
70791	GTCGCGATTCGATAAGGAAACAGGCTCTCGGTGCACC	TATTCACTCGTGGACAGAAAAGTTCACACAA	70858
70859	GTCGCGATTCGATAAGGAAACAGGCTCTTGGTGCACC		70895

BLAST spacers   Reset   Reverse selection
Find cas genes near this CRISPR
View CRISPRs with the same DR

Fig. 3. Output of a CRISPRdb query. (**a**) Each sequence contained in the queried strain is listed with its RefSeq number, GenBank id and the number of CRISPRs found. (**b**) The list of CRISPRs found in the selected sequence is shown, with its position and the DR consensus sequence. (**c**) The selected CRISPR is shown.

vicinity of at least one structure in strains harboring true CRISPRs (see Note 3).

4. Selected CRISPRs can be viewed by clicking on the submit button (see Fig. 3b). The schematic representation is similar to that of CRISPRfinder and tools are available to BLAST spacers against Genbank, search for clustered *cas* genes and view CRISPRs with the same DR (see Fig. 3c).
5. Direct Repeats and spacers lists of "confirmed" CRISPRs are built from the CRISPR database after each update. They are viewable and downloadable in "CRISPR utilities" and available for BLAST searches at the "BLAST CRISPRS" menu item.

3.4. My CRISPRdb

1. It is possible to run CRISPRfinder on submitted sequences and to store the results in a dedicated database hosted by the CRISPR server. To create a private account, an email address and a password must be provided.
2. Sequences can be submitted to CRISPRfinder using the same form as in the main CRISPRfinder page.
3. The identified CRISPR structures can be saved after providing a sample name.
4. To access the database click on the "Consult your private database button." The private data browser is similar to that of the main CRISPRdb. It is also possible to compare CRISPRs in both databases using the CRISPRcompar tool (see Subheading 3.5).

3.5. CRISPRs Comparison

CRISPR polymorphism is generally observed within a species and this feature may be used to compare strains and to provide phylogenetic information. The following tools help in identifying CRISPR loci and classifying spacers.

1. When the genome sequences of several strains are available for a given species, and when each strain possesses several CRISPRs, it is important to be able to classify the different CRISPRs, particularly as their position on the genome might vary. The CRISPRcompar tool is based on the presence of identical DRs and similar flanking sequences (100–70% mismatch threshold is accepted) (see Note 11). This function is particularly useful when several CRISPRs are present in a single genome.

   Three optional forms are available when connected to the CRISPRcompar page. The "View all the genomes harbouring CRISPRs" form lists all the sequences of the CRISPRdb that contain true CRISPRs in alphabetical order. Another way to compare CRISPR loci is to activate "View the strains taxonomy browser" which recovers from CRISPRdb all members of a genus containing a CRISPR and allows comparison of each

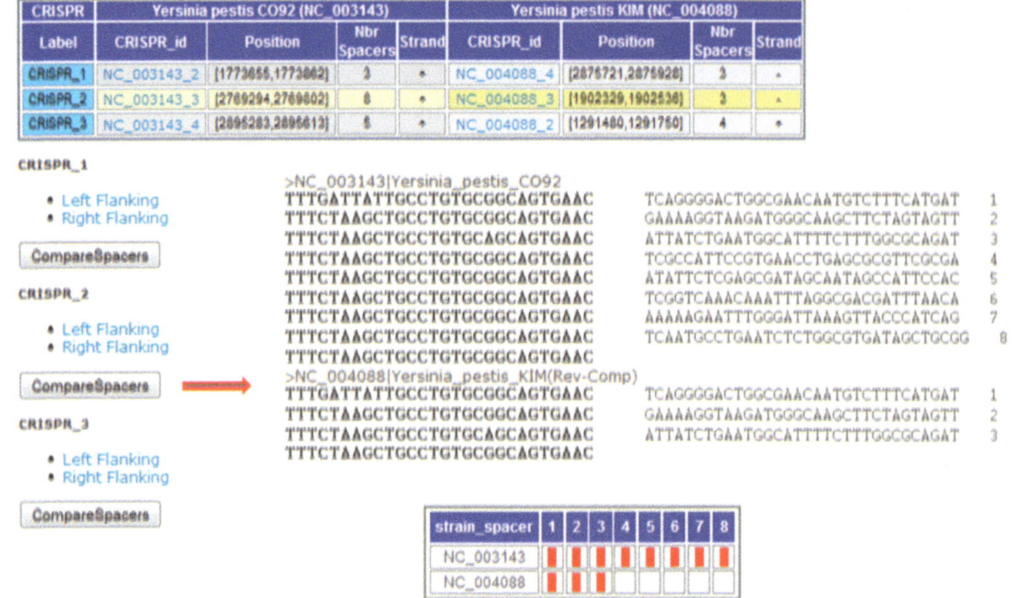

Fig. 4. Output of a CRISPRcompar query. The different CRISPRs of a given locus are compared by giving access to their flanking sequences and to their spacer's list.

of them. Additional strains can be added to the comparison. Finally logging into a private database leads to a page where it is possible to activate the link "CRISPRs comparison". Thereafter, the list of genomes present in the private database will be available together with the CRISPRdb list for comparing CRISPRs.

2. Select genomes, then activate the compare button. The resulting page shows the selected genomes, which can be retained or deleted from the comparison. Additional genomes can be added at that stage.
3. Select one of the genomes to serve as a reference for labeling the different CRISPR loci, and click the compare button. The resulting page shows the output of the comparison in a table in which CRISPR loci are listed under each sequence with their coordinates and number of spacers (see Fig. 4).
4. When two sequences possess a common CRISPR locus, their spacers can be compared by clicking on the "Compare Spacers" button. Activating this button will lead to the CRISPRtionary program in which spacers of submitted CRISPR sequences can be compared (see Subheading 3.6).
5. After getting the list of CRISPR loci in compared genomes, it is possible to view an alignment of sequences flanking the CRISPRs of a given locus by clicking on the "Left flanking" or

"Right flanking" links (see Note 11). Alignments between 5′ and 3′ sequences of CRISPRs are also directly available with the FlankAlign tool. In that case the user can choose the length of the regions to align (CRISPRdb sequences).

3.6. CRISPRtionary

1. After activating this page, upload or paste into the frame a list of sequences in FASTA format in order to create a repertoire of spacers, annotate them and order them (see Fig. 4). The spacers dictionary is an excel-formatted file with a specific format (examples are provided below the submission form, see Note 12). A dictionary containing the spacers identified in the submitted sequences will be created by the program if you do not specify any at this stage.
2. After activating the "Find CRISPRs" button click on continue if no dictionary was uploaded and the next page will show the CRISPRs identified in each of the submitted sequences.
3. To number the spacers it is necessary to select a DR sequence either from the "List of candidate DRs" or by introducing a sequence (see Note 13).
4. The "Find spacers" button leads to a page where each spacer has been given a number and different tables are available. Known spacers are numbered with the keys of the uploaded dictionary of spacers. New spacers will be given a new Id number and an updated dictionary will be created. The data is provided in the form of a tabulated text file, which can be stored and used for further analysis of new alleles.
5. Choose "see the html version" to see dictionaries as tables of spacers and strains.
6. The "Re-annotate Spacers" tool allows easy assessment of tentative phylogenetic relationships between alleles. The graphical representation of ordered spacers at the top of the page shows the spacer organization.

4. Notes

1. These structures have too low complexity to be identified de novo in sequences. However they are detected by comparison using the CRISPRtionary tool (as in *Yersinia pestis* Nepal 516, Accession number: NC_008149, coordinates: 233155-2331643).
2. Default parameters (as described in Grissa et al. (7) and in the web server "page manual" link) used in the regular CRISPRfinder are the following: In a first step, a search for two consecutive DRs with a length of 23–55 bp and an internal

a

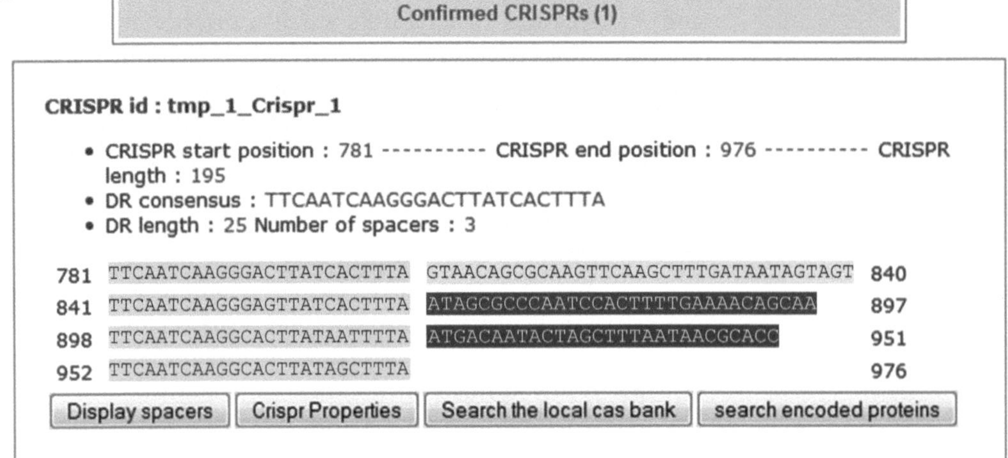

b

SYMNFKGSNIKVSGSSFKDDTDGGFSFSGNNNNSTISFNQTNFNQGTYHFSNSASSSF

DNSSFNQGSYHFNSAQSTFENSNFNQGTYNFNDNTSFNNDTFNQGTYSFNTSKVSFSG

Fig. 5. CRISPR-like sequence present in the Helpy-906 gene coding for cytotoxin VacA in *Helicobacter pylori*. (**a**) CRISPRfinder output using the all gene sequence. (**b**) Translation of the CRISPR-like region.

spacer size of 25–60 bp is performed. At this stage, one nucleotide mismatch is allowed between repeats. In the advanced version, these parameters are modifiable in the "Maximal repeat" section of the form. In a second step, the recovered sequences are clustered to build the final CRISPRs. At this stage, the spacer size should be from 0.6- to 2.5-fold the DR size. The default values of the allowed mismatch between DRs are 20% for the internal DRs and 33.3% for the degenerated one. The similarity between spacers should not exceed 60% in order to discard tandem repeats. These values are modifiable by the user in the "CRISPR properties" section of the advanced form.

3. Small CRISPRs harboring two or three DRs are classified as "Questionable." In such structures there are not enough DR copies to accurately define a consensus. Longer CRISPRs are annotated as "Confirmed." However a critical examination of the CRISPRfinder output is necessary because some structures may be misidentified as CRISPR if their characteristics fit all the CRISPRfinder program specifications. For example, portions of some genes consist of well-conserved sequences separated by more diverged sequences (see Fig. 5). A very low rate of mismatches between the internal DRs of a CRISPR is observed for most of the organisms analysed in the CRISPRdb.

4. A local database has been implemented that contains amino acid sequences of Cas proteins present in the Uniprot database (http://www.uniprot.org/). BLASTX searches against this database are made in order to find potential *cas* genes locations in DNA sequences. The presence of *cas*-like genes in cis highly validates CRISPRs-like structures but *cas* genes also act in trans in many cases.
5. CRISPR structures are not coding regions although they are made of gene fragments (the spacers) and repeats that may happen to produce an open reading frame. A putative CRISPR might in fact turn out to be part of a protein-coding sequence with repeated elements.
6. The sequence from which a spacer originated is called the protospacer. It is usually localized on a viral or plasmid genome although some spacers have been shown to originate from bacterial genes.
7. The CRISPR Id is made of the sequence NC number followed by a number corresponding to the CRISPR rank identified by the program in that sequence. Some CRISPR numbers may be missing in the database if they have been deleted by filters or during the curation process.
8. CRISPRdb stores the characteristics and location of CRISPRs identified de novo by the CRISPRfinder program in all the Bacteria and Archaea sequences (chromosome and associated plasmids) from the RefSeq database released at the ftp site of the NCBI (ftp://ftp.ncbi.nih.gov/genomes/Bacteria/). Several additional steps allow comparing candidates to CRISPRs already present in the database in order to convert some questionable structures into confirmed ones and to discard others. CRISPRdb is periodically updated and on this occasion each new CRISPR is critically observed. An administrator interface is dedicated to the sorting and to manual correction before final validation of the new version of the production database. A CRISPR harboring highly variable DRs is specifically examined and checked for the presence of long open reading frames. If a potential CRISPR corresponds to an ORF in Genbank, the CRISPR is deleted from the database. Indeed, some coding sequences possess exact repeats separated by sequences with a high degree of divergence. Similarly, the CRISPR is deleted when the spacer length is highly heterogenous, unless a set of *cas* genes is present in the vicinity.
9. Taxonomic information are those of PublicHouse, a publicly queryable relational databases (see http://biowarehouse.ai.sri.com/PublicHouseOverview.html for information). Genomes with missing information are not listed in the Taxonomy browser.

10. In the current version, 85% of archaea and 48% of bacteria possess one or several CRISPRs, and in 70% of all genomes possessing a CRISPR there is at least one *cas* gene referenced in the sequence annotations (chromosome or associated plasmids). When several CRISPRs are present in a single genome, a set of *cas* genes is generally clustered with at least one CRISPR. Up to three different CRISPR/Cas systems were observed in some genomes, such as in *Thermincola* sp.
11. The comparison program is performing alignments of the 200 bp regions flanking each CRISPR. The alignment is produced by the clustalW software (ktuple = 2, matrix = BLOSUM).
12. The dictionary must have three columns to be read by the program. The first column must contain numeric data and corresponds to the initial spacer keys that will appear in the "Spacers annotation" page at the right of each spacer. The second column, AnnotatedSpacer, must contain at least a string separated from a number by ":" (e.g., Sthermophilus:1). When the program is run, information about the strains that contain a given spacer and the position of that spacer in those strains are added in that field. The format of the resulting field will look like <strainA>:<spacer position in strainA>_<strainB>:spacer position in strain B>_<strainC> and so on. The third column contains the spacer sequences. When the re-annotate spacer program is launched, the spacer keys are changed to fit with the order of the spacers in the different strains CRISPRs.
13. The candidate DRs determined in each submitted sequence might differ at the last nucleotide, in particular when CRISPRs with a small number of motifs (a DR and a spacer) are analyzed. Therefore it is necessary to select a consensus DR, which is present in all the analyzed sequences. CRISPRtionary is particularly interesting to identify CRISPRs with a single spacer when one of the DR is degenerated. For example, the *Y. pestis* Nepal 516 strain possesses a CRISPR with one perfect DR and a truncated one separated by a unique spacer. That particular locus is allelic to a multi-DR locus in other *Y. pestis* strains. By imposing the use of the DR in CRISPRtionary it is possible to view the single spacer CRISPR.

References

1. Haft DH, Selengut J, Mongodin EF, Nelson KE (2005) A guild of 45 CRISPR-associated (Cas) protein families and multiple CRISPR/Cas subtypes exist in prokaryotic genomes. PLoS Comput Biol 1:e60
2. Horvath P, Barrangou R (2010) CRISPR/Cas, the immune system of bacteria and archaea. Science 327:167–170
3. Grissa I, Vergnaud G, Pourcel C (2007) The CRISPRdb database and tools to display

CRISPRs and to generate dictionaries of spacers and repeats. BMC Bioinformatics 8:172

4. Pourcel C, Salvignol G, Vergnaud G (2005) CRISPR elements in *Yersinia pestis* acquire new repeats by preferential uptake of bacteriophage DNA, and provide additional tools for evolutionary studies. Microbiology 151:653–663
5. Mojica FJ, Diez-Villasenor C, Garcia-Martinez J, Soria E (2005) Intervening sequences of regularly spaced prokaryotic repeats derive from foreign genetic elements. J Mol Evol 60:174–182
6. Lillestol RK, Redder P, Garrett RA, Brugger K (2006) A putative viral defence mechanism in archaeal cells. Archaea 2:59–72
7. Grissa I, Vergnaud G, Pourcel C (2007) CRISPRFinder: a web tool to identify clustered regularly interspaced short palindromic repeats. Nucleic Acids Res 35:W52–W57

Chapter 3

A Strategy for Identifying Noncoding RNAs Using Whole-Genome Tiling Arrays

Stephen G. Landt and Eduardo Abeliuk

Abstract

Whole-genome tiling arrays are powerful tools for detecting and characterizing novel RNA transcripts. Here, we describe a complete method combining elements of molecular and computational biology to identify small noncoding RNA (sRNA) transcripts. We focus on the key features of this approach, which include size-fractionation of input RNA, direct detection of array hybridization with antibodies that recognize RNA:DNA hybrids, and correlation-based computational methods for automated sRNA identification and boundary determination.

Key words: Small noncoding RNAs, High-density tiling arrays, Microarrays, Affymetrix, Transcription

1. Introduction

In recent years, a variety of strategies have been used to successfully identify prokaryotic small noncoding RNAs (sRNAs) (1). Here, we describe the use of whole-genome tiling microarrays for sRNA identification. The microarray approach has several specific advantages that recommend it: (1) it does not require the availability of a large number of sequenced genomes at various evolutionary distances, as do conservation-based approaches; (2) it is unbiased with respect to annotated genome features (promoter and terminators) which have been used for sRNA detection; and (3) in addition to being useful for sRNA detection, the arrays are generally useful for quantifying global gene expression and thus, when purchased at scale, can be cost-competitive with emerging sequencing technologies.

Kenneth C. Keiler (ed.), *Bacterial Regulatory RNA: Methods and Protocols*, Methods in Molecular Biology, vol. 905, DOI 10.1007/978-1-61779-949-5_3, © Springer Science+Business Media, LLC 2012

Several elements are important for a successful array-based sRNA annotation experiment and are detailed in the following protocol and notes. First, because virtually all known prokaryotic sRNAs are less than 500 nucleotides in length (2), it is advantageous to size-fractionate the input RNA population in order to eliminate the interfering signal resulting from rRNA and mRNA populations. Second, detection sensitivity and RNA boundary resolution are greatly affected by the array tiling density. Previous sRNA identification efforts have used arrays with tiling densities as high as every 5 nucleotides in intergenic regions. As array densities have improved, extremely high densities with high coverage of noncoding regions are becoming reasonable and the analysis of antisense transcription within coding regions has become possible. Third, hybridizing RNA directly to the array with detection by an antibody that recognizes RNA:DNA hybrids has been demonstrated to substantially improve detection sensitivity when compared to hybridizing cDNA generated from RNA (3, 4). Finally, optimal analytical methods for sRNA discovery are important. We present here a method for using signal correlation over multiple array hybridizations as an effective filter for sRNA identification and boundary determination.

2. Materials

2.1. RNA Extraction Components

1. RNase-free water (DEPC-treated or commercial). All solutions and reactions should be made with this grade of water.
2. Total RNA (see Notes 1 and 2).
3. TBE buffer: For 1 L of a 10× stock, mix 108 g of Tris base, 55 g of boric acid, 9.3 g of sodium-EDTA and bring volume to 1 L with water.
4. 5% PAGE 1× TBE gels: For 50 mL, mix 25 g of urea, 7.5 mL of 40% acrylamide/bisacrylamide (19:1), 5 mL of 10× TBE, 17.5 mL of water. Microwave to dissolve urea and let cool until it is warm to the touch. Add 250 μL of 10% ammonium persulfate and 25 μL of TEMED. Mix and pour immediately.
5. RNA size standards that include products in the 50–500 nucleotide range (such as the Fermentas Low Range RNA ladder).
6. Ethidium bromide.
7. Electroelution apparatus (for instance, the Elutrap system by Schliecher and Schuell).
8. 10× RNA fragmentation buffer: 500 mM Tris (pH 7.9), 1 M NaCl, 100 mM $MgCl_2$.
9. Ethanol.

10. Glycogen.
11. 3 M Sodium acetate, pH 5.2.

2.2. Array Hybridization Components

1. High-density tiling array (Affymetrix or other manufacturer).
2. 1× Hybridization buffer: 100 mM MES pH 6.6, 1 M NaCl, 20 mM EDTA, 0.01% Tween 20, 0.1 mg/mL herring sperm DNA, 0.5 mg/mL BSA, 50 pM Affymetrix control oligo B2.
3. Antibody recognizing RNA:DNA hybrids (4).
4. 1× Staining buffer: 100 mM MES, pH 6.6, 1 M NaCl, 0.05% Tween 20, 2 mg/mL BSA.
5. Biotin-labeled rabbit anti-mouse IgG.
6. Rabbit IgG (Sigma, St Louis, MO).
7. R-Phycoerythrin Streptavidin (Molecular Probes).
8. Wash buffer A: 6× SSPE (900 mM NaCl, 60 mM NaH_2PO_4, 6 mM EDTA, pH 7.4), 0.01% Tween 20.
9. Wash buffer B: 100 mM MES pH 6.6, 1 M NaCl, 0.01% Tween 20.
10. Array processing and detection equipment (usually through core facility).

2.3. Software Tools

1. Affymetrix GCOS software.
2. R/Bioconductor software.

3. Methods

3.1. RNA Isolation

1. Pour a 1 mm thick 5% denaturing PAGE (19:1) TBE gel. Pre-run the gel until it reaches a temperature of ~50–55°C. Use a syringe to thoroughly rinse the wells immediately prior to loading RNA samples.
2. Combine RNA with an equal volume of deionized formamide and denature for 5 min at 85°C (see Note 3). Similarly denature 1 μg of RNA size standards. At least one lane must contain the bromophenol blue dye to a final concentration of 0.006% (this can be included in the loading dye for the RNA samples or loaded in separate lanes).
3. Run the gel until the dye has migrated ~2/3 the length of the gel.
4. Cut the lanes containing RNA markers and transfer them to a glass dish containing 1 μg/mL ethidium bromide (SYBR-green or other appropriate stains may also be used).

5. Stain for 10 min followed by 10 min of destaining in deionized water.
6. Visualize the marker lanes by UV transillumination and mark the location of the maximum size RNA to be purified.
7. Realign stained marker lanes with the rest of the gel and use the marker bands and the bromophenol blue band to identify the desired region of the gel. Use a clean scalpel to cut the desired size range from each lane (see Note 4).
8. Isolate the RNA from the gel fragments by electroelution. If using the Elutrap, electroelution is done in prechilled 1× TBE buffer for 6 h at 200 V at 4°C. Buffer should be added to the sample chamber to cover the gel pieces. The elution chamber should contain 800 μL 1× TBE. After elution, the polarity of the apparatus is reversed for 20 s to dislodge eluted materials from the boundary membranes (see Note 5).
9. Precipitate the eluted RNA in eppendorf tubes by adding 0.1 volume 3 M sodium acetate (pH 5.2) and 2.5 volumes ethanol.
10. Mix thoroughly and store at least 1 h at −80°C.
11. Centrifuge at maximum speed in a tabletop centrifuge.
12. Wash pellet once with 70% ethanol.
13. Allow to air dry for 10 min and resuspend in 50–100 μL of RNase-free water.
14. Fragment size-fractionated RNA (at least 2 μg) in a volume of 100 μL by incubation for 30 min at 95°C in 1× RNA fragmentation buffer.
15. Ethanol precipitate the RNA as in steps 10–14 and resuspend in 50 μL of RNase-free water (see Note 6).
16. Determine RNA concentration by standard spectrophotometric or fluorescent dye-binding measurements.

3.2. Array Hybridization

1. Pre-warm microarrays to room temperature.
2. For each array, suspend 2 μg of RNA in 200 μL of 1× Hybridization buffer, heat to 99°C for 5 min and then 50°C for 5 min. Add to the array and hybridize at 50°C for 16 h (see Note 7).
3. Wash 10 cycles, 25°C, with 2 mixes/cycle in Wash buffer A.
4. Wash 4 cycles, 50°C, with 15 mixes/cycle in Wash buffer B.
5. Add primary antibody recognizing RNA:DNA hybrids to the array at 0.02 mg/mL in 1× staining buffer. Staining is for 60 min at 25°C (see Note 8).
6. Wash 10 cycles, 25°C, with 4 mixes/cycle in Wash buffer A.

7. Add the biotin-labeled rabbit anti-mouse secondary antibody at a concentration of 0.02 mg/mL along with 0.4 mg/mL rabbit IgG in 1× staining buffer.
8. Wash 10 cycles, 25°C, with 4 mixes/cycle in Wash buffer A.
9. Add the R-Phycoerythrin Streptavidin at a concentration of 10 μg/mL in 1× staining buffer.
10. Wash 15 cycles, 30°C, with 4 mixes/cycle in Wash buffer A.
11. Hold at 25°C prior to scanning.

3.3. Data Preprocessing

1. Remove Individual probe signals that are non-uniformly distributed across the spot ("feature") on the microarray using the Affymetrix GCOS software. These probes are flagged as part of the standard output from GCOS (see Note 9).
2. To measure the RNA signal at each intergenic position, calculate a signal given by the PM-MM* where PM and MM are the corresponding "perfect match" and "mismatch" probes associated with the intergenic position. MM* is an ideal mismatch value that guarantees that PM > MM (see Note 10).
3. Remove microarray background noise using a method such as the MAS 5.0 background adjustment method to remove intra-chip noise as implemented in the Bioconductor package *affy* (5) (http://www.bioconductor.org/packages/2.0/bioc/html/affy.html). It is recommended that no inter-chip normalization be performed (see Note 11).
4. To ensure uniform sampling of probes across intergenic regions (or other regions of interest), the number of probes is either upsampled or downsampled and their corresponding values are interpolated in order to guarantee a uniform probe density. Generally, this is determined by the array tiling density (see Note 12).

3.4. Statistical Scoring and Identification of Putative sRNAs

1. For each intergenic region, compute the matrix of Pearson correlation coefficients between each pair of probe signal vectors (see Fig. 1 and Note 13).
2. For every possible set of consecutive probes within an intergenic region, count the number of probe pairs with a correlation above $t = 0.5$ (see Note 14).
3. Compute p-values for all possible combinations of consecutive probes (defining a genomic segment of reasonably minimum size, see Note 15) and rank all these segments according to the p-values. The highest-ranking segment (with corresponding boundaries, see Note 16) in each intergenic region should be considered for further analysis (see Notes 17 and 18).
4. Identify differentially expressed sRNAs by first normalizing all microarrays to the same mean expression value (see Note 19).

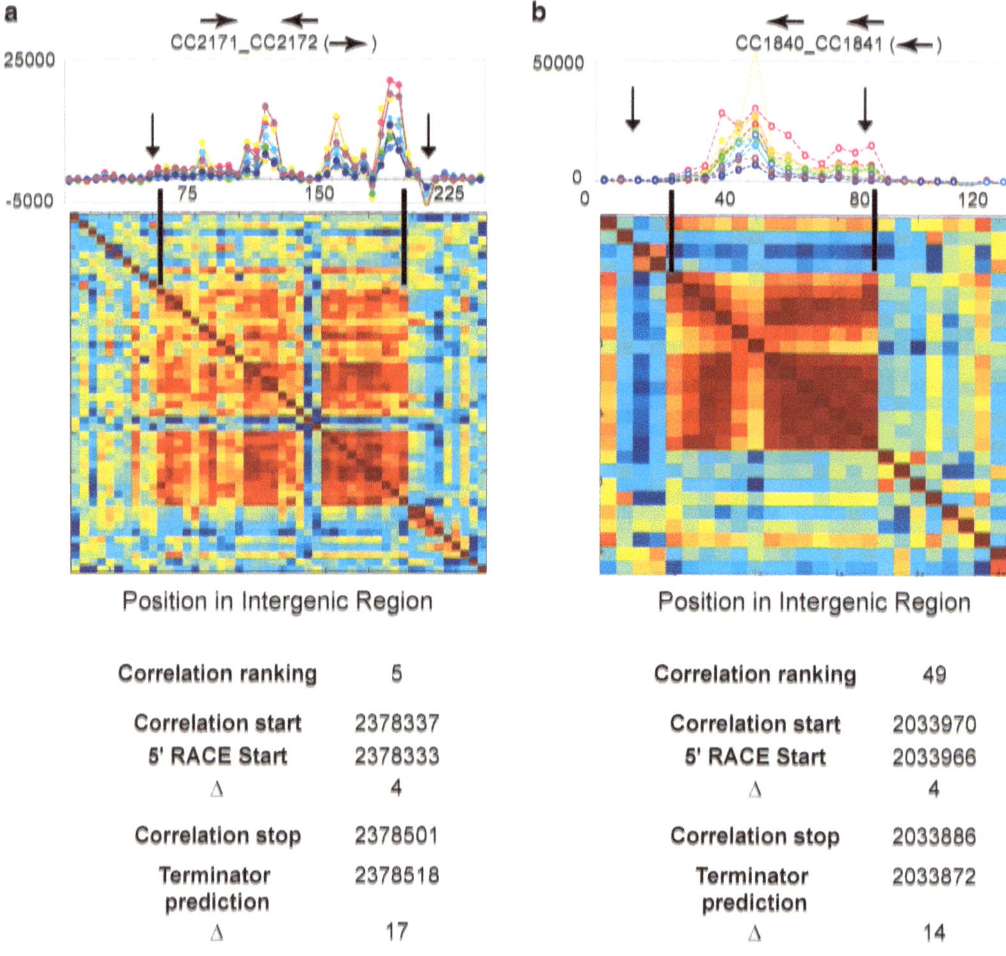

Fig. 1. Novel small noncoding RNAs (sRNA) transcripts identified using probe correlation analysis of whole-genome tiling microarray data. CauloHI1 whole-genome tiling arrays were used to measure levels of RNA prepared as described in the text from the *Caulobacter crescentus* genome at 12 different time points following cell-cycle synchronization. Analysis of the intergenic regions between genes CC2171 and CC2172 and genes CC1840 and CC1841, each of which was shown to contain a previously unannotated sRNA (6), are shown in (**a**) and (**b**). The *upper plot* in each panel shows raw (unnormalized) levels of RNA across the intergenic region, with each color representing a different point in the time series. The *x*-axis gives the location in the intergenic region relative to the end of the annotated upstream ORF. The *lower plot* in each panel shows Pearson correlation values between all probes in the intergenic region, calculated using un-normalized data. *Red* denotes high positive correlation and *blue* denotes high negative correlation. *Vertical black bars* indicate boundaries identified by the correlation analysis for the most significant transcript predicted in the intergenic region. *Arrows* indicate independently confirmed sRNA boundaries. Values at the *bottom* of each panel compare sRNA boundary estimates from the probe correlation analysis with those from 5′ RACE experiments or transcriptional terminator predictions. *Numbers* indicate positions in the *C. crescentus* CB15N genome.

Using these normalized values and the estimated sRNA boundaries, calculate expression values for each sRNA at each time point/experimental condition by averaging all the probe signals that target the corresponding sRNA. Analyze these values for significant variation over the course of the experiment.

5. It is recommended that expression from each intergenic region be visually inspected (see Note 20).

3.5. Additional Bioinformatics

1. This approach will identify many small RNA species that are not derived from independent, noncoding transcription units (6). To better identify genuine sRNAs, it is useful to filter other classes of RNA identified by this approach. These include:
   - Short mRNAs or degradation products comprised of mRNA 5′ and 3′ untranslated regions (UTRs). These can often be identified by continuous microarray signal leading up to the boundary of the adjacent open reading frame (ORF) or the presence of predicted transcriptional regulatory elements (transcription factor binding sites or transcriptional terminators) that link the expressed region and the adjacent ORF. An empirical separation threshold (for instance 50 nucleotides) may be applied to all putative sRNAs that are on the same strand as annotated ORFs to remove likely UTR elements.
   - Transcribed regulatory regions (such as riboswitches). These are often highly conserved and often may be identified by searching databases such as RFam (7) (http://rfam.janelia.org/).
   - Unannotated ORFs. These are identifiable by the presence of discrete coding potential within the transcript and may be found in the updated outputs from ORF finding algorithms such as Glimmer (8) (http://www.cbcb.umd.edu/software/glimmer/) and GeneMark.hmm (9) (http://opal.biology.gatech.edu/GeneMark/prokaryotes_database/index.cgi).
2. To validate predicted sRNAs, it is useful to search for properly oriented transcriptional regulatory elements that suggest the existence of independent transcription. These include rho-independent terminators (identified by TransTerm (10) (http://transterm.cbcb.umd.edu/cgi-bin/transterm/predictions.pl)) located at the extreme 3′-end of the putative sRNA and known transcription factor binding motifs, upstream of the putative sRNA (e.g., see ref. (6)). Identification of these elements provides additional confidence that a transcript is transcribed independently of adjacent ORFs.

4. Notes

1. We generally isolate total RNA using the Trizol reagent (Invitrogen) according to the manufacturer's instructions. We normally recover ~33 μg of total RNA/optical density unit of cells. To recover the 2 μg of size-fractionated, fragmented

RNA desired for array hybridization, an average of ~35 μg of total RNA was required. Ideally, ~80 μg of total RNA will be available to provide adequate material for additional array hybridizations, should they be required.

2. A minimum of four to five samples is recommended to provide enough data points to use for later correlation analyses.
3. We load up to 40 μg of RNA in each lane of the gel.
4. In previous sRNA identification efforts, we have isolated RNA ranging from ~35 to 500 nucleotides (6). As the bromophenol blue migrates similarly to a 35 base oligonucleotide in a 5% denaturing PAGE gel, the location of this dye can be used to determine the lower cut point.
5. We recover ~3–10% of the RNA originally loaded on the gel in this fraction.
6. On average, we recover ~75% of input RNA from the fragmentation step.
7. Our microarray-based sRNA discovery work has been performed using the custom designed Affymetrix CauloHI1 microarray (11), http://www.stanford.edu/group/caulobacter/CauloHI1. For all hybridization, washing, and detection steps (except incubations with the primary antibody recognizing RNA:DNA hybrids and the biotinylated secondary antibody), reagents and conditions were as described in the Affymetrix GeneChip Expression Analysis Technical Manual (http://www.affymetrix.com, P/N 702232). These two steps replace the first two staining steps (streptavidin staining followed by biotinylated anti-streptavidin antibody staining) in the standard protocol. The 50°C hybridization temperature was chosen during the array design to account for the high GC-content of the *Caulobacter crescentus* genome and should be adjusted for the array to be used according to the design. All array-handling steps are generally performed at a core facility using Affymetrix instrumentation. If other microarray platforms are used for detection, it should be possible to adapt hybridization and wash conditions and reagents from standard protocols, but this should be determined in consultation with the manufacturer.
8. The antibody against RNA:DNA hybrids used in our original protocol (6) was obtained from the Digene Corporation. This antibody is no longer available. An alternate protocol, with publicly available antibodies, has recently been published by Hu et al. (4). We recommend the use of one of these antibodies and the staining conditions they use.
9. We observed this phenomenon more with antibody based detection of RNA hybridization as compared to hybridization of labeled cDNA, perhaps due to localized antibody precipitation

on the arrays. It is important to exclude these outlier values from further analysis.

10. The MAS 5.0 method for processing microarrays relies on PM and MM probe pairs. However, recent methods that rely solely on PM probes, such as the RMA method, can outperform MAS 5.0 (12) and therefore mismatch probes can be ignored if this package is used in the analysis.
11. Correlation analysis of this unnormalized data takes advantage of signal variation caused by variation in sample preparation (experimental variation) to identify sRNAs, even if there is no biological variation in expression of an sRNA over the conditions sampled in the experiment (as might be expected for the majority of sRNAs). A formal analysis of error models that include both experimental and biological variations present in microarray signals can be found in (13, 14).
12. Uniform probe density is an assumption of the statistical scoring method to be performed. On the CauloHI1 array, intergenic probes were generally spaced every 5 nucleotides (11), http://www.stanford.edu/group/caulobacter/CauloHI1. Where probes were missing (often because they contained repeated sequence), or tiling density was higher, interpolation was used to maintain 5 nucleotide spacing between probes.
13. The underlying assumption is that signals for all probes measuring a sRNA will be highly correlated, either due to experimental variation or due to biological variation in sRNA expression.
14. Given the null hypothesis that a set of consecutive probes are all uncorrelated, the probability of counting k pair of probes with correlation above t is given by a Binomial distribution $B(N,q)$, where N is the total number of probe pairs in the set and q is the probability of an uncorrelated pair having a correlation above t. The parameter q can be estimated by computing $P(\text{correlation} > t)$ from a background distribution of correlation coefficients. In practice, this background distribution is generated from a distribution of correlation coefficients derived by correlating intergenic probes targeting distant locations within a chromosome.
15. A minimal number of probes required to suggest a sRNA should be set (for instance, at least three probes that overlap less than 50% with each other). Where we have used the CauloHI1 array with 5 nucleotide tiling spacing, we required at least eight consecutive highly correlated probes corresponding to an sRNA of at least ~60 nucleotides.
16. Boundaries are simply the endpoints of the regions with the most significant p-values. Since the intergenic probes are spaced every 5 base pairs, there is an inherent error of ±5 base pairs in

the estimate of the boundary positions within intergenic regions.

17. To simplify the calculation, one can consider only the highest score within each intergenic region. However, a consideration of non-overlapping regions can identify additional independent transcripts in a region (although this has not been widely observed) or alternate boundaries for transcripts (perhaps indicating alternately processed transcripts).
18. As the majority of intergenic regions are unlikely to contain sRNAs, most expressed sRNAs should rank highly on this list. In our identification of *C. crescentus* sRNAs, ~70% of sRNAs identified in regularly tiled regions ranked in the top 200 of the correlation rankings, while most of the rest were poorly identified by correlation (6) (see Note 19).
19. A number of methods exist for this, ranging from simple normalization to a common mean to more rigorous methods. The RMA method for normalizing multiple arrays is commonly used for this procedure (15).
20. We observed that some highly expressed sRNAs were poorly detected by this method and we speculate that hybridization signal is saturated for these sRNAs. Thus, information about experimental variation is lost. We therefore recommend visual inspection of the signal for all intergenic regions to identify obvious sRNAs that were missed in the correlation analysis. Note: In our analysis using the CauloHI1 microarray, ~20% of identified sRNAs fell into this category (6).

Acknowledgements

This work was supported by NIH grants GM51426 and GM32506, by DOE grants DE-FG03ER63219-A001 and DE-FG02ER63219, and also supported by the Stanford NIH Genome Training Grant (SGL).

References

1. Sharma CM, Vogel J (2009) Experimental approaches for the discovery and characterization of regulatory sRNA. Curr Opin Microbiol 12:536–546
2. Waters LS, Storz G (2009) Regulatory RNAs in bacteria. Cell 136:615–628
3. Zhang A, Wassarman KM, Rosenow C, Tjaden BC, Storz G, Gottesman S (2003) Global analysis of small RNA and mRNA targets of Hfq. Mol Microbiol 50:1111–1124
4. Hu Z, Zhang A, Storz G, Gottesman S, Leppla SH (2006) An antibody-based microarray assay for small RNA detection. Nucleic Acids Res 34:e52
5. Gautier L, Cope L, Bolstad BM, Irizarry RA (2004) affy—analysis of Affymetrix GeneChip

data at the probe level. Bioinformatics 20:307–315

6. Landt SG, Abeliuk E, McGrath PT, Lesley JA, McAdams HH, Shapiro L (2008) Small non-coding RNAs in *Caulobacter crescentus*. Mol Microbiol 68:600–614
7. Griffiths-Jones S, Moxon S, Marshall M, Khanna A, Eddy SR, Bateman A (2005) Rfam: annotating non-coding RNAs in complete genomes. Nucleic Acids Res 33:D121–D124
8. Delcher AL, Bratke KA, Powers EC, Salzberg SL (2007) Identifying bacterial genes and endo-symbiont DNA with Glimmer. Bioinformatics 23:673–679
9. Lukashin A, Borodovsky M (1998) GeneMark. hmm: new solutions for gene finding. Nucleic Acids Res 26:1107–1115
10. Kingsford K, Ayanbule K, Salzberg SL (2007) Rapid, accurate, computational discovery of Rho-independent transcription terminators illuminates their relationship to DNA uptake. Genome Biol 8:R22
11. McGrath PT, Lee H, Zhang L, Iniesta AA, Hottes AK, Tan MH et al (2007) High-throughput identification of transcription start sites, conserved promoter motifs and predicted regulons. Nat Biotechnol 25:584–592
12. Irizarry RA, Bolstad BM, Collin F, Cope LM, Hobbs B, Speed TP (2003) Summaries of Affymetrix GeneChip probe level data. Nucleic Acids Res 31:e15
13. Ideker T, Thorsson V, Siegel AF, Hood LE (2001) Testing for differentially expressed genes by maximum-likelihood analysis of microarray data. J Comput Biol 7:805–818
14. Li C, Wong WH (2001) Model-based analysis of oligonucleotide arrays: expression index computation and outlier detection. Proc Natl Acad Sci U S A 98:31–36
15. Bolstad BM, Irizarry RA, Astrand M, Speed TP (2003) A comparison of normalization methods for high density oligonucleotide array data based on variance and bias. Bioinformatics 19:185–193

Chapter 4

Genetic Screens to Identify Bacterial sRNA Regulators

Pierre Mandin

Abstract

Small regulatory RNAs (sRNAs) are versatile regulators that have been shown to be involved in the gene regulation of a growing number of biological pathways in bacteria. While finding the targets of a given sRNA has been the focus of many studies, fewer methods have been described to uncover which, if any, sRNAs regulate a given gene. Here I present two genetic screens that are designed to search for sRNAs regulating a gene of interest. Before the screens are performed, a translational fusion is made between the gene of interest and *lacZ*, designed so that mostly post-transcriptional effects on the gene's expression can be analyzed. I describe here a simple and rapid way to obtain this fusion, even when the transcriptional start site is unknown, by combining PCR or 5′RACE with recombination in the chromosome of a special strain of *Escherichia coli*. The first genetic screen uses a genomic multicopy library to find regulator genes that, when overexpressed, affect the expression of the fusion. While this technique is a classical genetic screen, particular attention is paid to how it can be used to specifically find sRNAs. A second screen is described that takes advantage of a specific library of sRNAs of *E. coli* that provides an easier and more rapid way to look for sRNA regulation. The library is transformed into the fusion containing strain using a serial transformation protocol developed in microtiter plates. The transformants can then be directly assayed for effects on the beta-galactosidase activity of the fusion in liquid, providing a precise and rapid way to evaluate sRNA regulation. Use of one or both of these screens should help uncover new pathways of regulation by sRNAs.

Key words: Small RNAs, noncoding RNAs, Regulatory RNAs, Hfq, Genetic screen, Recombineering, Beta-galactosidase assays

1. Introduction

In recent years, noncoding regulatory RNAs have emerged as major regulators in a wide variety of physiological processes (1). To date, slightly more than a 100 of these regulatory RNAs have been identified in *Escherichia coli*. In this chapter, I will use the term "sRNA" to refer to a particular class of bacterial regulatory RNAs that act by base pairing to mRNA targets encoded in trans in the genome (reviewed in refs. (1–3)). In Gram-negative

Kenneth C. Keiler (ed.), *Bacterial Regulatory RNA: Methods and Protocols*, Methods in Molecular Biology, vol. 905, DOI 10.1007/978-1-61779-949-5_4, © Springer Science+Business Media, LLC 2012

bacteria, and in particular in *E. coli*, most of these small regulatory RNAs (sRNAs) have been shown to require the RNA chaperone protein Hfq for their action (4). By binding to the sRNA, it is proposed that Hfq protects the sRNA from RNase degradation and helps it find and base pair to the mRNA targets. The mRNA targets are also bound by Hfq. Consequently, immunoprecipitation of Hfq has proven to be an invaluable tool to find not only sRNAs, but also their target mRNAs (5, 6). To date, about 35 sRNAs have been shown to bind Hfq in *E. coli* (5, 7, 8). However, the number of mRNA targets regulated by sRNAs is probably much higher as many, if not all, sRNAs have multiple targets (9). Given the variety of pathways employing sRNAs and the growing number of genes regulated by sRNAs, designing genetic screens that identify sRNA regulators of a given gene is of obvious importance.

The first question one may ask is "is my favourite gene subject to sRNA regulation"? A variety of characteristics may indicate sRNA regulation of a given gene. For example, genes that have a well-conserved sequence or RNA structure in the 5′ untranslated region are likely subject to some sort of post-transcriptional regulation, possibly by sRNAs. Genes that are known to be regulated at the post-transcriptional level, but have no obvious regulatory protein, are also candidates for sRNA regulation. The chances are high that an sRNA is involved in regulating genes that are differentially expressed in an *hfq* mutant as compared to a wild-type strain. In these cases, the use of a reporter fusion like those described below may be necessary to determine if the sRNA is affecting gene transcription (presumably indirectly) or acting post-transcriptionally. Finally, genes for which the mRNA is co-immunoprecipitated with Hfq are highly likely to be directly regulated by an sRNA. While none of these types of evidence directly prove that sRNA regulation exists, they provide strong motivation for an sRNA genetic screen.

The screens described here are in their essence classical genetic screens that use differential expression of a fusion of the gene of interest to a reporter gene, but they have been adapted specifically towards the identification of sRNAs (10, 11). In both screens, the reporter construct is stably integrated into the chromosome and the regulatory genes to be screened are expressed from multicopy plasmids.

In a first step, a translational fusion is constructed between the gene of interest and the beta-galactosidase reporter gene *lacZ*. In this construct, the endogenous promoter is replaced by an inducible one, so that no promoter-related transcriptional control should be seen. Similarly, the translational fusion contains limited 3′ sequence to avoid undesired post-translational regulation related to stability of the fusion protein (see Note 1). Thus, the construct should report predominantly on post-transcriptional regulatory events, such as the control exerted by sRNAs. The method described in Subheading 3.1 details construction of the fusion in a particular strain of *E. coli*, named PM1205, that allows the rapid cloning of the gene of interest starting from its own transcriptional start site, even when unknown, by combining 5′RACE with gene recombineering (see Fig. 1) (10, 12, 13). First, a PCR product corresponding to the 5′ region of the gene, from the transcriptional start site to the 10th–20th codon, is produced by 5′RACE or PCR (Fig. 1a, b, respectively). This product is then inserted into the chromosome of the strain PM1205 by homologous recombination (Fig. 1c). Recombinants are selected both for loss of the *sacB* gene, which is toxic on minimal sucrose plates, and for obtention of the *lacZ* fusion using X-Gal. The resulting fusion

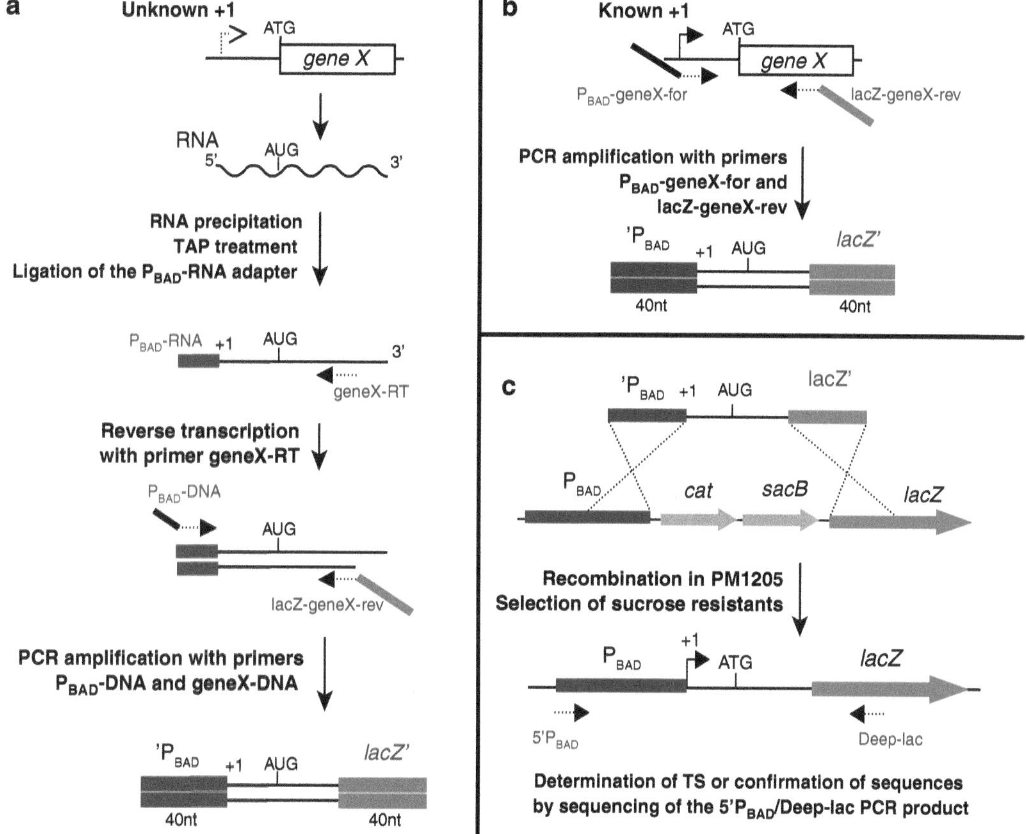

Fig. 1. Schematic representation of the translational fusion construction protocol. (**a**) When the transcriptional start site (+1) is unknown, a 5′RACE with modified oligonucleotides is performed to obtain the desired PCR fragment. RNA is extracted from the bacteria and RNA 5′ triphosphates are converted to 5′ monophosphates using tobacco acid pyrophosphatase (TAP). The P_{BAD}-RNA adapter is then ligated at the 5′ end of the mRNA. The RNA is then reverse-transcribed using the geneX-RT oligonucleotide. The resulting cDNA is amplified by PCR using DNA primers P_{BAD}-DNA and lacZ-geneX-rev. (**b**) When the transcriptional start site (+1) is known, the PCR product is obtained in one step by amplifying the gene with oligonucleotides P_{BAD}-geneX-for and lacZ-geneX-rev. (**c**) The PCR product obtained in (**a**) or (**b**) is inserted into the chromosome of the strain PM1205 by mini lambda-mediated recombination. Recombinants are selected for sucrose resistance and obtention of the *lacZ* fusion on minimal sucrose plates containing X-Gal. The resulting clones are sequenced using primers 5′P_{BAD} and Deep-lac.

is under the control of an arabinose-inducible P_{BAD} promoter that allows the control of the basal activity of the fusion as a function of arabinose concentration. This method not only has the advantage of being rapid and easy but also provides a strain that is optimized for the subsequent genetic screens (see Note 2).

Two types of screening methods are described using the resulting reporter strain. The first method, described in Subheading 3.2, makes use of a multicopy plasmid library to identify genes that, when overexpressed, will affect expression of the reporter at the post-transcriptional level (10). Basically, multicopy plasmids that contain fragments of the genome are transformed into the reporter strain and the overexpression phenotypes determined on MacConkey lactose plates. Clones that give a distinguishable phenotype as compared to the empty vector are then sequenced to identify which genes are responsible for the observed effects. One major advantage of this screen is that it makes no assumptions as to which regulators will be found, except that they must act post-transcriptionnally. Thus, regulatory proteins and indirect regulators (for instance, genes that turn on or off sRNAs) can be identified, in addition to sRNAs. Another advantage is that the screen can identify sRNAs that have not been found previously. However, setting up the screen can be tedious and analyzing the results can be time consuming. For example, multiple genes are often found on plasmids that give obvious phenotypes, and sub-cloning may be required to identify which gene is responsible for the observed effect. On the other hand, when a known or predicted sRNA is contained on the plasmid, the search can be relatively straightforward (10).

A drawback of genomic libraries comes from the fact that they can miss sRNAs for various reasons: the sRNA may be toxic when overexpressed, the sRNA may be under tight auto-repression, or the sRNA may simply be missing from the library because of cloning issues. It was partly to avoid this drawback that a library containing the Hfq-binding sRNAs from *E. coli* was constructed (11). The library is comprised of 26 plasmids, each carrying one of the *E. coli* sRNAs that binds Hfq, placed under the control of an inducible promoter (see ref. (11) for a list of the sRNAs). While not all the Hfq-binding sRNAs are yet cloned in this library, it is continuously being updated and available upon request. The approach using this sRNA library, which is described in Subheading 3.3, can greatly facilitate and enhance the screening. In a first step, the plasmid library, as well as the control empty plasmid, is transformed into the fusion-containing reporter strain. In addition to the classical observation of phenotypes on colour indicator plates, I describe a simple and efficient method to precisely evaluate the effects of the plasmids on beta-galactosidase activity from the reporter in microtiter plates (Fig. 2). This method

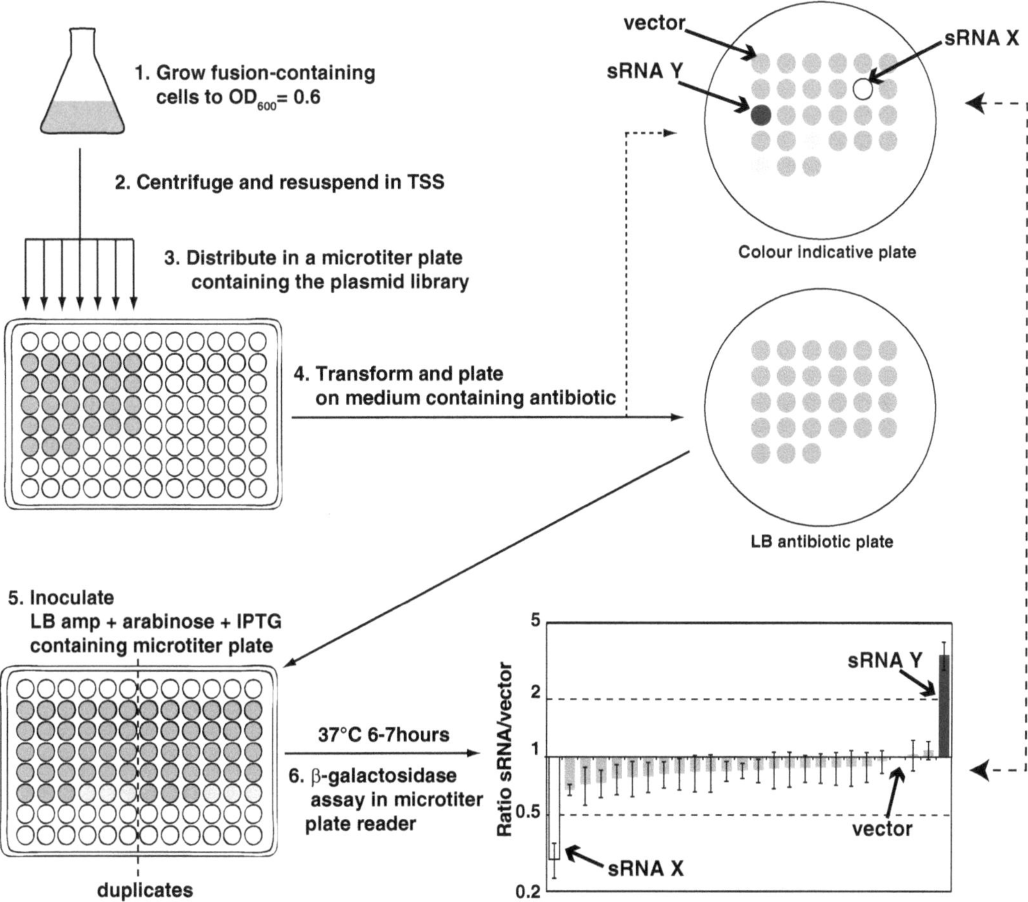

Fig. 2. Screening with the small regulatory RNAs (sRNA) library. The library plasmids are transformed into the reporter strain (see Subheading 3.3, steps 1–4), and the transformants are plated on LB plates and on colour indicator plates, such as MacConkey lactose. In this example, sRNA X represses expression of the translational fusion, resulting in a darker colony on the colour indicator plate, and the sRNA Y activates expression of the translational fusion, resulting in a lighter colony. The effect of overexpressing the sRNAs on the fusion can then be determined directly using beta-galactosidase assays in a microtiter plate reader (see Subheading 3.3, steps 5 and 6). Results are plotted as the ratio of specific activities obtained from the sRNA and the vector-only control (graph at *bottom right*). Thresholds of 0.5 and 2 are set for repression and activation by the sRNAs, respectively.

not only has the advantage of being very rapid and simple as compared to the overexpression screen, but it can also give important information on sRNA "competitiveness" for regulation (see Subheading 3.3, interpretation of the data).

Unfortunately such a short chapter on genetic screening cannot be exhaustive. A number of variations on the screens described herein could be used for particular observations. In particular, reporter genes other than *lacZ* may be tested; for example one could make use of a fluorescent reporter, such as GFP, to monitor gene expression. Use of such reporters has already proven to be successful to demonstrate sRNA regulation (14).

2. Materials

2.1. Translational Fusion Construction

1. *E. coli* strain PM1205: MG1655 *mal::lacI^q ΔaraBAD araC+ lacI′::P_{BAD}-cat-sacB-lacZ, mini λ tet^R*. The *P_{BAD}-cat-sacB* cassette is inserted at the *lac* site of this strain for the construction of translational fusions through homologous recombination and counter-selection on minimal sucrose plates (see Fig. 1c). The mini lambda allows homologous recombination when induced (13). This strain should always be grown at temperatures below 32°C to avoid induction and loss of the mini lambda. To ensure presence of the mini lambda, the strain can be checked for tetracycline resistance on LB-tet plates at 30°C. In addition, the strain contains a *lacI^q* allele to allow better control from P_{lac} promoters, such as the one driving expression of the sRNAs in the library, and is deleted for *araBAD* so that arabinose is used only as an inducer and not as a carbon source. Strain should be stored at −80°C upon arrival (see Note 3).
2. Oligonucleotides:
   (a) When the transcriptional start site (+1) from which transcription from the P_{BAD} promoter should start is known or can be arbitrarily determined (see Fig. 1b), obtain the following DNA oligonucleotides: "P_{BAD}-geneX-for" (5′-ACCTGACGCTTTTTATCGCAACTCTCT ACTGTTTCTCCAT-N(20)-3′), where N(20) represents the 5′ end sequence of the gene to amplify, the first base after the N being the transcriptional start site; "lacZ-geneX-rev" (5′-TAACGCCAGGGTTTTCCCAGTCACG ACGTTGTAAAACGAC-N(20)-3′), where N(20) represents the reverse complement 20 nucleotides in the 3′ end of the sequence of the gene to fuse to *lacZ*. This sequence should be in the ORF of the gene to clone. Make sure that the first N is also the third nucleotide of a codon of the protein to ensure that the fusion will be in frame with *lacZ* (see Note 1).

   (b) When the transcriptional start site is unknown, a 5′RACE is done with the following oligonucleotides (see Fig. 1a): RNA adapter oligonucleotide "P_{BAD}-RNA" (5′-ACUCU CUACUGUUUCUCCAU-3′); gene-specific DNA oligonucleotide "geneX-RT" for reverse transcription (5′-N(20)-3′); "lacZ-geneX-rev" as described in item 2a above—the annealing part of the sequence of this oligo should be downstream and overlapping the "geneX-RT" oligo (see Fig. 1a); 5′-adapter-specific DNA "P_{BAD}-DNA" for PCR amplification (5′-ACCTGACGCTTTT TATCGCAACTCTCTACTGTTTCTCCAT-3′).

   (c) DNA oligonucleotides "5′P_{BAD}" (5′-GCGCTTCAGCCA TACTTTTCATAC-3′) and "Deep-lac" (5′-CGGGCCTCT TCGCTA-3′) for colony PCR and sequencing (see Fig. 1c).

3. PCR and DNA analysis materials: DNA polymerase and supplied buffer, dNTPs, agarose, electrophoresis apparatus.
4. Electroporation materials: Electroporation cuvettes and electroporation apparatus such as the Micro Pulser (Biorad).
5. LB broth: To 1 L water, add 10 g bacto-tryptone, 5 g yeast extract, 10 g NaCl. Adjust pH to 7.5 with NaOH and sterilize by autoclaving.
6. LB agar: Add 15 g of agar to 1 L LB broth before autoclaving.
7. LB-cm: LB broth with 10 μg/mL chloramphenicol. Add antibiotic when medium is cooled after autoclaving.
8. LB-amp: LB broth with 100 μg/mL ampicillin. Add antibiotic when medium is cooled after autoclaving.
9. LB-tet: LB broth with 10 μg/mL tetracycline. Add antibiotic when medium is cooled after autoclaving.
10. 5× M63 (stock): Mix 10 g $(NH_4)_2SO_4$, 68 g KH_2PO_4, 2.5 mg $FeSO_4{\cdot}7H_2O$ and add water to 1 L. Adjust to pH 7 with KOH. Autoclave.
11. M63-sucrose-X-Gal plates: To 200 mL of 5× M63, add 50 g sucrose, 10 mL 20% glycerol, 250 μL 1% vitamin B1, 1 mL 1 M $MgSO_4{\cdot}7H_2O$, 1 mL X-Gal (20 mg/mL in DMF). Adjust to 500 mL with sterile water. Filter and mix with 500 mL of freshly autoclaved 3% agar in water and immediately pour the plates.

2.2. Multicopy Library Screening

1. Plasmid multicopy library: The library used here is a pBR322-based *E. coli* genomic library that was constructed as described briefly hereafter and as used previously (10, 15). In a first step, genomic DNA is digested to completion with one or two restriction enzymes so that the resulting fragments will be on average 1–5 kb in length. Alternatively, digestion may not be done to completion and the desired length fragments can be purified from an agarose gel. Restriction products are then ligated into a compatible vector digested with the same restriction enzymes. While this or similar libraries can be obtained from various laboratories, such libraries can be constructed relatively easily. Please refer to refs. (15, 16) for a detailed protocol. In most cases the control vector consists of a non-digested native plasmid (here pBR322).
2. MacConkey lactose plates. Use MacConkey lactose base (Difco or BBL). Prepare the media by suspending 50 g of powder in 1 L of purified water. Heat while stirring and let boil for 1 min. The powder should be completely dissolved. Autoclave. Let cool to 50–60°C before adding arabinose and/or antibiotics, when needed, and then pour the plates.

3. Ara-amp-MacConkey plates: MacConkey broth with 100 μg/mL ampicillin and the determined concentration of arabinose (see Subheading 3.2, step 1). Add antibiotic and arabinose when medium is cooled after autoclaving.
4. Arabinose (stock): Add 20 g of arabinose to 100 mL of water and autoclave. Store at room temperature.
5. Oligonucleotides for sequencing plasmid inserts: pBR-for (5′-CCTGACGTCTAAGAAACCATTATTATC-3′) and pBR-rev (5′-GCATTGTTAGATTTCATACACGG-3′).

2.3. sRNA Dedicated Library Screening

1. sRNA dedicated library: The sRNA library used in Subheading 3.3 is available from the author upon request (11). Upon arrival, transform the plasmids into an *E. coli* *lacI*^{q} strain (such as PM1205) and store the resulting transformants at –80°C. To use the library, prepare the plasmids individually and make aliquots containing 1 μg plasmid in 100 μL water and store them in a microtiter plate at –20°C.
2. Transformation and storage solution (TSS): LB broth (see above) with 10% PEG (molecular weight 3,350 or 8,000), 5% (vol/vol) DMSO, and 50 mM $MgCl_2$. Filter and store at 4°C.
3. IPTG (100 mM stock): Add 23.83 mg of IPTG to 1 mL water. Filter and make 100 μL aliquots. Store at –20°C.
4. Permeabilization buffer (stock): 100 mM Tris (pH 7.8), 32 mM $NaPO_4$, 8 mM 1,2-Cyclohexanediaminetetraacetic acid (CDTA chelator), 4% Triton X-100. Store at room temperature.
5. Polymixin B (stock): Add 20 mg polymixin B to 1 mL water and store at –20°C.
6. Permeabilization buffer (for one plate): Immediately before use, add 48 μL of 1 M fresh DTT and 60 μL of polymixin B stock to 6 mL of permeabilization buffer stock.
7. ONPG solution: Add 20 mL of 5× M63 (see Subheading 2.1, item 5) to 80 mL water, then and add 400 mg of ONPG and 200 μL of 1 M sodium citrate. Stir well. When ONPG is totally dissolved, filter and store as 6 or 12 mL aliquots (for one or two microtiter plates, respectively) at –20°C.
8. Microtiter plate reader: Microtiter plate reader that can read the OD at 600 nm and at 420 nm in 96-well plates, can warm the plate compartment to 28°C, and that allows kinetic assays. Microtiter plate readers such as Tecan's Infinite® series, in conjunction with their native software, are appropriate.
9. Microtiter plates: 96-well, sterile, flat bottom, transparent cell culture plates, with lids, such as Greiner Cellstar Cat-No.665 180.

3. Methods

3.1. Construction of a Translational Fusion of the Gene of Interest to lacZ

1. When the transcriptional start site of the gene is known, amplify the gene by PCR with a high fidelity polymerase and the oligonucleotides P_{BAD}-geneX-for and lacZ-geneX-rev, designed as described in Subheading 2.1, item 2a (see Fig. 1b). Check the purity and size of the PCR products on an agarose gel, purify DNA from the PCR reaction and proceed to step 3.
2. When the transcriptional start site of the gene to screen is unknown, perform a 5′RACE using the RNA adapter and oligonucleotides described in Subheading 2.1, item 2b (Fig. 1a; see refs. (10, 17) for a detailed protocol). Briefly, from a clean, DNase and TAP (tobacco acid pyrophosphatase) treated total RNA preparation, ligate the RNA adapter P_{BAD}-RNA using T4 RNA ligase. Reverse transcribe the ligated RNA using the gene specific oligonucleotide geneX-RT (10, 17). Finally, PCR amplify the desired transcript containing the desired 5′ end using oligonucleotides lacZ-geneX-rev and P_{BAD}-DNA (see Fig. 1a). Check the size and purity of the RACE products on an agarose gel. Purify bands from the gel and proceed to step 3 (see Note 4).
3. Add 60 μL of an overnight culture of strain PM1205 grown at 32°C in LB to 30 mL of LB broth in a well-aerated flask.
4. Grow with agitation at 32°C to an $OD_{600} \approx 0.6$ and then induce the mini lambda by placing the culture at 42°C for 15 min.
5. Immediately cool the culture by strongly agitating the flask in a mix of ice and water for 2 min. From here through the electroporation step bacteria should be kept on ice.
6. To obtain electrocompetent cells, centrifuge the bacteria at $6{,}000 \times g$ for 10 min at 4°C. Discard the supernatant and resuspend the cell pellet in an equal volume of cold (4°C), sterile, ultrapure water. Repeat the centrifugation and resuspension steps twice more. Do a fourth centrifugation and resuspend the pellet in 300 μL of a cold, sterile, 10% glycerol solution (see Note 5).
7. Mix 100 ng of the purified PCR product obtained in step 1 with 100 μL of electrocompetent cells. Include a negative control in which DNA is omitted from the electrocompetent cells. Transfer each mix to an electroporation cuvette and electroporate at 25 μFd, 200 Ω, 2.5 kV.
8. Add 1 mL of LB to the transformed cells and allow outgrowth for 1 h at 37°C, with agitation.
9. Plate 25 μL of the transformed cells on one minimal sucrose plate and 100 μL of the transformed cells on another minimal

sucrose plate. Centrifuge the rest of the transformed cells, remove the supernatant, resuspend the cells in 100 μL sterile M63, and plate them on a M63-sucrose-X-Gal plate.

10. Place the plates in a 37°C incubator for 36–48 h until colonies appear (see Notes 6 and 7). Purify 12–16 colonies on M63-X-Gal plates at 37°C for 24–36 h. Pick mostly blue colonies and some white ones (see Note 8).
11. For each of the clones obtained in step 10, re-streak the colonies on LB and LB-cm plates to check for chloramphenicol sensitivity (the correct recombinants should be chloramphenicol sensitive) (see Note 9). Include the parental strain (PM1205) as a control for chloramphenicol resistance. Incubate overnight at 37°C.
12. From the LB plate, resuspend one colony of each chloramphenicol-sensitive clone (i.e. that did not grow on the LB-cm plate) in 50 μL ultrapure water and boil for 10 min at 95°C. Use 0.5 μL of the boiled cells for a PCR reaction using oligonucleotides $5'P_{BAD}$ and Deep-lac (see Note 10).
13. Check the PCR products on an agarose gel. Purify the PCR and sequence two to four clones having the same size insert using oligonucleotides P_{BAD} and Deep-lac.
14. Store a couple of each clone that contains a translational fusion by growing them in LB to $OD_{600} \approx 0.6$ and mixing 1 mL of the culture with 1 mL of sterile 20% glycerol. Store at –80°C.

3.2. Screening the Reporter Strain Using a Multicopy Library

1. Determine the appropriate arabinose concentration for the screening. Streak colonies from the translational fusion-containing strain onto MacConkey lactose plates containing tenfold dilutions of arabinose, ranging from 0.2 to 0.00002%, or no arabinose. Grow overnight in a 30°C incubator.
2. Choose the arabinose concentration on which colonies have a relatively low red colour, neither dark red nor white, so you will be able to look for either activation or repression of expression of the fusion (see Notes 11 and 12) (16).
3. Add 100 μL of an overnight culture of the fusion-containing strain grown at 37°C in LB to 100 mL of LB broth in a well-aerated flask. Grow with agitation at 37°C to an $OD_{600} \approx 0.6$.
4. Prepare electrocompetent cells of the translational fusion-containing strain as described above (see Subheading 3.1, step 6).
5. Mix 100 ng of the plasmid library from Subheading 2.2 with 100 μL of electrocompetent cells. Include a positive control by mixing 100 μL of cells with 10 ng of the control vector (here pBR322), and a negative control without DNA added to the electrocompetent cells. Transfer each mix to an electroporation cuvette and electroporate at 25 μFd, 200 Ω, and 2.5 kV.

Store the remaining competent cells at −80°C for future rounds of screening.

6. Add 1 mL of LB to each transformed cell and move to a culture tube. Mix and incubate at 37°C for 1 h.
7. Plate 5 μL of cells + 120 μL of LB, 25 μL of cells + 100 μL of LB, and 125 μL of cells on three different MacConkey plates containing arabinose at the appropriate concentration (see steps 1 and 2) and ampicillin (100 μg/mL final concentration) (ara-amp-MacConkey plates). Incubate at 30°C overnight until colonies appear. Keep the rest of the transformed cells on ice overnight.
8. Determine which dilution of cells gives the most appropriate number of colonies (100–200 colonies/plate). Use this dilution to spread the rest of the cells transformed with the library onto 10–20 ara-amp-MacConkey plates.
9. Carefully look for colonies that have a different colour from the majority of the colonies, either light/white or dark/red ones (see Note 13). Light/white colonies indicate repression of the reporter, and dark/red colonies indicate activation of the reporter. Isolate these clones by streaking them on ara-amp-MacConkey plates and on LB-amp plates in parallel, including the strain containing only the empty vector for comparison. Incubate at 30°C overnight.
10. Choose the colonies that give the same phenotype as in step 10. Pick the corresponding clones from the LB-amp plate to inoculate 5 mL of LB with ampicillin and grow for 6–16 h at 37°C.
11. Prepare the plasmids from the cultures and determine the DNA concentration. Use 10 ng to transform the parental reporter strain as described in Subheading 3.2, steps 6 and 7. Plate the transformed cells on ara-amp-MacConkey plates and incubate overnight at 30°C.
12. Streak 8–16 transformants/plate, including one streak of cells transformed with the empty vector as a control, on ara-amp-MacConkey plates. Incubate overnight at 30°C.
13. Choose the clones that give a distinguishable phenotype after re-transformation in the parental strain. Those clones should contain plasmids that carry genes for which overexpression affects the activity of the reporter.
14. Prepare plasmid DNA and sequence the insert (i.e. the chromosomal DNA that has been cloned in the plasmid) of the plasmids using oligonucleotides pBR-for and pBR-rev.
15. To interpret the results, assemble the DNA sequences at the extremities of the inserts to find the boundaries of the chromosomal DNA that has been inserted in the plasmid (see Note 14).

If the screen is close to saturation, the same DNA fragments should be found in several plasmids. If this is not the case, then further rounds of screening may be done in parallel to the analysis of the plasmids to obtain more candidates (see Note 15). For this screening, start at step 7 with electrocompetent cells stored at −80°C. Note also that to confirm the phenotype, a liquid beta-galactosidase assay should be performed (see below for a protocol).

16. Check the DNA sequences to determine if a previously identified sRNA gene is present in one or several of the plasmids. If so, the sRNA is a strong candidate for a regulator of the fusion (see comments on demonstrating direct regulation of an mRNA by an sRNA in Subheading 3.3, step 18).
17. Interpretation of results: If no previously identified sRNA is included amongst the candidates, one can envision three hypotheses: (1) One or several plasmid(s) may contain a previously unidentified sRNA, presumably in one of its intergenic regions, that is responsible for the effects seen on the fusion. In this case, a minimal DNA sequence needed for regulation can be searched by sub-cloning the plasmids (e.g. cloning only its intergenic regions and determine the effects of their overexpression on the fusion individually). If a region with no previously annotated gene is found regulating the fusion, then it may be indicative of an sRNA that could be searched using diverse bio-informatics/molecular biology techniques (reviewed in ref. (9)). (2) One of the protein coding gene of the plasmid(s) is responsible for the effects seen, not an sRNA. This putative protein regulator could be directly acting on the fusion at the post-transcriptional level, or may be driving the expression of an sRNA located on the chromosome, itself responsible for the observed effect on the fusion. Again, further sub-cloning will be necessary to evaluate which gene overexpression is responsible for the phenotype. (3) The sRNA could not be found because the screen was not saturated or because it cannot be overexpressed from such a plasmid (e.g. toxicity of overexpression). In that case one may perform further rounds of screening and/or try another screen using a dedicated sRNA library such as described below (see Subheading 3.3).

3.3. Screening the Reporter Strain with an sRNA Library

1. Follow steps 1–3 of Subheading 3.2 to determine a suitable arabinose concentration for the expression of the translational fusion (see also Notes 11 and 12).
2. Inoculate 50 mL of LB with 50 μL of an overnight culture of the reporter strain. Grow in an incubator at 37°C with agitation to $OD_{600} \approx 0.5$.
3. Centrifuge the bacterial culture at 6,000 × *g* for 10 min. Discard the supernatant and keep the pellet on ice.

4. Resuspend the pellet in 5 mL of cold TSS solution. Keep on ice.
5. For each plasmid in the library, add 1 μL to a well in a microtiter plate (see Fig. 2). Keep on ice.
6. Add 100 μL of bacteria in TSS from step 4 to each well that contains plasmid DNA to begin the transformation. Keep on ice for 30 min.
7. Incubate at 37°C for 1 h with agitation.
8. With a multichannel pipettor, spot 4 μL of the transformants from the microtiter wells onto three LB-amp plates (see Note 16 for use of colour indicator plates). Incubate overnight at 32°C until colonies appear (see Note 17).
9. To all the wells of two sterile microtiter plates, add 100 μL LB-amp broth containing 10% of the arabinose concentration determined in step 1 (see Note 18).
10. Using a replicator (or sterile toothpicks), inoculate the wells with one colony of each plasmid-containing clone. Duplicates can be made on one microtiter plate (see Fig. 2).
11. Place in an incubator at 37°C with shaking for 6–7 h (see Note 19).
12. Carefully agitate the well to get rid of any cell debris in the wells. In a microtiter plate reader, read the OD_{600} of each well. Use wells containing only LB to determine the baseline (see Note 20).
13. Add 50 μL of permeabilization buffer/well. Keep at room temperature for 15 min.
14. Quickly add 50 μL of ONPG solution to each well and immediately proceed to step 15 (see Note 21).
15. In a microtiter plate reader, do a first round of agitation and then do a kinetic cycle to read the OD_{420} of the plate every 30 s for 2 h at 28°C.
16. For each well, calculate the V_{max} (OD_{420}/min) of the beta-galactosidase reaction. The V_{max} is the maximal velocity of the reaction, equal to the slope of the initial linear part of the curve when OD_{420} is plotted vs. time. Calculate the specific activity of each reaction, which equals 1,000 times the V_{max} of the reaction divided by the OD_{600} determined in step 13 (see Note 22).
17. Repeat steps 9–16 at least two more times using the remaining LB-amp plates containing transformants from step 8. Calculate the average of the specific activities (see Note 23).
18. To interpret the data, first calculate the ratio of the specific activity of each sRNA containing clone to that of the empty vector (see Note 24).

19. Calculate the average and standard deviation of these ratios for all the experiments. sRNAs that produce a ratio above 2 should be candidates for activators of expression of the fusion, and sRNAs that produce a ratio below 0.5 should be candidates for repressors. These thresholds are used to eliminate possible small indirect effects due to sRNA overexpression (see Fig. 2 and ref. (11) for an example) (see Notes 25 and 26).

4. Notes

1. To avoid post-translational effects, the last codon of the gene to be fused to *lacZ* should be relatively close to the beginning of the ORF. This is actually a necessity in the case of membrane and periplasmic proteins because beta-galactosidase is active only when expressed in the cytoplasm. We found that for most proteins, cloning at the 10th–20th codon gives satisfactory results. This location is in most cases acceptable because generally sRNA regulation occurs in the 5′ untranslated region of the mRNA, close to the ribosome binding site and/or in a region extending a few nucleotides (up to 100) upstream or downstream (18, 19). However, if sRNA regulation is suspected to occur downstream of this region, i.e. inside the ORF, one could use a fusion located further downstream. When a fusion in the ORF is used, special attention should be paid to discriminate potential post-translational regulators (regulated protease degradation, e.g.) from post-transcriptional/sRNA regulators.
2. In general, other classical cloning techniques could be used to obtain a similar strain with the gene of interest fused to *lacZ* under the control of P_{BAD} or another promoter. However, particular attention should be paid to the promoter used. In particular, to avoid false positives it is essential that the inducer does not serve as a metabolite for the cell (in PM1205, the *araBAD* genes for arabinose usage have been deleted). Also, the screening method described Subheading 3.3 requires a promoter compatible with P_{lac} (which is used to drive expression of the sRNAs in the library). Note also that PM1205 contains the *lacI^q* allele, which allows for tighter repression of genes under the control of a P_{lac} promoter in absence of inducer.
3. The strain PM1205 should be sucrose sensitive. However, if you encounter counter-selection problems in the fusion construction steps, you may want to check that the strain has kept its sucrose sensitivity. To do so, streak the PM1205 strain against a wild-type strain (MG1655, e.g.) on sucrose minimal plates (see recipe in Subheading 2.1, item 6) at 37°C overnight. Strain PM1205 should be unable to grow in such conditions.

4. Several bands may be observed on the gel in the case of multiple transcriptional start sites. In this case, each band should be extracted and purified from the gel and used separately for subsequent cloning in the procedure. Note that in this case you may get activity from putative promoters contained in the longer fragments.
5. Electrocompetent cells can be stored in 10% glycerol solution at −80°C for a few days, but may lose some competence efficiency. Make 100 μL aliquots of the cells and freeze them in a dry-ice ethanol bath before storage.
6. The number of colonies on the transformation plate is expected to be higher than the number on the negative control plate. However, we have noted that colonies may appear in the control strain electroporated without DNA. This is presumably due to the poor counter-selection given by the *sacB* gene on sucrose plates. Even in the case that there is not an apparent greater number of colonies in the DNA-electroporated cells as compared to the control, we recommend proceeding to Subheading 3.1, step 9 to check for chloramphenicol sensitivity of the obtained transformants.
7. Note that from this step to the obtention of the final strain, bacteria are grown at 37°C, implying that the mini lambda, which cannot replicate at temperatures ≥37°C, will be lost. If further genetic manipulations are envisioned in the desired strain, such as introduction of a mutation by homologous recombination, it may be useful to keep the mini lambda. In that case, grow the strains at a permissive temperature (i.e. 32°C) at this step and subsequently. Note however that growth will be slower, specially on minimal sucrose plates, and thus the selection process will take more time. Because the mini lambda can be lost during the induction step, if the mini lambda is desired check for its presence by streaking 12–16 clones on LB-tet plates at step 11 in Subheading 3.1. Mini lambda-containing clones should be tetracycline resistant.
8. In most cases, we found that even relatively weak fusions give a noticeable blue colour on X-Gal containing plates. Appearance of such colonies should be a sign of efficient recombination. However, it is possible that fusions with very weak activity will remain white on those plates. Thus, if no blue colonies are found, proceed as described with randomly picked colonies from the DNA-electroporated cells. Note also that rare dark colonies may be a sign of a mutation in the fusion. As noted above, recombination will ultimately be checked by loss of chloramphenicol resistance in Subheading 3.1, step 11, and the integrity of the fusion will be confirmed by sequencing.
9. While it may seem like a slower process, it is important that chloramphenicol sensitivity is checked after at least one

purification on minimal sucrose because otherwise chloramphenicol resistant contaminants may be present and give false negative results.

10. For relatively small fusions (less than 500 nucleotides from the +1 to *lacZ*), we found that colony PCR is sufficient to give confidently readable sequences. For larger fusions, however, we recommend to prepare chromosomal DNA from the clones and to do a PCR with a high fidelity enzyme before proceeding to Subheading 3.1, step 13.
11. It is possible that some fusions which have very high basal activity will appear red even at the lowest arabinose concentration. In this case, try to grow the cells at lower temperatures (25°C, e.g.) to minimize this effect. If the strain forms red colonies even in the absence of arabinose, it might be indicative of an additional promoter contained in the fusion. Conversely, some very weak fusions may remain pale or white even at high (0.2%) arabinose concentration, in which case incubation at higher temperatures (but not more than 37°C) could be used. If you cannot obtain a workable indicator colour for basal activity, you might consider other indicator media such as LB + X-Gal or tetrazolium lactose agar (see ref. (20) for details).
12. Since Hfq is involved in many sRNAs activity, activity of mRNA regulated by sRNAs is often changed in an *hfq* mutant. Thus, before proceeding to screening, you may want to check for possible variation of the activity of the fusion in an *hfq* mutant. To do so, P1 transduct an antibiotic marked *hfq* mutation into the translational fusion-containing strain (see ref. (16) for a detailed transduction protocol). Streak both the WT and *hfq* mutant strains on a MacConkey plate containing the above-determined arabinose concentration and incubate overnight at 30°C. If the *hfq* mutant looks lighter or darker than the WT, it is a good indication that an sRNA up- or down-regulates, respectively, the expression of the translational fusion. However, given that the *hfq* mutation has pleiotropic effects and a marked slow growth phenotype even in rich medium, shape and colour of colonies may look slightly different than that of the WT strain on a MacConkey plate, even if no sRNA regulates the fusion. Conversely, colour may look comparable even though sRNA regulation exists. Therefore, this experiment may be indicative only in the case of marked phenotypes. This is why we recommend in any case to proceed to subsequent screening steps to find possible sRNA regulators.
13. Note that some overexpression plasmids may induce phenotypes, such as slow growth or mucoidy, that may affect the shape and colour of the colony independently of beta-galactosidase activity. It is hard to discriminate such clones from those in which the reporter activity has changed. Most of the time these false

positives can be discarded after one or two rounds of purification on MacConkey lactose plates.

14. Note that depending on how DNA was cloned in the plasmids, it is possible that more than one piece of DNA may be inserted. These cases are easy to discriminate because in general, the sequence at the extremities are located at such a great distance (>10 kb) on the chromosome that it is not possible that the insert comes from only one fragment of DNA. Although it is possible to get the full sequence of the insert by primer walk sequencing, it is easier to discard these plasmids from further analysis if at all possible.
15. As a good starting point, at least three times the chromosome length should be screened when doing a multicopy library screening. For example, in the case described here, because the genome of *E. coli* is 4.7 Mb and the average size of the inserts in the plasmids is ~2 kb, the minimum number of transformants one should screen is given by 3×4.7 Mb/2 kb ≈7,000 clones. In the rare case where no candidates can be identified after several rounds of screening, it may mean that there is no regulator of the fusion or that the screen is not appropriate. Another screen such as the one described in Subheading 3.3 may be used.
16. Transformants can also be spotted on beta-galactosidase activity indicator plates such as MacConkey lactose and/or LB-X-Gal (containing IPTG at 100 μM) in parallel, to give an idea of the effect of sRNA overexpression from the plasmids (see Fig. 2). In both cases, arabinose at an appropriate concentration (see Subheading 3.2, step 2) and ampicillin should be included in the medium. Note however that results may be difficult to interpret, particularly as we found that some sRNAs are toxic to the cells when overexpressed (Spot42, DicF, OxyS, e.g.) and thus should not give colonies in presence of inducer.
17. Spotted cells should give 10–20 small, isolated colonies after an overnight incubation at 37°C. However, if colonies are too confluent, try transforming with less plasmid (1 ng, e.g.) and/or incubating at 30°C to obtain smaller, isolated colonies. Confluence of colonies should be avoided since this may result in poor selection and give cells that do not contain the plasmids.
18. In general for beta-galactosidase tests from liquid cultures such as this one, an arabinose concentration ten times lower than that found on MacConkey arabinose plates is the most efficient. For example, if you find that your fusion gives a convenient red colour at 0.002% arabinose concentration on a MacConkey plate, then use 0.02% arabinose in LB for the test.
19. After 6–7 h of growth, cells have entered stationary phase, resulting in a more consistent OD_{600}. Growing to stationary

phase therefore limits differences due to uneven growth of the clones. Growing longer than 7 h may result in slight evaporation from the wells that will increase noise in the results.

20. Given that the sample in the well does not have a path length of 1 cm, the OD_{600} determined in the microtiter plate is not equivalent to that of one read in a spectrophometer. For a cell culture of 100 μL, the OD_{600} read in a microtiter plate reader is about tenfold lower than that read by a spectrophotometer.
21. It is very important that ONPG solution is added as fast as possible to the cell extracts containing wells as reaction will start instantly after addition. Some plate readers have injectors that will automatically dispense the appropriate volume in the wells. If not available on your reader, we recommend that you use a multichannel pipettor/dispenser that will allow you to dispense the ONPG rapidly.
22. This calculated specific activity is expressed in arbitrary units that are approximately, but not strictly, equivalent to Miller units.
23. Given that one cannot choose an exact OD_{600} at which to stop the cultures, there is inevitably a certain degree of variation amongst the specific activities found from one experiment to another. This is why we recommend repeating the experiment at least three times to minimize the standard deviations of the values.
24. Alternatively, one could directly use the specific activities of the sRNA-containing clones to find which induce more variations on the specific activities. However we found that given the fair amount of variation from one experiment to another, calculating the ratio of specific activities of sRNA-containing clones to the vector-only control gives generally more consistent results.
25. If no sRNA is found to activate or repress expression of the fusion, it may either mean that no sRNA is involved in the regulation of the gene or that the sRNA is absent from the sRNA library. In such case one may try to find other regulators using an overexpression library such as described in Subheading 3.2. In the case where neither the screen in Subheading 3.2 nor the screen in Subheading 3.3 identifies an sRNA regulator of the fusion, biochemical approaches should be tried. In particular, one limit to the methods described here is that they assume the critical regions for regulation are in the 5′ untranslated region and 5′ end of the coding region, and are not dependent on sequences further downstream, since downstream sequences are not present in the fusion constructs (see also Note 1).
26. Note that all sRNAs changing activity of the fusion may not necessarily be direct regulators. In particular, it is possible that

some sRNAs act indirectly by either (1) inducing or repressing the expression of a direct sRNA regulator located on the chromosome or (2) by competing with a direct sRNA regulator for Hfq (11, 21, 22). The latter case should particularly be taken into account when working with overexpressed sRNAs, such as the ones used here. Therefore, it is important to subsequently confirm or infer direct regulation of the sRNA on the mRNA encoded by the fusion. Prediction of sRNA base pairing to the target mRNA can be searched using tools such as TargetRNA (http://snowwhite.wellesley.edu/targetRNA) (23) and confirmed by testing the effects of mutations in the sRNA/mRNA base pairing region (see refs. (11, 24–27) for examples of such experiments).

Acknowledgements

I greatly thank F. Barras for financial, space and intellectual support. I thank S. Gottesman for critical reading and helpful discussions on the manuscript. I also thank C. Beisel, Y. Duverger, M. Guillier, H.J. Lee, K. Moon, and B. Py for help and comments on the preparation of the manuscript.

References

1. Gottesman S, Storz G (2011) Bacterial small RNA regulators: versatile roles and rapidly evolving variations. Cold Spring Harb Perspect Biol 3(12, pii):a003798
2. Waters LS, Storz G (2009) Regulatory RNAs in bacteria. Cell 136:615–628
3. Gottesman S, McCullen CA, Guillier M, Vanderpool CK, Majdalani N, Benhammou J, Thompson KM, FitzGerald PC, Sowa NA, FitzGerald DJ (2006) Small RNA regulators and the bacterial response to stress. Cold Spring Harb Symp Quant Biol 71:1–11
4. Brennan RG, Link TM (2007) Hfq structure, function and ligand binding. Curr Opin Microbiol 10:125–133
5. Zhang A, Wassarman KM, Rosenow C, Tjaden BC, Storz G, Gottesman S (2003) Global analysis of small RNA and mRNA targets of Hfq. Mol Microbiol 50:1111–1124
6. Sittka A, Lucchini S, Papenfort K, Sharma CM, Rolle K, Binnewies TT, Hinton JC, Vogel J (2008) Deep sequencing analysis of small non-coding RNA and mRNA targets of the global post-transcriptional regulator, Hfq. PLoS Genet 4:e1000163
7. Wassarman KM, Repoila F, Rosenow C, Storz G, Gottesman S (2001) Identification of novel small RNAs using comparative genomics and microarrays. Genes Dev 15:1637–1651
8. Raghavan R, Groisman EA, Ochman H (2011) Genome-wide detection of novel regulatory RNAs in *E. coli*. Genome Res 21(9):1487–1497
9. Sharma CM, Vogel J (2009) Experimental approaches for the discovery and characterization of regulatory small RNA. Curr Opin Microbiol 12:536–546
10. Mandin P, Gottesman S (2009) A genetic approach for finding small RNAs regulators of genes of interest identifies RybC as regulating the DpiA/DpiB two-component system. Mol Microbiol 72:551–565
11. Mandin P, Gottesman S (2010) Integrating anaerobic/aerobic sensing and the general stress response through the ArcZ small RNA. EMBO J 29:3094–3107
12. Datsenko KA, Wanner BL (2000) One-step inactivation of chromosomal genes in *Escherichia coli* K-12 using PCR products. Proc Natl Acad Sci U S A 97:6640–6645
13. Court DL, Swaminathan S, Yu D, Wilson H, Baker T, Bubunenko M, Sawitzke J, Sharan SK (2003) Mini-lambda: a tractable system for chromosome and BAC engineering. Gene 315:63–69

14. Urban JH, Vogel J (2009) A green fluorescent protein (GFP)-based plasmid system to study post-transcriptional control of gene expression in vivo. Methods Mol Biol 540:301–319
15. Ulbrandt ND, Newitt JA, Bernstein HD (1997) The *E. coli* signal recognition particle is required for the insertion of a subset of inner membrane proteins. Cell 88:187–196
16. Sambrook J, Russel D (2001) Molecular cloning: a laboratory manual, 3rd edn. CSHL Press, Cold Spring Harbor, NY
17. Argaman L, Hershberg R, Vogel J, Bejerano G, Wagner EG, Margalit H, Altuvia S (2001) Novel small RNA-encoding genes in the intergenic regions of *Escherichia coli*. Curr Biol 11:941–950
18. Bouvier M, Sharma CM, Mika F, Nierhaus KH, Vogel J (2008) Small RNA binding to 5' mRNA coding region inhibits translational initiation. Mol Cell 32:827–837
19. Pfeiffer V, Papenfort K, Lucchini S, Hinton JC, Vogel J (2009) Coding sequence targeting by MicC RNA reveals bacterial mRNA silencing downstream of translational initiation. Nat Struct Mol Biol 16:840–846
20. Shuman HA, Silhavy TJ (2003) The art and design of genetic screens: *Escherichia coli*. Nat Rev Genet 4:419–431
21. Hussein R, Lim HN (2011) Disruption of small RNA signaling caused by competition for Hfq. Proc Natl Acad Sci USA 108: 1110–1115
22. Olejniczak M (2011) Despite similar binding to the Hfq protein regulatory RNAs widely differ in their competition performance. Biochemistry 50:4427–4440
23. Tjaden B, Goodwin SS, Opdyke JA, Guillier M, Fu DX, Gottesman S, Storz G (2006) Target prediction for small, noncoding RNAs in bacteria. Nucleic Acids Res 34: 2791–2802
24. Papenfort K, Bouvier M, Mika F, Sharma CM, Vogel J (2010) Evidence for an autonomous 5' target recognition domain in an Hfq-associated small RNA. Proc Natl Acad Sci USA 107:20435–20440
25. Figueroa-Bossi N, Valentini M, Malleret L, Bossi L (2009) Caught at its own game: regulatory small RNA inactivated by an inducible transcript mimicking its target. Genes Dev 23:2004–2015
26. Guillier M, Gottesman S (2008) The 5' end of two redundant sRNAs is involved in the regulation of multiple targets, including their own regulator. Nucleic Acids Res 36: 6781–6794
27. Majdalani N, Hernandez D, Gottesman S (2002) Regulation and mode of action of the second small RNA activator of RpoS translation, RprA. Mol Microbiol 46:813–826

Part II

Function and Expression of Regulatory RNAs

Chapter 5

Competition Assays Using Barcoded Deletion Strains to Gain Insight into Small RNA Function

Errett C. Hobbs and Gisela Storz

Abstract

Ordered collections of barcoded deletion mutants can be screened rapidly in mixed cultures to uncover resistance and sensitivity phenotypes associated with the loss of a gene. As such, they are invaluable tools for assigning gene function in the post-genomic era. In this protocol, we describe methodologies for creating and employing barcoded deletion mutants in competitive screens as well as how to analyze the ensuing results.

Key words: Competition assays, sRNA, Microarray

1. Introduction

Barcoded deletion mutagenesis has its roots in signature sequence-tagged mutagenesis (STM) experiments to identify genes involved in pathogenesis (reviewed in ref. (1)). In short, a series of transposons containing random 40-mer pieces of DNA (signature sequence tags) were used to randomly mutagenize bacteria. The mutagenized bacteria were then used to infect an animal and the abundance of each signature sequence tag before and after infection was determined by Southern blotting.

With the advent of DNA microarray technology, STM was expanded to delete almost every nonessential gene in *Saccharomyces cerevisiae* with DNA barcoded antibiotic resistance cassettes (2, 3). Barcodes are unique 20-mer DNA sequences that are incorporated upstream (UP) and downstream (DN) of an antibiotic resistance cassette by PCR. Instead of employing Southern blots to measure strain abundance after control or stress treatments, DNA barcodes can be amplified by PCR and hybridized to a TAG4 microarray (4). These barcodes were designed so that they would

Kenneth C. Keiler (ed.), *Bacterial Regulatory RNA: Methods and Protocols*, Methods in Molecular Biology, vol. 905, DOI 10.1007/978-1-61779-949-5_5,

not cross-hybridize with one another and would have the same optimal hybridization temperature on the DNA microarray. All told, there are 11,812 barcodes available that can be used in conjunction with the TAG4 microarray, which should be sufficient to delete all of the genes contained within most bacterial chromosomes (4). As of now, two sets of barcoded *Escherichia coli* mutants have been reported in the literature (5, 6). One set is deleted for a series of genes involved in DNA repair, and the other is deleted for over 100 genes that encode small proteins (≤50 aa) or small regulatory RNAs in addition to some genes involved in stress response.

Very detailed methodologies exist for employing and analyzing large barcoded deletion collections (4) (see Note 1). It should be further noted that in the future, barcoded deletion collections may be analyzed more efficiently by employing deep sequencing technologies. Nevertheless, the procedures described below are sufficient for analyzing relatively small collections containing around 100 strains in a time-efficient manner.

2. Materials

The solutions containing MES or phycoerythrin are photo sensitive and should be shielded from light, where indicated. Even though it may not be necessary, it was for this reason that the biotin staining mixture was always freshly prepared on the day of use from the premixed stocks described below.

1. Primers

Oligo name	Sequence (5′–3′)
kan-UP (5′-Bio)	BIOTIN-gaagcagctccagcctacac
kan-DN (5′-Bio)	BIOTIN-ggtcgacggatccccggaat
B213 (5′-Bio)	BIOTIN-ctgaacggtagcatcttgac
kan-UP	gaagcagctccagcctacac
kan-DN	ggtcgacggatccccggaat
Common UP	gatgtccacgaggtctct
Common DN	cggtgtcggtctcgtag
Complement—kan-UP	gtgtaggctggagctgcttc
Complement—kan-DN	attccggggatccgtcgacc
Complement—Common UP	agagacctcgtggacatc
Complement—Common DN	ctacgagaccgacaccg

2. Luria-Bertani (LB) medium: 10 g tryptone, 5 g yeast extract, and 10 g NaCl per liter of H_2O. Autoclave at 121°C, 15 psi for 15 min.

3. 12× MES (2-(*N*-morpholino)ethanesulfonic acid) stock (1.22 M MES, 0.89 M Na^+, pH 6.5–6.7): 0.70 g MES free acid monohydrate, 1.9 g MES sodium salt per 10 mL of dH_2O. Sterile filter. Store at 4°C, shielded from light.
4. 2× hybridization buffer (200 mM MES, 1 M Na^+, 20 mM EDTA, 0.01% Tween 20): To make 50 mL of 2× hybridization buffer, mix 8.3 mL of 12× MES stock, 17.7 mL of 5 M NaCl, 4.0 mL of 0.5 M EDTA, 0.1 mL of 10% Tween 20, and 19.9 mL of dH_2O. Store shielded from light at 4°C.
5. Mixed oligonucleotides (12.5 μM for each oligonucleotide): Combine 100 μL each of 100 μM stocks of the following oligonucleotides: kan-UP, kan-DN, Common UP primer, Common DN primer, Complement–kan-UP, Complement–kan-DN, Complement–Common UP, and Complement–Common DN.
6. Hybridization mix: Combine 75 μL of 2× hybridization buffer, 0.5 μL of B213 oligonucleotide, 12 μL of mixed oligonucleotides, and 3 μL of 50× Denhardt's solution (1% Ficoll (type 400), 1% polyvinylpyrrolidone, and 1% bovine serum albumin).
7. 20× SSPE (3 M NaCl, 200 mM NaH_2PO_4, 20 mM EDTA, pH 7.4): Dissolve 175.3 g of NaCl, 27.6 g of $NaH_2PO_4{\cdot}H_2O$, and 7.4 g of Na_2EDTA in 800 mL of dH_2O. Bring pH to 7.4 using 10 M NaOH. Add dH_2O to a final volume of 1 L and autoclave to sterilize.
8. Wash A (6× SSPE-T): Mix 300 mL of 20× SSPE and 1 mL of 10% Tween 20 with 699 mL of dH_2O.
9. Wash B (3× SSPE-T): Mix 150 mL of 20× SSPE and 1 mL of 10% Tween 20 with 849 mL of dH_2O.
10. Biotin staining mix: Combine 51.4 μL of 20× SSPE, 3.4 μL of 50× Denhardt's solution, 1.7 μL of 1% Tween 20, and 0.29 μL of 1 mg/mL streptavidin-phycoerythrin (Molecular Probes—cat. #S-866) with 114 μL of dH_2O. Store at 4°C, shielded from light. This solution should be prepared the day the microarray is stained.

3. Methods

3.1. Construction of Deletion Strains

The barcoded antibiotic resistance cassettes can be generated in a two-step PCR process as previously described (5) (Fig. 1). All of the barcoded antibiotic resistance cassettes are constructed separately.

1. In the first-round PCR, amplify the kanamycin antibiotic resistance cassette from pKD13 (7) by using upstream (P1)

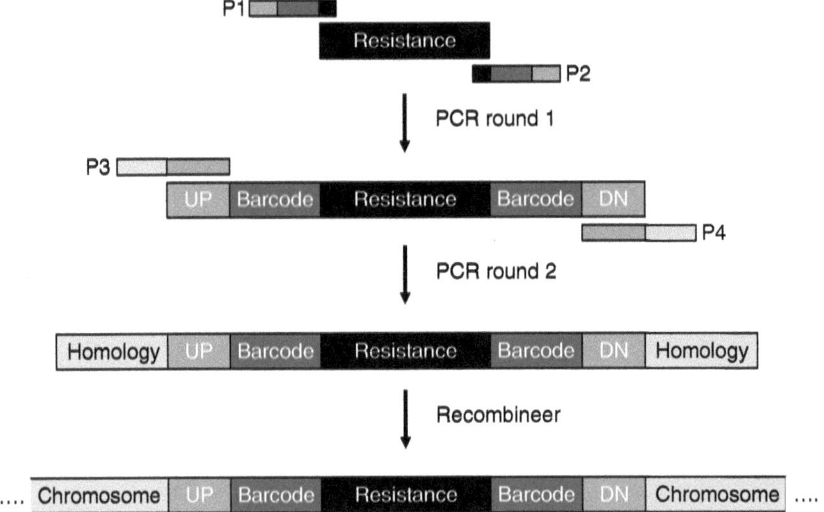

Fig. 1. Barcoded deletion cassettes are generated by a two-step PCR process and incorporated into the chromosome by recombineering. The P1 and P2 primers used in the first-round PCR are complementary to the antibiotic resistance cassette and are used to incorporate the barcodes and common regions. The P3 and P4 primers used in the second-round PCR are complementary to the common regions and are used to incorporate regions of homology for recombineering the cassette into the target locus on the chromosome.

and downstream (P2) primers to incorporate the upstream and downstream barcodes in addition to the common regions flanking the barcodes in every cassette.

2. Subject the first-round PCR products to electrophoresis at 120 V in a 1% TBE agarose gel until they are clearly resolved. Excise the amplified kanamycin cassette from the gel and purify using a Qiagen Gel Extraction Kit.
3. Use the purified first-round PCR products as a template in a second-round PCR with the P3 and P4 primers to incorporate upstream and downstream sequences homologous to the chromosomal regions flanking the locus to be deleted.
4. Purify the second-round PCR products by using a Qiagen PCR Purification Kit followed by ethanol precipitation.
5. Incorporate the barcoded antibiotic resistance cassettes into the genome by λ-Red mediated recombineering (8, 9).

3.2. Competition Assays

1. Streak all strains for single colonies on LB solid medium or pin from glycerol stocks in a 96-well plate to LB solid medium in a Nunc OmniTray. Incubate for 16 h at 30°C (see Note 2).
2. Inoculate 5 mL of liquid LB in a culture tube or 600 μL of liquid LB in a deep-well 96-well plate with individual colonies. 96-well plates should be sealed with a sterile air-permeable membrane. Incubate for 16 h at 30°C with 250 rpm shaking.

3. Mix 100 μL of each culture in a sterile 50 mL conical tube. The amounts of cells associated with each strain do not have to be equal, as the cell numbers for each strain in the experimental culture will eventually be compared to their number in the control condition.
4. Subject one aliquot of the mixed culture to a control condition and another to a stress condition. In general, the stress condition can take one of two forms; "acute" or "chronic" stress. Cells subjected to an "acute" stress are exposed to an adverse condition for a set period of time and then to a period of outgrowth under permissive conditions. If cell abundance is measured immediately after an acute stress, there will be no measurable difference in cell abundance between strains with and without a phenotype, as dead cells still have DNA from which barcodes can be amplified. Therefore, both the control and stress cultures must be diluted into fresh media and allowed to outgrow to equal optical densities. This will allow cells that are better suited to surviving the stress condition to divide and overwhelm those cells that are not able to survive or recover from the stress in question. We find that allowing for at least 12 generations of growth is enough to distinguish cells showing a phenotype from those that do not. In practice, this means diluting both the stress culture and the control culture 1:5,000 and allowing both cultures to achieve the optical density of the original, untreated culture.
5. In contrast to "acute" stress, cells subjected to "chronic" stress are diluted and allowed to grow under a stress condition that is minimally permissive to growth. In this instance, the mixed culture is diluted 1:2,000. The diluted culture is split into a control and a stress-treated culture. Both cultures are grown to the same OD_{600}. Not as much outgrowth seems to be required as for an "acute" stress. For example, 8–10 generations of outgrowth were sufficient to distinguish strains that had cell envelope stress phenotypes from those that did not (5).

3.3. Amplification of Barcodes from Genomic DNA

1. Harvest between 5 and 6 OD_{600} of cells. For example, if the final culture density is 5 OD_{600}, then centrifuge 1 mL of culture in a benchtop microfuge at maximum speed, dispose of the supernatant, and proceed with the protocol (if desired, the cell pellet can be frozen for later use). If the final culture density is 2.5 OD_{600}, then collect the pellet from 2 mL of culture.
2. Isolate genomic DNA from cells harvested from the control and stress cultures. The Wizard Genomic DNA Purification Kit (Promega) works well.
3. Amplify the "UP" and "DN" DNA barcodes in two separate sets of polymerase chain reactions. Approximately 200 ng of genomic DNA works best (see Note 3).

Reagents	1 × (μL)	3.5 × (μL)
UP barcodes		
10× buffer	10	35
10 mM dNTPs	2	7
50 mM $MgSO_4$	4	14
Platinum Taq HiFi	0.8	2.8
dH_2O	77.2	270.2
kan-UP (5′-Bio)	2	7
Common UP primer	2	7
Total	98	343
DN barcodes		
10× buffer	10	35
10 mM dNTPs	2	7
50 mM $MgSO_4$	4	14
Platinum Taq HiFi	0.8	2.8
dH_2O	77.2	270.2
kan-DN (5′-Bio)	2	7
Common DN primer	2	7
Total	98	343

Reaction	Template DNA	Reaction mix
1	Control culture	UP barcodes
2	Control culture	DN barcodes
3	Stress culture	UP barcodes
4	Stress culture	DN barcodes
5	No DNA	UP barcodes
6	No DNA	DN barcodes

Cycling parameters	
94°C	3 min
94°C	30 s
55°C	30 s
68°C	30 s
Go to step 2 34×	
4°C	∞

4. Subject 5 μL of each reaction product to electrophoresis in a 2% agarose gel to verify that the barcodes were amplified in reactions 1–4. There will be excess primer in each reaction, but it is important that no amplified products are present in reactions 5 and 6, which do not contain any DNA template.

3.4. Hybridizing and Scanning DNA Microarrays

1. Affymetrix arrays contain two gaskets through which liquids can be applied to the array. Simply place a 20–200 μL micropipettor tip into one gasket to allow air to evacuate from the array as liquids are added through the other gasket. Using this technique, fill each of two DNA microarrays (Affymetrix Genflex Tag 16 K Array v2) with 140 μL of 1× hybridization buffer (see Note 4).
2. Incubate the microarrays for 10 min in a hybridization oven set to 42°C, 20 rpm.
3. UP and DN tags from within each culture are measured on the same DNA microarray, but analyzed separately. To start, combine 30 μL each of reactions 1 and 2 with 90 μL of hybridization mix (CNTL). Also combine 30 μL each of reactions 3 and 4 with 90 μL of hybridization mix (STRESS).
4. Boil the CNTL and STRESS tubes for 2 min and then incubate them on ice for 2 min.
5. Remove the 1× hybridization buffer from the two arrays (labeled "CNTL" and "STRESS").
6. Add the contents of the CNTL and STRESS tubes to the appropriate array. A bubble might be visible in the array, which is normal.
7. Place a Tough Tag (Diversified Biotech) over each of the two gaskets on each array.
8. Incubate the arrays for 10–16 h in a hybridization oven set to 42°C and 20 rpm.
9. Aspirate the hybridization mix and wash as described below. One mixing cycle consists of pipetting 150 μL of wash solution in and out of the microarray. An automated fluidics station can also be used to perform all wash cycles.

Solution (temperature)	Solution changes	Mixes per solution change
Wash A (room temperature)	2	4
Wash B (42°C)	6	4
Wash A (room temperature)	1	2

10. Aspirate Wash A from the microarrays and replace it with 140 μL of biotin staining mix.
11. Incubate for 10 min in a 42°C hybridization oven set to 20 rpm.
12. Aspirate the biotin staining mix and wash each array (pipette in and out four times) with six 150 μL changes of Wash A at room temperature.

13. Carefully fill the arrays with fresh Wash A. At this stage, no bubbles can be left on the array. If necessary, the arrays can be overfilled with Wash A to flush any bubbles that remain behind.
14. Remove all micropipettor tips from the arrays, dry the gaskets with a Kimwipe or other lint-free paper towel, and apply Tough Tags to all gaskets. Here, the Tough Tags are applied to prevent any liquids from spilling into the microarray scanner.
15. Scan each array using an Affymetrix GeneArray Scanner set to an emission wavelength of 560 nm.

3.5. Analysis of Results

More sophisticated protocols and computer programs exist to analyze larger data sets (4). However, the methods below describe a robust and easy way to analyze the results of small barcoded deletion mutant collections.

1. Every barcode is queried by five probes on the microarray. Extract the arbitrary fluorescence intensities associated with each probe. UP and DN barcodes should be analyzed separately, as they were amplified separately.
2. Average the five arbitrary fluorescence intensities associated with each barcode to obtain the mean fluorescence intensity.
3. Determine the background fluorescence for each microarray by averaging the probe intensity values associated with the barcodes that are not present in any strain.
4. Subtract the background fluorescence from all mean fluorescence intensities associated with strains in the test set.
5. Exclude barcodes exhibiting background-corrected mean fluorescence intensities of less than 200 arbitrary fluorescence units in the control array.
6. The signal for individual probes in DNA microarrays saturates, and the relationship between strain abundance and barcode signal becomes nonlinear. As a result, more abundant barcodes tend to be underrepresented when measured with a DNA microarray. To correct for probe saturation, multiply every background-corrected mean fluorescence intensity by $e^{(0.00031*\text{mean barcode intensity})}$.
7. Divide the corrected fluorescence intensity obtained for each barcode in the mock-treatment sample by its corresponding fluorescence intensity in the stress-treatment sample to obtain a relative abundance value (R.A.).
8. Arrange the R.A. values in descending order. Sensitive strains will have R.A. values above 1, and resistant strains will have R.A. values below 1. In our experience, these analyses are most reliable when the competitive screens are performed in triplicate. There are two barcodes associated with each deletion mutant,

so performing the screens in triplicate results in six R.A. measurements per strain. There will be variability in the absolute R.A. values obtained between different trials, but the rank order of the most sensitive and most resistant strains tends to remain relatively constant. In a pool ~100 deletion mutants, our experience indicates that strains consistently appearing among the top 20 and bottom 5 R.A. values in at least four of the six series of R.A. values are reliably sensitive or resistant.

4. Notes

1. The methods described in this chapter are also detailed by the Davis group at the following website: http://chemogenomics.stanford.edu/supplements/04tag/analysis.
2. Until the deletion mutants in a collection have been well characterized, we find it best to assume that they will all have a temperature sensitivity phenotype and to incubate them at a lower temperature for all strain maintenance steps until execution of the experiment.
3. Using too little genomic DNA can introduce sampling errors. Our group employed Platinum Taq HiFi polymerase (Invitrogen), but any DNA polymerase should work well in principle. Set up the reactions as designated above. It is essential that the kan-UP (5′-Bio) and kan-DN (5′-Bio) oligonucleotides are biotinylated at their 5′-ends, as the biotin groups are required to measure hybridization of the barcodes to the array.
4. The hybridization mix contains oligonucleotides complementary to common regions used to amplify all of the UP and DN barcodes within the population. Including the complementary oligonucleotides limits cross-hybridization between barcodes. The B213 (5′-Bio) oligonucleotide hybridizes to landmarks within the microarray that facilitate recognition of the proper chip orientation by the scanning software.

Acknowledgements

This research was supported by the Intramural Research Program of the *Eunice Kennedy Shriver* National Institute of Child Health and Human Development (E.C.H. and G.S.) and by the Pharmacology Research Associate Program of the National Institute of General Medical Sciences (E.C.H.).

References

1. Mazurkiewicz P, Tang CM, Boone C, Holden DW (2006) Signature-tagged mutagenesis: barcoding mutants for genome-wide screens. Nat Rev Genet 930:929–939
2. Shoemaker DD, Lashkari DA, Morris D, Mittmann M, Davis RW (1996) Quantitative phenotypic analysis of yeast deletion mutants using a highly parallel molecular bar-coding strategy. Nat Genet 14:450–456
3. Winzeler EA, Shoemaker DD, Astromoff A, Liang H, Anderson K, Andre B, Bangham R, Benito R et al (1999) Functional characterization of the *Saccharomyces cerevisiae* genome by gene deletion and parallel analysis. Science 285:901–906
4. Pierce SE, Fung EL, Jaramillo DF, Chu AM, Davis RW, Nislow C, Giaever G (2006) A unique and universal molecular barcode array. Nat Methods 3:601–603
5. Hobbs EC, Astarita JL, Storz G (2010) Small RNAs and small proteins involved in resistance to cell envelope stress and acid shock in *Escherichia coli*: analysis of a bar-coded mutant collection. J Bacteriol 192:59–67
6. Rooney JP, Patil A, Zappala MR, Conklin DS, Cunningham RP, Begley TJ (2008) A molecular bar-coded DNA repair resource for pooled toxicogenomic screens. DNA Repair (Amst) 7:1855–1868
7. Cherepanov PP, Wackernagel W (1995) Gene disruption in *Escherichia coli*: Tc^R and Km^R cassettes with the option of Flp-catalyzed excision of the antibiotic-resistance determinant. Gene 158:9–14
8. Datsenko KA, Wanner BL (2000) One-step inactivation of chromosomal genes in *Escherichia coli* K-12 using PCR products. Proc Natl Acad Sci U S A 97:6640–6645
9. Yu D, Ellis HM, Lee EC, Jenkins NA, Copeland NG, Court DL (2000) An efficient recombination system for chromosome engineering in *Escherichia coli*. Proc Natl Acad Sci U S A 97:5978–5983

Chapter 6

Use of Semi-quantitative Northern Blot Analysis to Determine Relative Quantities of Bacterial CRISPR Transcripts

Ksenia Pougach and Konstantin Severinov

Abstract

The Northern blot technique is widely used to study RNA. This relatively old method allows one to detect RNA molecules ranging in size from ~20 to thousands of nucleotides and simultaneously estimate the size of an RNA and detect its degradation/processing products. The method does not rely on enzymes such as reverse transcriptases or RNA ligases used in most advanced RNA detection methods, which can be advantageous since biases in detection of individual RNAs can be avoided. We used this approach to the transcripts of Clustered Regularly Interspaced Palindromic Repeats (CRISPR) phage defense loci in *Escherichia coli*. CRISPR loci are transcribed into a single long pre-crRNA, which is then processed at multiple sites to generate ~60 nt fragments (crRNA) each able to mount defense against a specific phage. The Northern blot technique allowed us to estimate the abundance of individual crRNAs and determine stabilities of both pre-crRNA and crRNA. The procedures described in this chapter can be used with very minor modifications to monitor the abundance and stabilities of transcripts of various lengths from many bacterial sources.

Key words: CRISPR, RNA, Northern blot, Short RNAs, Capillary transfer, Electrophoretic transfer

1. Introduction

The existence of short miRNAs and siRNAs in Eukaryotes is a relatively recent discovery, whereas examples of short regulatory RNA molecules in bacteria (sRNAs, ranging from ~50 to 500 nt, see ref. (1) for an excellent review) were known for years (2–4). However, the frequency with which bacteria rely on sRNAs for regulation of diverse processes has not been appreciated until quite recently. New examples of sRNAs in diverse bacteria are discovered on an almost daily basis. A vast majority of these sRNAs are discovered through computational analyses or global approaches such as

Kenneth C. Keiler (ed.), *Bacterial Regulatory RNA: Methods and Protocols*, Methods in Molecular Biology, vol. 905,
DOI 10.1007/978-1-61779-949-5_6, © Springer Science+Business Media, LLC 2012

microarrays or deep sequencing; their functions, thus, remain hypothetical (see ref. (5) for a review). Nevertheless, clear cases of sRNA involvement in control of a wide range of cellular processes in bacteria—from regulation of mobile elements to oxidative stress response are well documented (see refs. (1, 5) for review).

The recently discovered clustered regularly interspaced palindromic repeats (CRISPR)/Cas defense systems of bacteria and archaea function through small RNAs called crRNAs that recognize foreign nucleic acids through complementary interactions and lead to their destruction through mechanisms that remain to be defined (see refs. (6, 7) for review). CRISPR/Cas loci consist of a CRISPR cassette (a segment of DNA containing multiple identical non-perfect palindromic repeats separated by spacers of identical length but variable sequence) and *cas* (for CRISPR associated) genes. The *cas* genes and CRISPR cassettes are transcribed separately, and the CRISPR transcript (pre-crRNA) is processed by one of the *cas* gene products. Processing occurs at sites located within pre-crRNA repeat sequences, generating short crRNAs each containing a spacer sequence bound by fragments of the repeat sequence. Individual crRNAs, in complex with Cas proteins, can specifically target foreign DNA.

Escherichia coli contains several CRISPR cassettes. Popular laboratory K12 strains contain one cassette, CRISPR 1, with 13 repeats (12 spacers), a second cassette, CRISPR 2, with 7 repeats. A third small cassette with 3 repeats is located close to CRISPR2. The repeat sequences in these cassettes are almost identical, and crRNAs generated from pre-crRNAs transcribed from each cassette are ~60 nt long. In our recent study, we wanted to estimate the abundance of crRNAs from different cassettes as well as to determine the relative abundance of individual crRNAs generated from the same pre-crRNA (8). The latter determination was important since it was expected that RNA polymerase (RNAP) may terminate transcription while synthesizing long untranslated pre-crRNA containing multiple palindromic repeats.

The small size made it impossible to estimate individual crRNA quantities using RT-PCR or Real-Time PCR methods. Therefore, we used Northern blots to quantify the amount of each crRNAs by determining hybridization signals with radioactively labeled probes corresponding to various spacers. To allow semi-quantitative estimates of individual crRNAs abundance, we compared intensities of crRNA hybridization signals with signals originating from known amounts of DNA oligonucleotide markers complementary to hybridization probes. The oligonucleotide markers were subjected to electrophoretic separation and blotting along with *E. coli* total RNA samples. To estimate the amounts of long pre-crRNA transcripts, we compared intensities of pre-crRNA hybridization signals with signals originating from known amounts of in vitro transcribed pre-crRNA fragments. In the end, this strategy allowed us to determine intracellular levels of pre-crRNA and individual crRNAs, to put forward a model of CRISPR-cassette transcription and processing, and to predict the existence of a yet-unidentified nuclease that is involved in control of pre-crRNA abundance (8).

2. Materials

2.1. Reagents

1. RNA samples in formamide loading buffer: Dissolve 5–10 μg of total *E. coli* RNA for each sample in Gel loading buffer II (Ambion).
2. Radiolabeled DNA oligonucleotide probe.
3. T4 polynucleotide kinase (PNK) and 10× PNK buffer.
4. [γ-^{32}P]ATP (7,000 Ci/mmol, 150 mCi/mL).
5. Radiolabeled short RNA marker (such as Decade marker system, Ambion—range 10–150 nt).
6. Long RNA marker (such as RiboRuler Low range RNA ladder, Fermentas—range 100–1,000 nt).
7. In vitro transcription: T7 RNAP and 10× T7 RNAP buffer (from New England Biolabs, for example); 0.1 M DDT; 100 mg/mL BSA; NTPs (10 mM each); RNaseOUT (Invitrogen).
8. TBE 5× buffer: 89 mM Tris-borate, 2 mM EDTA, pH 8.3; mix 54 g of Tris base, 27.5 g of boric acid, 20 mL of EDTA (from 0.5 M, pH 8.0 stock), and H_2O (double-distilled) up to 1 L. Store at room temperature.
9. 20% Acrylamide, 7 M urea, TBE 1× solution: mix 19 g of acrylamide, 1 g of methylenebisacrylamide, 42 g of urea, 20 mL of TBE 5×, and H_2O (double-distilled) up to 100 mL. Filter through a 0.45-μm Corning filter and store at 4°C.
10. 7 M urea, TBE 1× solution: mix 42 g of urea, 20 mL of TBE 5×, and H_2O (double-distilled) up to 100 mL.
11. High percentage (12%) denaturing PAAG with 7 M urea: mix 3 mL of 20% acrylamide, 7 M urea, TBE 1× solution, 2 mL of 7 M urea, TBE 1× solution, 50 μL SPS (sodium persulfate, from 10% w/v stock), and 5 μL TEMED (*N*,*N*,*N'*,*N'*-Tetramethylethylenediamine). 5 mL of final solution is sufficient to cast one (7×10×0.085 cm) gel that can be used in Bio-Rad Mini-Protean Tetra cell.
12. Agarose.
13. Formaldehyde.
14. Formaldehyde gel-running buffer 10×: 0.2 M MOPS (pH 7.0), 80 mM $NaCH_3COO$, 10 mM EDTA; prepare with

DEPC (diethylpyrocarbonate)-treated water (see Notes 1 and 2). Dissolve 41.8 g of MOPS in 700 mL of H_2O and adjust pH to 7.0 with NaOH. Then add 26.7 mL of $NaCH_3COO$ (from 3 M, pH 5.2 stock), 20 mL of EDTA (from 0.5 M, pH 8.0 stock), and H_2O (DEPC-treated) up to 1 L.

15. SSC 20×: 3 M NaCl, 0.3 M sodium citrate, pH 7.0. Mix 175.3 g of NaCl, 88.2 g of $Na_3C_6H_5O_7{\cdot}2H_2O$ (sodium citrate), and H_2O (DEPC-treated) up to 1 L. Adjust pH to 7.0 with NaOH.
16. Hybridization buffer: we used ExpessHyb from Clontech but Church buffer (below) works perfectly well too: mix 10 g of BSA, 2 mL of EDTA (from 0.5 M, pH 8.0 stock), 500 mL of sodium phosphate buffer (from 1 M, pH 7.2 stock), 70 g of SDS, and H_2O (DEPC-treated) up to 1 L.
17. Wash solution #1 (SSC 2×, SDS 0.05%, see Note 3): 100 mL SSC (from SSC 20× stock), 5 mL SDS (from 10% stock), and H_2O (double-distilled) up to 1 L.
18. Wash solution #2 (SSC 0.1×, SDS 0.1%): 5 mL of SSC (from SSC 20× stock), 10 mL of SDS (from 10% stock), and H_2O (double-distilled) up to 1 L.
19. Positively charged transfer membrane (such as Hybond XL membrane, GE Healthcare).
20. Whatman 3MM papers (six sheets the size of the gel).

2.2. Equipment

1. Electrophoresis apparatus (we used Mini-Protean Tetra cell, Bio-Rad).
2. Electroblot chamber (such as Mini Trans-Blot Electrophoretic Transfer Cell, Bio-Rad).
3. Crosslinker (such as BioLink DNA Crosslinker, Biometra).
4. Hybridization oven with rotating rack.
5. Glass hybridization tubes.
6. Phosphorimager with phosphor storage cassettes.

3. Methods

3.1. Isolation of RNA and Sample Preparation

Different RNA isolation protocols must be used to prepare samples for long and short RNA Northern blots. When RNAs of interest are expected to be longer than 200 nt (like pre-crRNA, which is about 800–1,100 nt long), the RNeasy (Qiagen) kit can be used. Shorter RNA molecules cannot be isolated with this kit. To purify crRNAs (~60 nt-long) the following protocol with TRIzol reagent (Invitrogen) should be used.

1. Spin down up to 3 mL of *E. coli* culture (OD_{600} = 0.3–0.6), (see Note 4).
2. Add 10 μL of EB buffer (Qiagen) and 5 μL of 20 mg/mL lysozyme water solution.
3. Shake samples well to resuspend cell pellets and let them stand at room temperature for 5 min.
4. Add 0.5 mL of TRIzol reagent to each sample, vortex tubes for about 30 s, and let them stand for 10 min.
5. Add 100–200 μL of cold chloroform. Vortex tubes for about 30 s. Place tubes on ice and incubate until the aqueous and organic phases separate (~10–15 min).
6. Centrifuge at maximum speed for 5–10 min and transfer the upper (aqueous) phase (~300 μL) to a new microcentrifuge tube.
7. Add 5 μL of sodium acetate (pH 5.2) and 900 μL of 96% EtOH, mix well, and leave at −20°C overnight (or at −70°C for 1 h). The amount of sodium acetate used in this protocol is less than 1/10 of the total volume because TRIzol reagent already contains some.
8. Centrifuge tubes for 20 min at maximum speed. Remove and discard the supernatant.
9. Air-dry the pellet, then add 10–15 μL of loading buffer and dissolve the pellet by vortexing and/or pipeting. The pellets are difficult to dissolve. Sometimes heating at 60–70°C for a few minutes is necessary to dissolve them completely.
10. Briefly spin tubes in a microcentrifuge and store them on ice or, for longer storage (up to several months), place at −70°C.

The amount of crRNA in some samples (like in wild-type *E. coli* K12 cells) can be so insignificant that an additional enrichment procedure employing mirVana miRNA isolation kit (Ambion) has to be implemented starting with total *E. coli* RNA prepared from 15 to 20 mL of *E. coli* culture using the TRIzol procedure (above).

11. Mix 50–100 μg of total RNA with 5 volumes of Lysis/Binding buffer.
12. Add 1/10 volume of miRNA Homogenate Additive to the RNA mixture from step 1, and mix well by vortexing.
13. Leave the mixture on ice for 10 min.
14. Add 1/3 volume of 96–100% ethanol, mix well.
15. Place a Filter Cartridge into one of the Collection Tubes supplied with the kit.
16. Pipet the RNA/ethanol mixture from step 4 onto the Filter Cartridge (up to 700 mL).
17. Centrifuge for 1 min at 5,000 × *g*.

18. Collect the *flow-through* and transfer it to a fresh tube.
19. Add 2/3 volume of room temperature 96–100% ethanol to flow-through.
20. Place a Filter Cartridge into one of the Collection Tubes supplied.
21. Pipet the RNA/ethanol mixture from step 9 onto the Filter Cartridge (up to 700 μL).
22. Centrifuge for 1 min at 5,000 × *g*.
23. Discard the flow-through.
24. Apply 700 μL of miRNA Wash Solution 1 to the Filter Cartridge and centrifuge for 1 min at 5,000 × *g*. Discard the flow-through.
25. Apply 500 μL of Wash Solution 2/3 to the Filter Cartridge and centrifuge for 1 min at 5,000 × *g*. Discard the flow-through.
26. Repeat step 15.
27. Centrifuge for an additional 1 min to remove residual ethanol from the filter.
28. Put Filter Cartridge from step 17 into a fresh Collection tube. Apply 50 μL of preheated (95°C) Elution Solution on the membrane. Incubate for 2 min.
29. Centrifuge for 1 min at 10,000 × *g*.
30. Repeat the elution step.
31. Add 10 μL of 3 M sodium acetate (pH 5.2) and 300 μL of 96% EtOH, mix well, and leave at −20°C overnight (or for 1 h at −70°C).
32. Centrifuge for 20 min at maximum speed. Remove and discard the supernatant.
33. Air-dry the pellets, then add 10–15 μL of loading buffer and dissolve the pellet by robust vortexing and/or pipeting.

3.2. Preparation of the Radioactively Labeled Probe

We used CRISPR spacer-specific oligos, which were at least 30 nt in length; shorter oligos led to messier hybridization signals and blots of poorer quality. 5′-end labeling of oligonucleotides was carried out using the following protocol:

1. Mix 0.5 μL DNA oligonucleotide (100 pmol/μL), 2 μL 10× PNK buffer, 4–5 μL [γ-^{32}P]ATP (7,000 Ci/mmol, 150 mCi/mL), 1 μL PNK, H_2O up to 20 μL.
2. Incubate at 37°C for 30 min and then at 65°C for 20 min to inactivate the enzyme.
3. Purify oligos from unincorporated [γ-^{32}P]ATP with Bio-Spin 6 gel-filtration columns (Bio-Rad) using manufacturer's instructions. After completing purification, check the radioactivity of

the eluted sample and remaining radioactivity on the column using a Geiger counter. The residual radioactivity on the column shall be two to three times less than that of the probe (efficiency of labeling at least 70%) (see Note 5).

To prepare radiolabeled Decade marker (Ambion)

1. Mix the following components provided with the marker system: 1 μL of Decade Marker RNA (100 ng), 6 μL of nuclease-free H_2O, 1 μL of kinase reaction buffer, 1 μL of [γ-^{32}P]ATP (7,000 Ci/mmol, 150 mCi/mL), 1 μL of T4 PNK.
2. Incubate at 37°C for 1 h, then add 8 μL of nuclease-free H_2O, 2 μL of 10× cleavage reagent.
3. Incubate at room temperature for 5 min. Add 180 μL Gel Loading Buffer II. Decade Marker can now be loaded on a denaturing polyacrylamide gel.

3.3. Preparation for Quantification of RNA in the Samples

To estimate the quantity of individual crRNAs we used sequentially diluted (e.g., 5, 1 ng/μL, etc.) unlabeled CRISPR spacer-specific DNA oligos complementary to radiolabeled oligonucleotide probes (we used oligos of the same size as radiolabeled oligo probes) which were loaded on the gel in parallel with *E. coli* RNA samples. We compared the intensities of hybridization signals for individual crRNAs with signals from known amounts of DNA oligos.

To estimate the quantity of long pre-crRNA transcripts we used serial dilutions of in vitro synthesized pre-crRNA fragments. A PCR-fragment encompassing part of the *E. coli* CRISPR cassette was prepared and purified with QIAquick Gel Extraction kit (Qiagen). One of the PCR primers was designed to contain a sequence of the T7 RNAP promoter on its 5′ end, so the resulting PCR product also had the T7 RNAP promoter sequence (see Note 6). The amplified DNA fragment was transcribed in vitro with T7 RNAP using the following protocol.

1. Mix in the microcentrifuge tube: 10 μL of 10× T7 RNAP buffer, 10 μL of DDT (0.1 M), 1 μL of BSA (100 mg/mL), 10 μL of NTPs (10 mM each), 5–20 pmol of PCR-product (purified), 2.5 μL of RNaseOUT, 5 μL of T7 RNAP, H_2O (DEPC-treated) up to 100 μL.
2. Incubate at 37°C for 1 h. Prior to purification, DNase I treatment is recommended to get rid of the DNA template.
3. Purify in vitro transcribed pre-crRNA fragment with the BioSpin 6 gel-filtration columns using manufacturer's instructions.
4. Estimate the quantity of purified pre-crRNA product with Nanodrop or any other detection system. Prepare serial dilutions of in vitro transcribed RNA to be loaded on the gel.

3.4. Denaturing PAAG Electrophoresis of crRNAs

Our procedure is based on the "Northern Blots for Small RNA and MicroRNAs Protocol" described in Hannon et al. "RNA a laboratory manual" (9) with some adjustments and modifications.

1. Cast a 12% denaturing PAAG gel.
2. Heat total RNA samples for 5 min at 95°C, then chill on ice for 1 min. Briefly spin tubes in a microcentrifuge.
3. Load samples on the gel in the following order: RNA samples under investigation, serial dilutions of complementary oligo, RNA ladder marker.
4. Run the gel in 1× TBE at constant voltage (200–250 V) until the bromophenol blue dye almost reaches the bottom of the gel.

3.5. Electrophoretic Transfer for PAAG Gel

All procedures are carried out at room temperature.

1. Separate glass plates and transfer the gel on a Whatman 3MM paper pre-wetted in 0.5× TBE.
2. Open the transfer cassette. Soak both sponges and remaining five Whatman 3MM papers in 0.5× TBE. Assemble cassette in the following order (Fig. 1): negative pole (cathode), sponge, three Whatman 3MM papers, gel, transfer membrane (after placing the membrane on the gel carefully roll it over with a plastic pipette to remove any bubbles between the membrane and the gel) (see also Note 7), three Whatman 3MM papers, sponge, positive pole (anode).
3. Clamp the cassette closed.
4. Place cassette into transfer chamber filled with 0.5× TBE and turn the current on for 1.5 h at 200 mA.
5. Proceed to Subheading 3.8. All procedures following the transfer stage are identical for the short and long RNA detection methods.

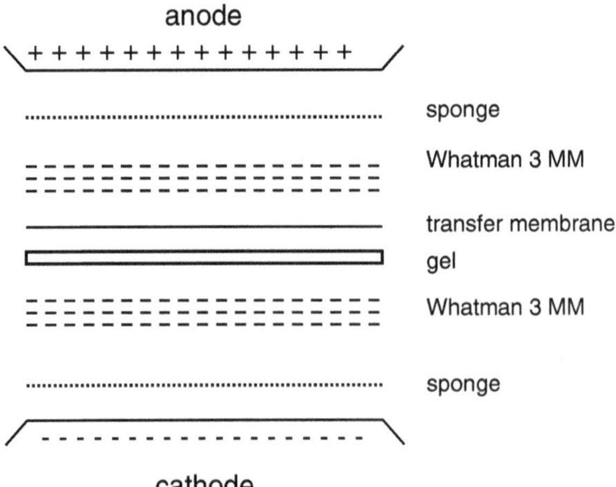

Fig. 1. A scheme of cassette assembly for electrophoretic transfer of short RNAs.

3.6. Formaldehyde-Agarose Gel Electrophoresis of Pre-crRNA

We used classical formaldehyde-agarose protocol for Northern blots described in Sambrook et al. (10) with only minor modifications.

1. To prepare an agarose gel containing formaldehyde, melt 1 g of high-grade low-melting agarose in 70 mL of DEPC-treated H_2O and cool the solution to 55–65°C. Add 22 mL of formaldehyde gel-running buffer 5× and 20 mL of 37% formaldehyde. Use a chemical hood because formaldehyde is highly vaporous, especially when warm. The resulting solution suffices to cast one 11×13×0.5 cm gel.
2. To prepare the samples mix the following in a microcentrifuge tube: 4.5 μL of RNA analyzed (5–15 μg), 2 μL of 5× formaldehyde gel-running buffer, 3.5 μL of formaldehyde, 10 μL of formamide. Long RNA marker is prepared in the same way.
3. Heat samples at 70°C for 10 min then chill on ice for 1 min. Spin down the fluid condensed on cap and walls of the tube in a microcentrifuge.
4. Add 5 μL of 5× GelPilot loading dye (Qiagen) (or any agarose gel loading buffer).
5. Prerun the gel for 5 min at 5 V/cm and immediately load the samples and the marker on a gel.
6. Run the gel at 30–50 V until the bromphenol blue migrates ~2/3 of the length of the gel.
7. Cut off the marker lane—it will be stained with ethidium bromide (or GelStar, Lonza) and then used to estimate molecular weights of RNA products (see Note 8).

3.7. Capillary Transfer for Formaldehyde-Agarose Gels

All procedures are carried out at room temperature.

1. Minimize the size of the gel by cutting off edges that do not contain sample lanes.
2. Take a glass baking dish of appropriate size and put a piece of Plexiglass (we use an agarose gel casting tray from the same gel turned upside down) in it. Pour the transfer buffer (10× SSC) as shown on Fig. 2.
3. Place a piece of rectangular Whatman 3MM paper pre-wetted with 10× SSC on the Plexiglass piece such that the right and left sides of the paper sheet are submerged in transfer buffer.
4. Place the gel atop of the paper and then cover it with transfer membrane pre-wetted in 10× SSC. Make sure that there are no air bubbles left between the gel and the membrane.
5. Cover the membrane with two pre-wetted Whatman 3MM sheets of the same size as the gel.

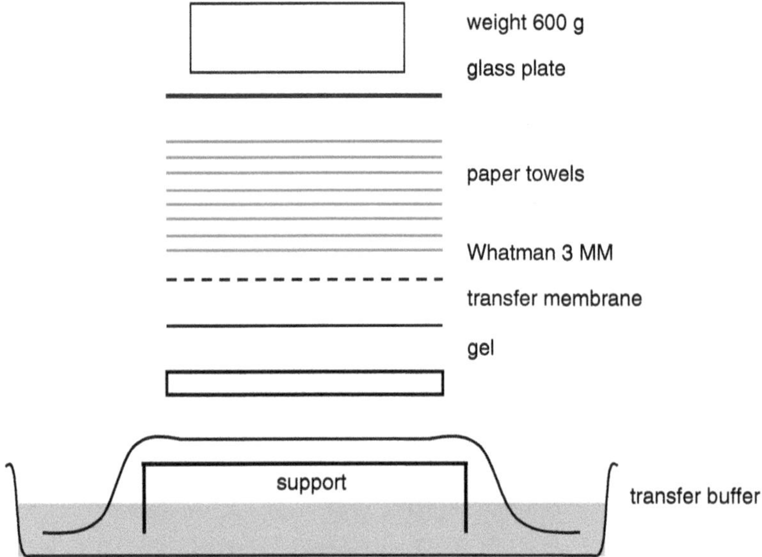

Fig. 2. A scheme of capillary transfer of long RNA from agarose gels.

6. Mount a stack of dry paper towels (about 5–10 cm high) cut to the size of just a little bit smaller than Whatman 3MM sheets atop of gel assembly.
7. Cover the stack with a glass plate and put it under weight (about 600 g).
8. Leave at room temperature for 4 h to allow the transfer. Replace paper towels as they get wet.
9. Proceed to Subheading 3.8. All procedures following the transfer stage are identical for the short and long RNA detection methods.

3.8. Cross-Linking

1. Disassemble the transfer unit and open cassette. Mark gently the side of the membrane that was facing the gel with a pencil.
2. Completely dry the membrane at room temperature.
3. Place the dried membrane marked side up in the cross-linker and expose to UV light (254 nm, 1,250 mJ (optimal cross-link settings in our machine) (see Note 9).

3.9. Prehybridization and Hybridization

1. Put the membrane into hybridization tube (the marked side facing the inside of the tube).
2. Add 5 mL of ExpressHyb (or 15–20 mL of Church buffer). Gently remove any bubbles between the membrane and the wall of the tube with forceps. Try not to damage the membrane.
3. Close the tube and prehybridize by placing the tube for 30 min in an oven at medium rotation speed at 37°C for

30 min (if using ExpressHyb buffer) or for 1.5 h at 42°C in the case of Church buffer.

4. Open the tube, discard hybridization solution, and then add 3 mL of ExpressHyb buffer or 5 mL of Church buffer. Add radiolabeled oligonucleotide (50–100 pmol of purified labeled oligo in 20–40 μL) (see Note 10).
5. Hybridize at 37°C for 2 h (in case of ExpressHyb buffer) or overnight at 42°C in the case of Church buffer.

3.10. Washing

Performed at room temperature with ExpressHyb buffer or at 37°C with Church buffer.

1. Dispose of radioactive hybridization solution.
2. Rinse the tube with 5–10 mL of Wash solution #1.
3. Add 30 mL of Wash solution #1. Close the tube, return it in the oven and leave it for 20 min at appropriate temperature (see above) at high rotation speed. Repeat this step one more time.
4. Rinse the tube with Wash solution #2.
5. Add 30 mL of Wash solution #2 and wash for 20 min as above.
6. Withdraw the membrane and wrap it with plastic wrap. Do not dry the membrane if you plan to reuse it with a different oligonucleotide probe—see below).
7. Expose the membrane to a phosphor screen for at least 6 h. Analyze the image on the Phosphorimager (see Note 11).

A representative result obtained with a crRNA probe specific for a CRISPR 1 cassette spacer is presented in Fig. 3. A representative result obtained with CRISPR 1 cassette pre-crRNA (a total length of ~800 nt) is presented in Fig. 4.

3.11. Reusing the Membrane

Though reusing the membrane is generally not recommended since some RNA will be lost during washing, repeated use for abundant RNAs (for example, 5S rRNA, which can be used as a loading control) is possible (see Note 11).

1. Place the membrane in the hybridization tube. Add 50 mL of 0.5% SDS preheated to 95°C.
2. Place the tube into hybridization oven rack and rotate for 10 min at 95°C.
3. Let the tube cool down and take the membrane out. Now it can be reused with a different oligonucleotide probe (it is a good practice to first expose the washed membrane to a phosphor screen to confirm that no traces of prior hybridization are left).

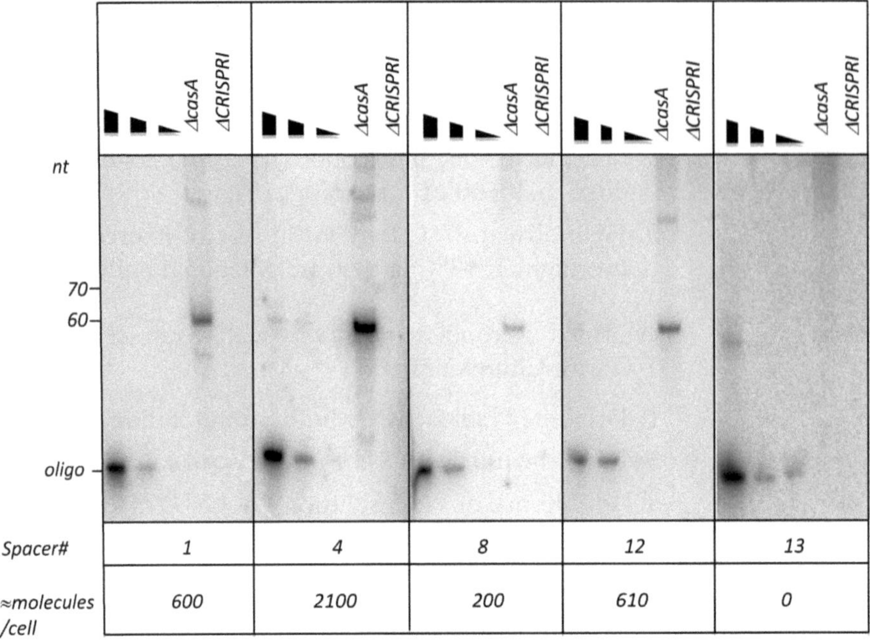

Fig. 3. An example of semi-quantitative Northern blots with short RNA. Total RNA purified from *Escherichia coli* cells was used for Northern blotting with probes specific for CRISPR I cassette spacers indicated. In each panel, the first three lanes contained increasing amounts of DNA oligonucleotides complementary to probes used for blotting. Approximate numbers of individual processed crRNA per cell are shown below.

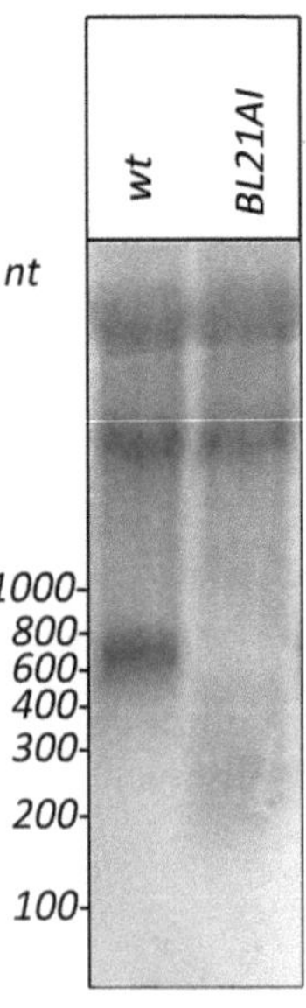

Fig. 4. A Northern blot showing the results of pre-crRNA analysis. A repeat-specific probe was used for hybridization. The pre-crRNA is indicated with an *arrow* (the BL21A1 *E. coli* cells were used as control; they naturally lack CRISPR cassettes (11)). Additional labeled bands present in both types of cells are due to nonspecific hybridization with 16S and 23S ribosomal RNA.

4. Notes

1. DEPC-treated (and therefore RNase-free) water is used to reduce the risk of RNA being degraded by RNases. To prepare DEPC-treated water add 100 μL of DEPC for every 100 mL of water. Incubate at 37°C with agitation for at least 1 h (better overnight) and then inactivate DEPC by autoclaving for 30 min. Inactivation leads to degradation of DEPC to CO_2, H_2O, EtOH.
2. Formaldehyde gel-running buffer may turn yellow with age, when exposed to light, or autoclaved. For best results, it is recommended to use fresh or slightly yellow buffer solution.
3. SDS precipitates if mixed directly with SSC or cold water, so it is recommended to mix warm water (at least room temperature) with SSC first, and then add SDS.
4. If working with cells from stationary phase use no more than 0.5 mL of culture for every 500 μL of TRIzol reagent.
5. Purified labeled oligos shall be used within a few days. Radiolabeled probes stored for longer periods of time give blots of poorer quality.
6. When choosing which of the PCR primers (forward or reverse) to add the T7 RNAP promoter sequence, keep in mind the orientation of the radiolabeled probe used for the Northern blot. We used a probe complementary to CRISPR repeat sequence, and in vitro transcribed RNA was identical to pre-crRNA and thus complementary to hybridization probe. It is also recommended to add 4–5 extra bases upstream of the T7 RNAP promoter when designing the primer, which increases the overall efficiency of in vitro transcription.
7. Transfer membrane should be first pre-wetted with double-distilled water and then with 0.5× TBE. Disposable plastic pipettes can be used to roll out air bubbles.
8. Stain a fragment of the gel containing the marker lane in ethidium bromide solution (~1 μg/mL) or GelStar (Lonza) with gentle agitation for about 10–15 min. Align with a fluorescent ruler on the surface of a transilluminator, photograph, and print the image. Compare appropriately sized print of the marker lane and fluorescent ruler with the blot to estimate molecular sizes of hybridization products.
9. A membrane with cross-linked RNA molecules can be stored at room temperature for a few (1–3) days without significant decrease in the quality or resulting blots.
10. 5S rRNA from total RNA can sometimes unspecifically hybridize with radiolabeled probe thus obscuring the specific signal

and reducing the overall quality of the blot. In such cases, 20–50 pmol of an unlabeled ~30-nt long oligo complementary to any part of the 5S rRNA molecule can be added to hybridization tube together with radiolabeled probe.

11. The results of Northern blot analysis report the steady-state levels of transcripts, which are determined by both the rate of transcript synthesis and the rate of transcript degradation. The latter rate can be estimated using antibiotic rifampicin, which is known to block transcription initiation by bacterial RNAP. Since cells treated with rifampicin are unable to synthesize new RNA, only preexisting molecules that can be detected. By following the decrease in hybridization signal over time after the addition of rifampicin, the first-order decay rate constant can be obtained. With *E. coli*, rifampicin concentrations of ~150 μg/1 mL of bacterial culture works well (8). With other bacteria, preliminary experiments to determine which rifampicin concentration blocks transcription might be needed. Aliquots of cells taken at different time points after rifampicin treatment are analyzed by Northern blotting using the procedures described above and half-life times of hybridized RNAs are estimated. This in turn allows one to estimate the transcription rate.

Acknowledgement

This work was supported by an NIH grant GM59295 to KS.

References

1. Walters LS, Storz G (2009) Regulatory RNAs in bacteria. Cell 136:615–628
2. Tomizawa J, Itoh T, Selzer G, Som T (1981) Inhibition of ColE1 RNA primer formation by a plasmid-specified small RNA. Proc Natl Acad Sci U S A 78:6096–6100
3. Stougaard P, Molin S, Nordström K (1981) RNAs involved in copy-number control and incompatibility of plasmid R1. Proc Natl Acad Sci U S A 78:6008–6012
4. Simons RW, Kleckner N (1983) Translational control of IS10 transposition. Cell 34:683–691
5. Thomason MK, Storz G (2010) Bacterial antisense RNAs: how many are there and what are they doing? Annu Rev Genet 44:167–188
6. Al-Attar S et al (2011) Clustered regularly interspaced short palindromic repeats (CRISPRs): the hallmark of an ingenious antiviral defense mechanism in prokaryotes. Biol Chem 392: 277–289
7. Marraffini LA, Sontheimer EJ (2010) CRISPR interference: RNA-directed adaptive immunity in bacteria and archaea. Nat Rev Genet 11: 181–190
8. Pougach K et al (2010) Transcription, processing and function of CRISPR cassettes in *Escherichia coli*. Mol Microbiol 77(6):1367–1379
9. Rio DC, Ares M Jr, Hannon GJ, Nilsen TW (2011) RNA: a laboratory manual. Cold Spring Harbor Laboratory Press, Cold Spring Harbor
10. Sambrook J, Russel DW (1989) Molecular cloning. A laboratory manual, 3rd edn. Cold Spring Harbor Laboratory Press, Cold Spring Harbor
11. Jeong H et al (2009) Genome sequences of *Escherichia coli* B strain REL606 and BL21(DE3). J Mol Biol 394(4):644–652

Chapter 7

RNA Visualization in Bacteria by Fluorescence In Situ Hybridization

Jay H. Russell and Kenneth C. Keiler

Abstract

Detecting localized RNA in bacteria is difficult due to the properties of RNA and the small size of the cell. Fluorescence in situ hybridization (FISH) has been an invaluable method for detecting and imaging RNA. In FISH, RNA is fixed in its native subcellular position through chemical cross-linking. An oligonucleotide probe conjugated to a fluorophore is annealed to the target RNA, and the target RNA/probe hybrid is visualized using fluorescence microscopy. This chapter describes the use of FISH to visualize tmRNA, a regulatory RNA required for *trans*-translation. The method can be adapted to visualize the localization of other regulatory and messenger RNAs as well.

Key words: RNA localization, FISH, Fluorescence microscopy, RNA, Regulatory RNA, tmRNA

1. Introduction

Studies of RNA dynamics in bacteria have demonstrated that model RNA molecules can diffuse throughout the cell after they are released from RNA polymerase (1, 2). Nevertheless, some natural RNAs are localized to a particular subcellular site or structure within bacteria (3, 4). RNA localization has been proposed to play a role in controlling gene expression, RNA turnover, *trans*-translation, and protein secretion (5). Two types of techniques have been used to visualize RNAs in bacterial cells: fluorescence in situ hybridization (FISH), and indirect labeling using fluorescent fusion proteins (3, 4, 6–8). These are complementary techniques, with different advantages and drawbacks. In FISH, a fluorescently labeled oligonucleotide probe is annealed to the RNA of interest in fixed cells. FISH does not require any modifications to the native RNA, but cell fixation can produce artifacts and prevents studies on live cells. Indirect labeling approaches use an RNA binding

Kenneth C. Keiler (ed.), *Bacterial Regulatory RNA: Methods and Protocols*, Methods in Molecular Biology, vol. 905, DOI 10.1007/978-1-61779-949-5_7, © Springer Science+Business Media, LLC 2012

protein fused to GFP or other fluorescent protein. For RNAs that do not have a specific protein binding partner, the RNA sequence must be modified to include a protein-binding motif such as an MS2 aptamer or an eIF4A-binding sequence (2, 9). The RNA can then be observed by producing MS2-GFP or eIF4A-GFP. Indirect labeling can be used in live or fixed cells, allowing time-lapse studies of individual cells, but modification of the RNA and binding of an exogenous protein could alter the localization. Ideally, FISH and indirect labeling techniques should be used in tandem to take advantage of their complementary strengths.

FISH was first adapted for RNA localization studies in bacteria to visualize the location of tmRNA in *Caulobacter crescentus* (3). tmRNA and its protein binding partner, SmpB, are the key molecules in *trans*-translation, a pathway for releasing stalled ribosomes that is found in all bacteria (10, 11). In *C. crescentus*, tmRNA was localized in a helix-like pattern that changed as a function of the cell cycle (3). This tmRNA localization pattern was verified in live cells by indirect labeling using an SmpB-GFP fusion protein. In addition, a combination of FISH and immunofluorescence (IF) was used to simultaneously visualize tmRNA and SmpB in the same cells, confirming their co-localization (3).

To perform FISH, the bacterial cells are first fixed and the RNA is partially denatured. A fluorescent oligonucleotide probe is added and allowed to anneal with the RNA, excess probe is washed away, and the cells are visualized using fluorescence microscopy. The protocol for visualizing tmRNA is described below, and similar procedures can be used for other RNAs in other bacterial species. However, several steps are variable and will need to be optimized for different systems. For example, appropriate fixation times will depend on the bacterial species and growth conditions. Likewise, the probe sequence depends on the RNA that will be visualized, and hybridization and washing procedures will vary according to the affinity and specificity of the probe for the target RNA. These variables are outlined in the Subheading 4.

2. Materials

All solutions should be filter-sterilized and prepared using ultrapure RNase-free water (prepared by purifying deionized water to attain a sensitivity of 18 Mω × cm at 25°C) and analytical grade reagents. Prepare and store all solutions at 25°C unless otherwise indicated.

2.1. Solutions for Cell Fixation

1. M2 minimal salts media with glucose (M2G): Weigh 0.865 g of dibasic anhydrous sodium phosphate, 0.530 g of monobasic potassium phosphate, 0.497 g of ammonium chloride, and

0.060 g of magnesium sulfate. Add 989.0 mL of ddH_2O and mix. While mixing, add 1.0 mL of ferrous sulfate [EDTA chelate, Sigma Chemical Company]. Add 10 mL of 20% glucose solution, for a final glucose concentration of 0.2% and mix.

2. 1× PBS (pH 7.4): Weigh 8.18 g of sodium chloride, 0.224 g of potassium chloride, 1.14 g of dibasic anhydrous sodium phosphate, 0.204 g of monobasic anhydrous potassium phosphate. Transfer chemicals to a 1-L glass beaker and mix with 100 mL of ddH_2O. Make solution up to 1 L with water and check pH using a pH meter, buffer should be approximately pH 7.4.
3. 5× Fix solution: Weigh 0.216 g of monobasic sodium phosphate and mix with 2.0 mL of ddH_2O. Adjust pH to 7.5 with sodium hydroxide. Add 10.0 mL of 16% formaldehyde and mix. Store at 4°C.

2.2. Solutions for Cell Attachment

1. Glucose Tris-EDTA (pH 7.5) (GTE): Weigh 0.02 g of tris, 0.9 g of glucose and mix with 80.0 mL of ddH_2O. Add 10.0 mL of 0.5 M EDTA (pH 8.0) and mix. Adjust pH to 7.5 with HCl. Raise final volume to 100.0 mL with ddH_2O.
2. 0.1% Polylysine solution: Weigh 0.1 g of polylysine and mix with 10 mL of ddH_2O. Make 1.0 mL aliquots and freeze at −20°C.
3. Lysozyme solution: Weigh 50 μg of lysozyme (chicken egg white). Transfer to Eppendorf tube with 1.0 mL of GTE and mix.
4. Microscopy slides and coverslips.
5. 7 N HCl.
6. 0.1% Triton X-100: Add 50 μL of 100% Triton X-100 into 49.95 mL of ddH_2O.

2.3. Solutions and Materials for In Situ Hybridization

1. 2× SSCT (pH 7.0): Weigh 0.876 g of sodium chloride and 0.435 g of sodium citrate and mix with 10 mL of ddH_2O. Add 50 μL of Tween-20. Adjust the volume to 50 mL with ddH_2O and adjust pH to 7.0 with sodium hydroxide. Final concentrations should be 300 mM sodium chloride, 30 mM sodium citrate, and 0.1% Tween-20. Make 1.0 mL aliquots and freeze at −20°C.
2. 2× SSCT, 25% formamide (pH 7.0): Prepare the 2× SSCT as previously described with the following exception. Add 12.5 mL of formamide after mixing sodium chloride, sodium citrate, and Tween-20. Adjust the volume 50 mL with ddH_2O and adjust pH to 7.0 with sodium hydroxide. Make 1.0 mL aliquots and freeze at −20°C.
3. 2× SSCT, 50% formamide (pH 7.0): Prepare the 2× SSCT as previously described with the following exception. Add 25 mL of formamide after mixing sodium chloride, sodium citrate, and Tween-20. Adjust the volume 50 mL with ddH_2O and

adjust pH to 7.0 with sodium hydroxide. Make 1.0 mL aliquots and freeze at −20°C.

4. 1× Hybridization solution: Weigh 1.75 g of sodium chloride, 0.882 g of sodium citrate, and 1 g of dextran sulfate. Mix reagents in a 15-mL conical tube containing 5.5 mL of formamide. Adjust the volume to 10 mL with ddH_2O. Make 1.0 mL aliquots and freeze at −20°C.
5. Glass or plastic dish with tight-fitting lid.
6. Whatman filter paper cut to fit the base of the glass or plastic dish.
7. Plastic wrap.
8. Heat block set at 95°C.
9. Cy-3 fluorophore conjugated tmRNA probe (20 ng/μL) in hybridization solution (see Notes 1 and 2).

2.4. Solutions and Materials for Post Hybridization

1. 2× SSCT, 50% formamide (pH 7.0); 2× SSCT, 25% formamide (pH 7.0); 1× PBS (pH 7.4) prepared as described in Subheading 2.3.
2. 2× SSCT containing 100 μg/mL of DAPI: Prepare 2× SSCT as previously described in Subheading 2.3. Add 10 μL of 10 mg/mL DAPI (see Note 2).
3. Mounting media, such as Cytoseal Mounting media.
4. Epi-fluorescence or confocal microscope equipped with a CCD binning black and white camera (see Note 3).

3. Methods

3.1. Cell Fixation

1. Pick a single colony from a plate containing the appropriate antibiotic and grow overnight in minimal salts medium, such as M2G for *C. crescentus* (see Note 4).
2. Dilute the overnight culture 1:500 into 5 mL of fresh M2G containing the appropriate antibiotic and grow to mid-logarithmic phase or to the desired physiological condition.
3. Add 1/4 volume of 5× Fix solution and incubate for 15 min with shaking at desired temperature (see Notes 5 and 6).
4. Quickly move the culture to ice and incubate for 30 min.

3.2. Slide Preparation

Slide preparation steps can be carried out during cell fixation.

1. Wash standard microscopy slides and coverslips with 7 N HCl for 5 min while rocking (see Note 7).

2. Immediately wash the slides and coverslips three times with ddH_2O for 5 min while rocking. Set aside to air dry. After slides and coverslips have dried, they can be stored in a covered container for later use.
3. Move the coverslips to a flat service and spot 100 μL of 0.1% polylysine solution and incubate for 5 min.
4. Aspirate off the polylysine solution and spot 100 μL of ddH_2O on the same area of the coverslip.
5. Incubate the coverslips with ddH_2O for 5 min and aspirate off the liquid.
6. Allow the coverslips to dry completely.

3.3. Hybridization Chamber Set-Up

1. Cut two pieces of Whatman paper to fit the bottom of a glass dish (a plastic container can be used instead).
2. Saturate the Whatman paper with ddH_2O.
3. Place a piece of plastic wrap over the glass dish.
4. Cover the sides and bottom of the glass dish with aluminum foil so no light can enter when the dish is covered.
5. Place the lid of glass dish over the plastic wrap and pre-warm at 37°C to produce a humid chamber (Fig. 1).

3.4. Prehybridization and Hybridization

When setting up slides for hybridization, make a mock control slide that contains no probe that can be used to compare with slides containing probe.

1. Incubate the fixed cells from Subheading 3.1 have incubated on ice for 30 min, centrifuge 0.5 mL of the fixed cells for 2 min at 16,000 ×*g* and aspirate off the supernatant (see Note 8).
2. Wash the cells by resuspending the pellet in 1.0 mL of PBS, centrifuging for 2 min at 16,000 ×*g*, and aspirating off the supernatant. Repeat this step three times.
3. After the final wash, resuspend the cells in 100 μL of GTE.
4. Add 1.0 μL of Lysozyme Solution and incubate for 4 min at 25°C.

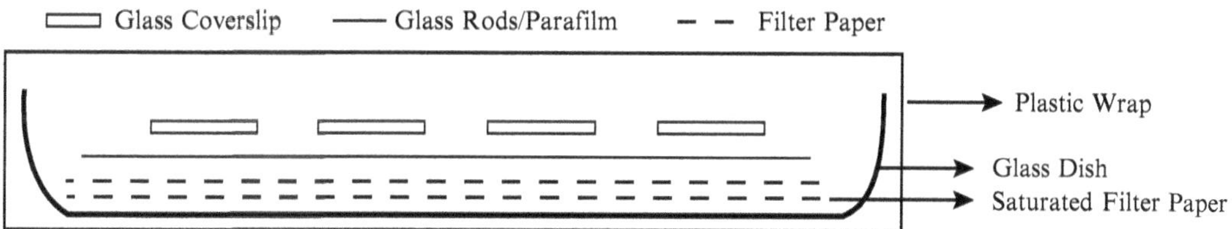

Fig. 1. Hybridization chamber set-up for FISH. Whatman filter paper is placed in a glass or plastic dish and saturated with RNase-free ddH_2O. Glass rods are placed on top of the saturated filter paper to act as a platform and barrier for coverslips. Coverslips with fixed cells are placed on top of the glass rods, and the glass or plastic dish is sealed with plastic wrap.

5. Centrifuge for 2 min at 3,341 × *g* (see Notes 9–11).
6. Gently resuspend the cells in 100 mL of GTE and spot 100 μL onto polylysine-treated coverslips (see Note 12).
7. After incubating for 5 min, aspirate off most of the excess liquid leaving only a thin liquid film (see Note 13).
8. Allow the cells to air dry completely on the coverslip.
9. Spot 100 μL of 0.1% Triton X-100 on the cells and incubate for 5 min at 25°C. Aspirate off most of the excess liquid and allow to air dry completely.
10. Wash by spotting 100 μL of 2× SSCT on the cells, incubating the coverslips for 5 min at 25°C in the humid chamber, and aspirating off the liquid. Repeat this wash step twice.
11. Spot 100 μL of 2× SSCT, 50% formamide on the cells, and incubate the coverslips in the humid chamber at 37°C for 30 min (see Notes 14 and 15).
12. During this incubation, prepare the probe solution: Add probe to a final concentration of 20 ng/μL in 100 μL of hybridization. Keep the probe solution wrapped in aluminum foil and on ice (see Note 16).
13. Heat the probe solution at 95°C for 5 min, immediately centrifuge for 1 min at 16,000 × *g*, and place on ice (see Note 17).
14. After the incubation in step 11 is complete, aspirate off the liquid removing as much as possible without disrupting the fixed cells (see Note 13).
15. Heat the probe and target RNA by spotting 25 μL of the probe solution on the fixed cells and placing the coverslips directly on a 20-well 95°C heat block with wells filled with water for 4 min, shielded from light (see Note 18).
16. Place coverslips in the humid chamber, seal the humid chamber with plastic wrap and lid, and place the sealed humid chamber at 42°C for 8–12 h (see Notes 19 and 20).

3.5. Post Hybridization

1. Keeping the coverslips in the humid chamber, aspirate off the probe solution (see Note 13).
2. Wash out probe that has not annealed by spotting 100 μL of 2× SSCT, 25% formamide on the cells, sealing the humid chamber with plastic wrap, and incubating the humid chamber at 37°C for 30 min.
3. Aspirate off the liquid.
4. Stain with DAPI by spotting 100 μL of 2× SSCT containing 100 μg/mL of DAPI on the fixed cells, and incubating coverslips in the humid chamber at 25°C with rocking for 5 min (see Notes 2 and 21).

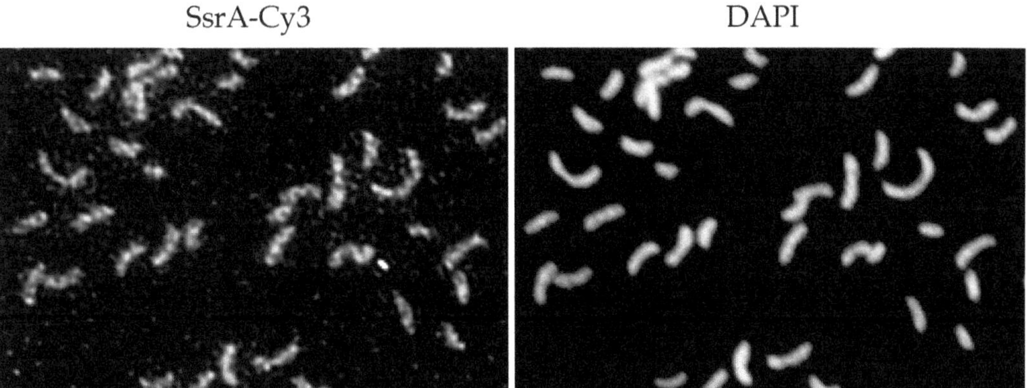

Fig. 2. Localization of tmRNA in *Caulobacter crescentus*. *C. crescentus* cells were probed for tmRNA using FISH with the fluorescently labeled probe SsrA-Cy3 (*left*). The same cells stained with DAPI are shown on the *right*

5. Aspirate off the liquid, spot 100 μL of 2× SSCT on the fixed cells, and incubate at 25°C with rocking for 5 min. Repeat this step three times.
6. After the final 2× SSCT wash, aspirate off the liquid, spot 100 μL of PBS on the fixed cells, and incubate the coverslips in the humid chamber with rocking at 25°C for 5 min. Repeat this step 5–10 times (see Note 22).
7. After the final PBS wash, aspirate off the liquid and air dry the coverslips.
8. Apply a single drop of mounting media to the coverslips with fixed cells, and place the coverslips on an acid washed microscopy slide.
9. Gently press down on the coverslips and wipe away excess mounting media with a tissue (see Note 23).
10. The slide is ready to be viewed, captured, and analyzed using fluorescence microscopy (Fig. 2). Exposing the sample to too much light will photobleach the fluorophore, therefore when viewing samples it is best to use short exposures to limit photobleaching (see Note 24).

4. Notes

1. The DNA or RNA oligonucleotide probe design should be the reverse complement sequence from an unstructured region of the target RNA. Ideally the probe should have a GC content of ~50%, and anneal to the target with a high melting temperature. These factors are important for constant hybridization kinetics. The oligonucleotide probe can be ordered with a fluorophore

or dye conjugated to it. An ideal fluorophore/dye to have conjugated to the probe should have a high quantum yield (emission efficiency) and longer half-life.

2. The emission and excitation properties of each fluorophore used should be checked to ensure there is no spectral overlap. In case of spectral overlap between fluorophores, there are several fluorescent nucleic acid stains with a wide range of emission and excitation properties.
3. CCD binning increases pixel size and detection sensitivity.
4. Minimal salts medium reduces background fluorescence when analyzing samples.
5. The fixation step may vary depending on growth rate of the bacterium or the stage of growth (lag, exponential, or stationary phase). Be Fast!!! You want to "capture" the RNA before the cell senses any environmental changes.
6. Our best fixation results came from diluting fresh 5× Fix solution. The 5× Fix solution was kept at 4°C for no longer than 1 week.
7. Washing with HCl removes any dust, smudges, etc. that may interfere with the probe and fluorophores.
8. The supernatant contains paraformaldehyde that will need to be disposed of in the proper chemical waste.
9. This protocol was optimized for *C. crescentus*. For other species, the lysozyme concentration and incubation times will need to be empirically determined.
10. The lysozyme treatment helps to weaken the cell wall of the bacteria.
11. The pellet should appear opaque and diffuse. Cells are centrifuged at low speeds to reduce cell lysis.
12. At this stage gently resuspend pellet. The cells are fragile because most of the cell wall has been removed.
13. Do not let the tip of the aspirator touch the coverslip. The ideal technique is to hold the aspirator at a 45° angle where the tip of the aspirator only touches the edge of the liquid. If the tip of the aspirator touches the coverslip the force from the vacuum will pull apart cells that have adhered.
14. Formamide will help to reduce secondary structure in RNA.
15. If an incubator with a rocking platform is available, rock the humid chamber during heated incubations.
16. Always shield fluorophores from direct light. Light can cause fluorophores to decay rapidly.
17. Heating the probe and snap chilling it on ice will help reduce secondary structure and facilitate annealing to the target RNA.

18. Heating the fixed cells and probe solution will further denature cellular RNA, enabling the probe to hybridize to the target.
19. This hybridization protocol was optimized for *C. crescentus* tmRNA. The hybridization temperature will need to be empirically determined for different RNAs.
20. Glass rods and parafilm act as a good barrier to keep coverslips containing fixed cells with hybridization solutions from direct contact with wet Whatman paper. Glass rods also provide a platform for the coverslips that will make handling the coverslips easier in future steps.
21. DAPI is a nucleic acid stain and allows visualization of the cells by staining the nucleoid.
22. PBS washing helps to remove concentrated salts from SSCT solutions as well as unhybridized probe.
23. Mounting Media adheres the coverslip and slide. Check to make sure the mounting media contains an anti-fade reagent. This will preserve the life span of the fluorophores.
24. Confocal microscopy is recommended because it provides a sharper image. Alternatively, the combination of epi-fluorescence microscopy and deblurring software can be used.

References

1. Valencia-Burton M, Shah A, Sutin J, Borogovac A, McCullough RM, Cantor CR, Meller A, Broude NE (2009) Spatiotemporal patterns and transcription kinetics of induced RNA in single bacterial cells. Proc Natl Acad Sci U S A 106:16399–16404
2. Golding I, Cox EC (2004) RNA dynamics in live *Escherichia coli* cells. Proc Natl Acad Sci U S A 101:11310–11315
3. Russell JH, Keiler KC (2009) Subcellular localization of a bacterial regulatory RNA. Proc Natl Acad Sci U S A 106:16405–16409
4. Nevo-Dinur K, Nussbaum-Shochat A, Ben-Yehuda S, Amster-Choder O (2011) Translation-independent localization of mRNA in *E. coli*. Science 331:1081–1084
5. Keiler KC (2011) RNA localization in bacteria. Curr Opin Microbiol 14:155–159
6. Pilhofer M, Pavlekovic M, Lee NM, Ludwig W, Schleifer KH (2009) Fluorescence in situ hybridization for intracellular localization of nifH mRNA. Syst Appl Microbiol 32: 186–192
7. Montero Llopis P, Jackson AF, Sliusarenko O, Surovtsev I, Heinritz J, Emonet T, Jacobs-Wagner C (2010) Spatial organization of the flow of genetic information in bacteria. Nature 466:77–81
8. Block KF, Puerta-Fernandez E, Wallace JG, Breaker RR (2011) Association of OLE RNA with bacterial membranes via an RNA-protein interaction. Mol Microbiol 79:21–34
9. Valencia-Burton M, McCullough RM, Cantor CR, Broude NE (2007) RNA visualization in live bacterial cells using fluorescent protein complementation. Nat Methods 4:421–427
10. Hong SJ, Tran QA, Keiler KC (2005) Cell cycle-regulated degradation of tmRNA is controlled by RNase R and SmpB. Mol Microbiol 57:565–575
11. Keiler KC (2008) Biology of trans-translation. Annu Rev Microbiol 62:133–151

Part III

RNA Structure and Biophysical Measurements

Chapter 8

RNA Structure Prediction: An Overview of Methods

Matthew G. Seetin and David H. Mathews

Abstract

RNA is now appreciated to serve numerous cellular roles, and understanding RNA structure is important for understanding a mechanism of action. This contribution discusses the methods available for predicting RNA structure. Secondary structure is the set of the canonical base pairs, and secondary structure can be accurately determined by comparative sequence analysis. Secondary structure can also be predicted. The most commonly used method is free energy minimization. The accuracy of structure prediction is improved either by using experimental mapping data or by predicting a structure conserved in a set of homologous sequences. Additionally, tertiary structure, the three-dimensional arrangement of atoms, can be modeled with guidance from comparative analysis and experimental techniques. New approaches are also available for predicting tertiary structure.

Key words: Statistical mechanics, Comparative sequence analysis, Gibbs free energy, Dynamic programming algorithm, RNA secondary structure

1. Introduction

In the first formulation of the Central Dogma of Molecular Biology, RNA was merely a passive carrier of genetic information between DNA and proteins (1). Recent discoveries, however, reveal an active role for RNA in many aspects of biology, including roles in catalysis (2, 3), genome maintenance (4), regulation (5), and protein synthesis (6–8). Each of these discoveries was awarded a Nobel Prize because it changed our understanding of cellular function.

The catalytic roles of RNA in RNase P and in protein synthesis in the ribosome are common to all kingdoms of life (2, 6–8). RNA sequences play additional roles particular to bacteria. Small RNA regulators (sRNAs) modulate the activities and stabilities of their one or more target mRNAs (9). Riboswitches are found in the untranslated regions of bacteria mRNA and usually control gene expression in response to binding small molecules (10, 11). This is

Kenneth C. Keiler (ed.), *Bacterial Regulatory RNA: Methods and Protocols*, Methods in Molecular Biology, vol. 905, DOI 10.1007/978-1-61779-949-5_8,

accomplished without the aid of proteins. RNA also plays a role in bacterial immunity against phages (12).

The functions of many of these RNA sequences that function without being translated into protein, so-called "noncoding RNAs" (ncRNAs), are dependent on their structures. RNA structure is hierarchical, beginning with the primary sequence, then the secondary structure, i.e., the set of canonical pairs, and ultimately the tertiary structure, i.e., the full three-dimensional structure. RNA secondary structure tends to form on a faster time scale and with larger free energy changes than the contacts that mediate tertiary structure, enabling the prediction of the secondary structure independent of knowledge about the tertiary structure (13).

RNA structure prediction plays two important roles. One, it can help in the interpretation of experiments relating to the mechanism of RNA function (14, 15). Two, it can help propose new experiments to probe function (16). An understanding of even the secondary structure alone can assist both of these. Accurate structure prediction is therefore a valuable tool for experimentalists to use in their pursuits.

This chapter reviews the available methods for predicting RNA structure, with an emphasis on secondary structure prediction and on methods based on thermodynamics. It puts the methods in the context of their limitations and also provides suggestions for choosing an appropriate method given the available data.

2. Comparative Sequence Analysis

Comparative sequence analysis is the most accurate means of RNA secondary structure prediction, and it is the first choice approach when determining the secondary structure of a novel RNA. It is based on the principle that structure is conserved by evolution to a greater extent than sequences. Therefore, a large set of homologous sequences provides sufficient information to determine structure.

Comparative sequence analysis was first used to solve the secondary structure of tRNA (17, 18). This analysis was completed without the aid of modern algorithms for structure prediction, but the model was shown to be correct when tRNA crystal structures were solved (19). An analysis of the secondary structures of ribosomal RNAs determined using comparative sequence analysis found that over 97% of predicted pairs were in subsequently solved crystal structures, and almost all of the predicted tertiary interactions and non-canonical pairs were found to be correct (20). No other means of secondary structure prediction can offer anything close to this level of accuracy, particularly for longer RNAs, nor do they offer the same insight to higher order contacts that may also be of structural and/or functional significance. Comparative sequence analysis is generally the standard by which structure prediction algorithms are evaluated, as only a limited number of sequences of certain RNA families have been crystallized.

Comparative sequence analysis is necessarily a manual undertaking. The goal of the analysis is to align the structures encoded by related RNA sequences, not strictly the sequence of nucleotides. This requires both considerable insight from the researcher and a relatively large number of sequences. Typically the sequences are homologs from multiple organisms, but sets of in vitro evolved sequences or homologs from a single organism can also be used. Identifying regions with coordinated mutations that do not preserve nucleotide identities, but instead preserve base pairs, is a strong evidence of a conserved and functionally important underlying structure. Two-nucleotide changes in sequence that preserve base pairing are called compensating base pair changes. For example, a G-C base pair in a conserved structure of one sequence is more likely to mutate into another canonical pair (AU, CG, UA, GU, or UG), even though such a change may involve changing two nucleotides instead of one, than it is to mutate into a non-canonical pair or undergo a single-nucleotide deletion of one of the pairing partners. Base paired regions of homologous RNAs from distant species may have low pairwise sequence identity but perfectly conserved secondary structures, as each of the changes that diverges the sequences conserves the structure. The importance of aligning structure conservation instead of sequence conservation is illustrated in Fig. 1. Other available examples of structures with divergent sequences solved by comparative sequence analysis are found in the seed alignments available in the Rfam database (21).

There is currently no automated means of performing comparative sequence analysis. No algorithm to date is as accurate or as capable as a skilled human. Many of the algorithms discussed below, however, can generate useful hypotheses to aid comparative analysis. These may quickly and easily show regions where pairing is possible or likely and how two or more RNA sequences could form a common structure, which can significantly reduce the amount of time the researcher spends constructing the alignment. When choosing an algorithm to aid in comparative sequence analysis, important features include the use of two or more sequences to identify conservation, the generation of multiple structures instead of a single output so as to have multiple hypotheses, and the prediction of base pair probabilities to identify regions that were likely predicted accurately vs. regions where more human insight will be needed. It is important not to let the algorithmic predictions dominate the comparative sequence analysis, but rather to let them illuminate regions of the RNA where a conserved structure is likely so that covarying base pairs can be more quickly identified.

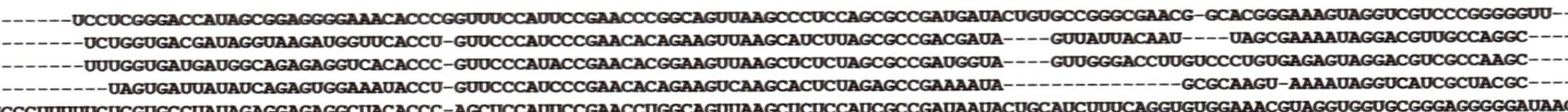

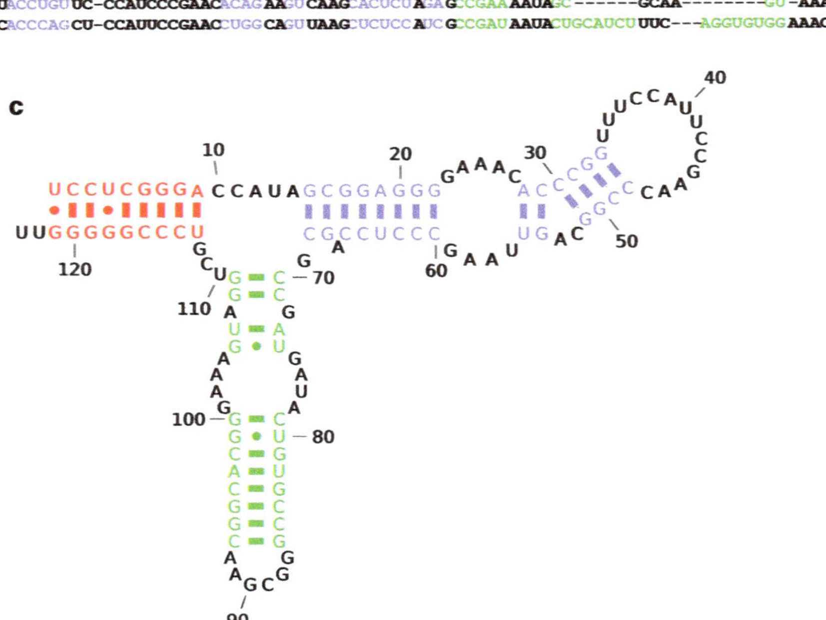

Fig. 1. (**a**) A sequence alignment of five bacterial 5S rRNA sequences from the 5S ribosomal RNA database (127, 128) calculated by MUSCLE 3.6 (129). (**b**) The correct structural alignment from the 5S Ribosomal RNA Database (127). The columns that make up the three conserved helices are color-coded in *red*, *green*, and *purple*. (**c**) The secondary structure of the first sequence. The colors of the helices match the colors of the corresponding columns from the structural alignment. While the sequence alignment algorithm aligns many nucleotides correctly, the errors, even mistaking a single gap for the wrong location, mask important structural features of the RNA.

3. Single-Sequence Structure Prediction Methods

3.1. Free Energy Minimization

Comparative sequence analysis is not always possible, as in the case when there is no known homolog of a sequence. The most popular method of predicting a secondary structure from a single sequence is free energy minimization. When a structure i is at equilibrium with the unpaired state, the equilibrium is described by the equilibrium constant K_i as in Eq. 1 below:

$$K_i = \frac{[\text{Structure } i]}{[\text{Unpaired state}]} = e^{-\Delta G_i^0/RT} \tag{1}$$

The Gibbs free energy change for structure i, ΔG_i^0, quantifies the favorability of a given secondary structure compared to the unpaired state. Similarly, the equilibrium between two structures i and j is described as in Eq. 2 below:

$$\frac{[\text{Structure } i]}{[\text{Structure } j]} = \frac{K_i}{K_j} = e^{-(\Delta G_j^0 - \Delta G_i^0)/RT} \tag{2}$$

Thus, the Gibbs free energy change quantifies the favorability of a structure at a given temperature. The structure with the lowest Gibbs free energy change will be the most prevalent in solution at equilibrium.

Free energy changes can be estimated using a nearest neighbor model. The nearest neighbor model assumes that the free energy change for forming a base pair depends only on the sequence identities of that pair and the immediately neighboring base pairs. Free energy changes associated with forming loop regions and other structural motifs are likewise assumed to not depend on anything outside of the loop sequence and the sequence of the bounding base pairs. Thus, the free energy change for a given structure can be computed simply by adding up all the energies associated with forming all the base pair stacks and all the other structural motifs. These parameters were determined from optical melting experiments (22–24). The parameters and tutorials on their use can be found on the Nearest Neighbor Database website, http://rna.urmc.rochester.edu/NNDB (25).

The structure predicted to have minimum free energy can be computed using a dynamic programming algorithm. Dynamic programing algorithms implicitly consider all possible structures and guarantee the optimal solution while maintaining efficient computation times. Several implementations of dynamic programming algorithms that find the minimum free energy structure have been reviewed (26, 27). Formally, this algorithm scales $O(N^3)$ in time, where N is the length of the RNA sequence, meaning that doubling the length of an input sequence results in a computational cost increase of a factor of approximately $2^3 = 8$. In practice, calculations on typical RNA sequences are fast and often will run in times on the order of seconds on modern computers. For example, the RNase P RNA from *M. tuberculosis*, a sequence of 434 nucleotides, takes 8.8 s to compute its minimum free energy structure on a single core of a system with a 2.4 GHz Intel Core 2 Quad processor using RNAstructure (28).

Secondary structure prediction algorithms are typically evaluated using the metrics sensitivity and positive predictive value (PPV). Sensitivity is defined as:

$$\text{Sensitivity} = \frac{\text{True positive pairs}}{\text{True positive pairs} + \text{False negative pairs}} \quad (3)$$

A true positive is a predicted base pair that is found in the known structure. A false negative is a base pair in the known structure that was not predicted. A false positive is a predicted base pair not found in the known structure. Thus, sensitivity measures the fraction of base pairs in the known structure that were predicted. PPV is defined in Eq. 4:

$$\text{PPV} = \frac{\text{True positive pairs}}{\text{True positive pairs} + \text{False positive pairs}} \quad (4)$$

A false positive pair is a pair predicted to exist that does not exist in the accepted structure. Thus, PPV measures the fraction of predicted base pairs that are correct. Both metrics are necessary to evaluate the accuracy of a structure prediction method. Structures predicted for sequences of fewer than 800 nucleotides by free energy minimization have an average sensitivity of 74% and PPV of 66% for a variety of RNA families (29). For longer sequences, this accuracy is considerably worse. For example, for full length small and large subunit rRNA, average sensitivities are 47.1% and 56.2%, respectively (30).

3.2. Predicting Base Pair Probabilities

Structure prediction by free energy minimization calculations assume that the RNA is at equilibrium, that the RNA has a single conformation, and that the nearest neighbor parameters are without error. The last assumption is known to not be true because some non-nearest neighbor effects have been observed (31–33), and some sequence-specific effects have been averaged (24). The first two assumptions may not be true for all RNA sequences.

On average, a predicted secondary structure will have correctly predicted pairs and incorrectly predicted pairs. Supplementing a free energy minimization calculation with a partition function calculation can suggest those pairs that are more likely to be correct. The partition function, Q, is defined as the sum of the equilibrium constants, K_i, for all possible structures.

$$Q = \sum_{\text{all structures}} K_i = \sum_{\text{all structures}} e^{-\Delta G_i^0 / RT} \quad (5)$$

Thus, the probability of a particular structure i being found in solution can be calculated according to Eq. 6 below:

$$P_i = \frac{e^{-\Delta G_i^0/\mathrm{RT}}}{Q} \tag{6}$$

Furthermore, the probability of a base pair $i-j$ can be calculated by summing the equilibrium constants for structures that contain that pair and dividing by the partition function.

$$P(i-j \text{ base pair}) = \sum_k \frac{e^{-\Delta G_k^0/\mathrm{RT}}}{Q} \tag{7}$$

In Eq. 7, the sum on the index k is taken over all structures that have the base pair $i-j$.

An algorithm to compute the partition function using a dynamic programming algorithm has been designed (34). Most significantly, the base pairs predicted to be most probable according to the partition function are also the base pairs most likely to be correctly predicted (29). For example, base pairs in the minimum free energy structure predicted to have 99% pairing probability are accurate 91% of the time, compared with 66% of all predicted pairs being accurate. The partition function calculation also scales $O(N^3)$ in time and it also tends to take on the order of seconds to compute for typical sequences on modern computers. For the same 434 nucleotide sequence used above to benchmark the computer cost of free energy minimization, the partition function calculation takes 12.7 s. Partition function calculations can be combined with free energy minimization calculations to annotate the predicted minimum free energy structure with base pair probabilities from the partition function. Color annotation can indicate regions likely to be accurately predicted and those that are not, as well as whether the overall structure has likely been well predicted or poorly predicted. Therefore, it is strongly recommended that any single-sequence free energy minimization calculation be supplemented with a partition function calculation to annotate the predicted structure(s) with base pair probabilities. This feature is available in both the RNAstructure (28) and Vienna packages (35).

3.3. Predicted Maximum Expected Accuracy Structures

Given that the base pairs predicted to be the most probable tend to be the most accurate, an alternative approach for structure prediction is to assemble a structure composed of the most probable base pairs. This approach was pioneered using knowledge-based potentials to predict pair probabilities (36, 37). Expected accuracy is defined by (36):

$$\text{Expected accuracy}(S) = \gamma \sum_{(i,j)\in \mathrm{BP}} 2P_{\mathrm{BP}}(i,j) + \sum_{k\in \mathrm{SS}} P_{\mathrm{SS}}(k) \tag{8}$$

where $P_{BP}(i,j)$ is the probability of a base pair (BP) between a nucleotide at position i and a nucleotide at position j, $P_{SS}(k)$ is the probability a nucleotide at position k is single-stranded (SS). The two sums are taken over all base pairs and over all single-stranded nucleotides in a structure, respectively. The factor γ is a weight factor between the two sums.

Maximum expected accuracy structures can be assembled from base pair probabilities computed by the partition function using a program called MaxExpect (38). Over a range of RNA families, MaxExpect has approximately the same sensitivity as free energy minimization (73%), but an improved PPV (66–68% compared to 66% for free energy minimization) (38). The structure of the RNase P RNA from *M. tuberculosis* as predicted by MaxExpect and color-annotated with base pair probabilities is shown in Fig. 2. MaxExpect also formally scales $O(N^3)$ in time and runs in times similar to a partition function calculation.

3.4. Suboptimal Structure Prediction

Suboptimal structures are structures that are similar in score to the structure that is predicted to have the best score. For example, in free energy minimization, suboptimal structures are structures that have low free energy change. These structures represent important alternative hypotheses about the structure of the RNA in addition to the predicted optimal structure. This is because minimum free energy or maximum expected accuracy structures are not always accurate, so suboptimal structures may be closer to the accepted structures. Additionally, some RNA sequences may have multiple secondary structures, such as riboswitches, which change secondary structure upon ligand binding (10, 39). An optimal structure alone cannot contain all the important structural information in this case. There are several available methods for generating suboptimal structures with low free energy. These methods were reviewed in detail previously and are summarized here (40).

One way to generate these suboptimal structures is a heuristic that computes representative alternative structures (41, 42). While the average sensitivity of the minimum free energy structure is about 73%, the best suboptimal structure in this set will have an average sensitivity of 87%, and when considering all the base pairs in any suboptimal structure, the sensitivity of the entire set of suboptimal structures is on average 97% (22). The set of suboptimal structures can contain nearly all base pairs in the real structure and is worth reviewing, especially if base pair probability information, intuition, or experimental data can guide the researcher toward those suboptimal structures, or those features in the suboptimal structures, that are likely to be correct.

It is also possible to generate all possible secondary structures within a specified energy increment of the minimum free energy structure (43). This approach is computationally expensive because the number of possible suboptimal structures increases exponentially

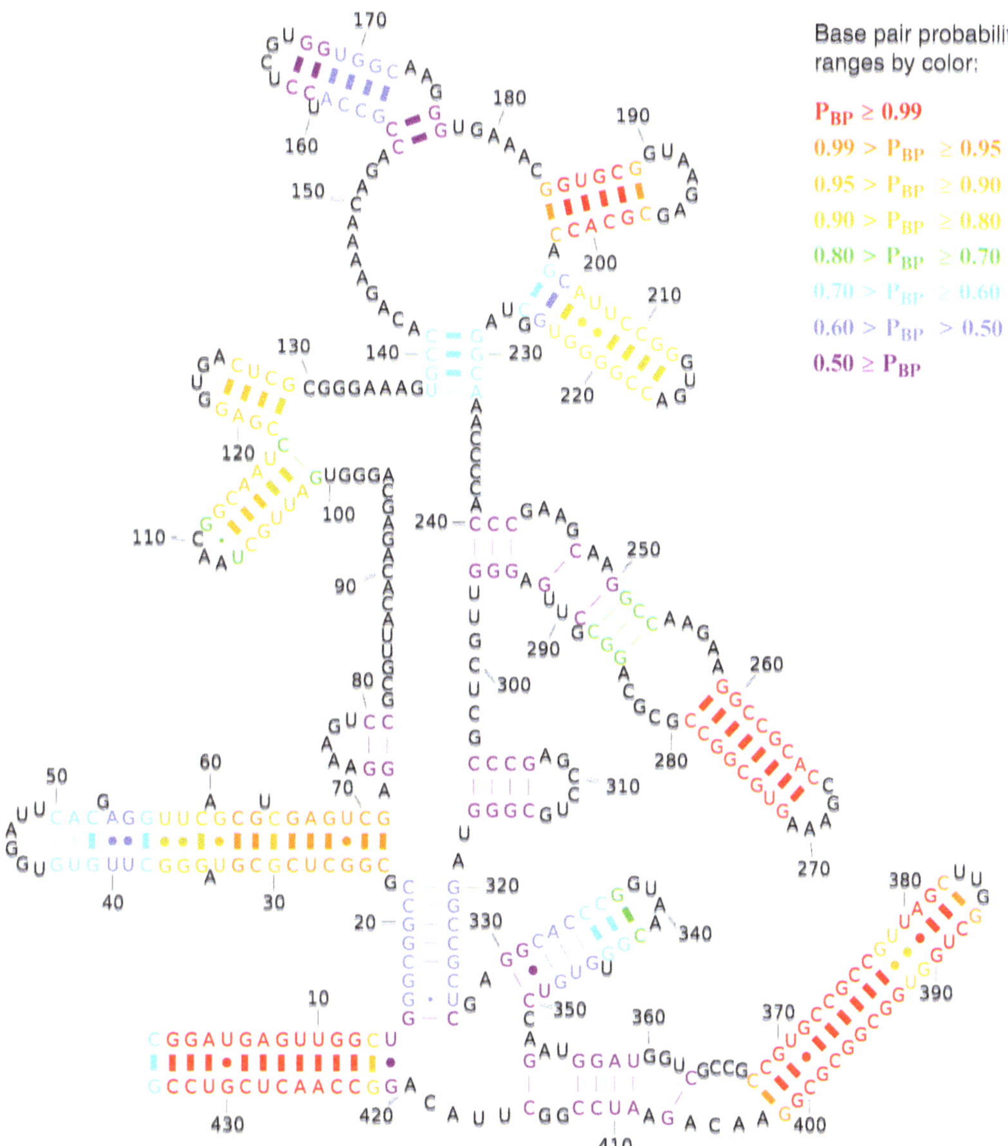

Fig. 2. The predicted secondary structure of the *M. tuberculosis* RNase P RNA as predicted by the MaxExpect algorithm. Correctly predicted base pairs are shown with *thick lines* or *circles*, and incorrect base pairs are shown with *thin lines* or *circles*. Base pairs are color-annotated with their computed pairing probabilities according to the partition function. In this example, all base pairs predicted to have 80% pair probability or higher are correct. Regions of dubious accuracy typically involve base pairs predicted to have below 60% probability, although some such base pairs are correct.

with the energy increment. The complete enumeration of suboptimal structures can be useful in the case where data from experiment can be used to choose reasonable structures from the set of predicted structures.

Another approach to generating suboptimal structures is to sample structures according to their probability in the Boltzmann ensemble (44, 45). The approach, implemented in the Sfold program (46), RNAstructure, and the Vienna Package, may reveal one or more structure clusters of closely related structures. A representative structure from the largest cluster, called a centroid structure, may be a more accurate representation of the correct structure than the lowest free energy structure (47). That is, the base pairs found in common in the set of highly probable structures may be both distinct and more accurate compared to the base pairs in the minimum free energy structure.

3.5. Pseudoknots and Pseudoknot Prediction

Pseudoknots represent a particular problem for sequence prediction algorithms. Pseudoknots are formed by non-nested base pairs. Formally, a pseudoknot is defined by at least two base pairs $i-j$ and $i'-j'$ such that in the sequence nucleotide i comes before i', i' before j, and j before j'. A simple pseudoknot is shown in Fig. 3. Finding the predicted lowest free energy structure including pseudoknots of any topology is an NP-hard problem, meaning that as sequence length increases, the computational expense increases exponentially (48) and, to further complicate things, energy parameters for pseudoknots have not been experimentally determined. There are, however, separate parameter sets determined using a polymer model (49), a lattice model (50), and an empirical approach that optimizes structure prediction accuracy (51).

All of the algorithms described above exclude pseudoknots from consideration in the calculation. Pseudoknots, however, are often found near functionally important regions of RNAs, such as the active site of the group I intron (52). Algorithms have been developed that attempt to predict these difficult to identify, but important motifs (53). A first approach is to use a dynamic programming algorithm that is capable of predicting pseudoknots of certain, limited topologies (51, 54–57). The pseudoknot topologies these algorithms are capable of predicting have been described and classified (58). A second approach uses an iterative approach to assemble structures from algorithms that are not capable of predicting pseudoknots in a single iteration (59–61). A third approach is to simulate a folding pathway or, in a related fashion, sample a stepwise addition of helices (62–66). Recently, a novel means of assembling maximum expected accuracy structures that can contain pseudoknots of any topology was published (30). Overall, however, the accuracy with which these algorithms predict pseudoknotted base pairs is low, and this remains an ongoing area of research (30).

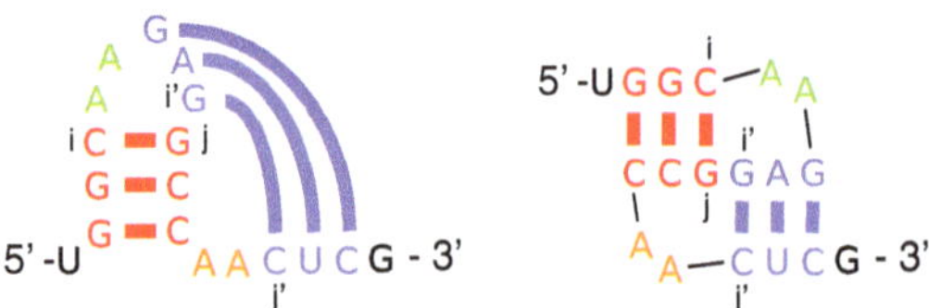

Fig. 3. Two different representations of the same pseudoknot. The base pairs between nucleotides i and j and between i' and j' are pseudoknotted. That is, in the sequence, $i<i'<j<j'$.

4. Experimental Mapping

Data from experimental mapping techniques can guide structure prediction. This guidance can dramatically improve the structure prediction accuracy in some cases. The first experiment to probe RNA structure in solution was using the double-strand-specific RNase V1 on tRNA (67, 68). This ribonuclease specifically cleaves RNA in regions that are helical, indicating where the RNA is base paired. Structure can be probed in much the same way using RNases specific for unpaired regions, such as RNase S1 or T2 (69, 70). Recent publications demonstrate that enzymatic probing of structure can be performed on whole cell transcripts using high-throughput sequencing (71, 72). Enzymatic cleavage has also been used to probe tertiary structure (73).

Chemical modification can also be used to probe structure. Probes that react with one or more of the nucleobases tend to react when the nucleotide is unpaired and are less reactive when the nucleotide is base paired. The results can be analyzed by reverse transcription, which will stop at the site of a modification. Common reagents are kethoxal, which reacts with the N1 or N2 of guanine (74), dimethyl sulfate, which reacts with the N1 of adenine and the N3 of cytosine (75), and 1-cyclohexyl-(2-morpholinoethyl) carbodiimide metho-*p*-toluene sulfonate (CMCT), which modifies the N3 of uracil and the N1 of guanine (76–78). The results of these experiments can be combined and used with data from the enzymatic cleavage techniques (77). Hydroxyl radical footprinting tends to be useful for identifying completely buried nucleotides, as the hydroxyl radical reacts with and break the sugar-phosphate backbone regardless of whether the base is paired or not (79).

More recently, a technique called Selective 2′-Hydroxyl acylation Analyzed by Primer Extension, or SHAPE, was developed (80). SHAPE uses the reagents *N*-methylisotoic anhydride (NMIA) or 1-methyl-7-nitroisatoic anhydride (1M7), which react with the RNA backbone, but with a preference for flexible regions, which tend to be single-stranded (80, 81). Due to reacting with the backbone regardless of nucleotide identity, a single SHAPE experiment yields data on the backbone flexibility at every position in the RNA. SHAPE has been used on RNA sequences as long as the entire HIV genome (82, 83) and can also be applied to complex mixtures using high-throughput sequencing (84).

These experimental techniques can be combined with the structure prediction approaches discussed above by either constraining or restraining the calculations. When considering enzymatic cleavage data, a particular nucleotide can be constrained to be either single-stranded or base-paired if enzymes with the same specificity cleave on both sides of that nucleotide (23). Chemical modification data can be directly incorporated into a dynamic programming algorithm, including a partition function calculation (22). Likewise, SHAPE data can be directly incorporated into a dynamic programming algorithm and can give highly accurate results (85). The best available evidence suggests that SHAPE mapping, coupled with quantification by capillary electrophoresis, should be used as the mapping technique of choice if a single technique is used to probe secondary structure (85, 86). Chemical modification and SHAPE constraints and restraints, respectively, are both available in the RNAstructure software package (28).

5. Multiple-Sequence Secondary Structure Prediction

Multiple-sequence secondary structure prediction attempts to emulate comparative sequence analysis by predicting structures conserved in two or more sequences. While these approaches are not yet as accurate as manual comparative sequence analysis, they can offer significant improvements in accuracy over single-sequence methods (87). Many of these strategies are more computationally expensive than single-sequence methods. Typically, however, calculation times and memory requirements for the algorithms discussed below remain modest on modern desktop computers.

5.1. Algorithms That Fold and Align Simultaneously

The first approach to folding homologous RNAs was proposed by David Sankoff (88). It considers the alignment and folding of any number RNA sequences simultaneously in a single calculation. Formally, the algorithm scales $O(N^{3s})$ in time and $O(N^{2s})$ in storage, i.e., memory use, for s sequences of length N. Like most single-sequence structure prediction methods, this algorithm cannot predict pseudoknots. This algorithm is expensive, particularly for more than two sequences. Restricting the alignment space to exclude biologically unlikely alignments and pairs offers significant improvements in time cost, however.

The first limited implementation of the Sankoff algorithm was FOLDALIGN (89). This algorithm used base pair maximization rather than the nearest neighbor model and free energy minimization. It also excluded from consideration branched structures, reducing the algorithm to $O(N^4)$ in time, but at the cost of excluding a common and important motif in RNA structure. Recent updates to FOLDALIGN have added support for the prediction of branched structures, a free energy model, and a pruning heuristic to accelerate the calculation, significantly increasing the accuracy of the algorithm (90).

Another implementation of the Sankoff algorithm, Dynalign, uses thermodynamics from the nearest neighbor model and allows for branched structures (91). It uses a simple alignment model that only counts gaps, and the gap penalty was optimized as a pseudo-free energy change to weigh the alignment with the folding. Dynalign significantly improves the quality of secondary structure prediction compared to single-sequence free energy minimization (91). For example, the accuracy of structure predictions for the small subunit rRNA improves from 47.4 to 73.3% in sensitivity and from 47.5 to 73.1% in PPV compared to single-sequence free energy minimization (87). Dynalign currently uses constraints on the allowed nucleotide alignments and on the allowed pairs (92) to exclude unlikely structures and alignments from consideration, further reducing the computational cost and increasing the accuracy (93). Dynalign also features the capability of generating suboptimal conserved structures (94).

A third approach to the Sankoff algorithm is LocARNA (95). LocARNA, rather than minimizing energy, maximizes the sum of pair probabilities for both sequences that are determined from separate single-sequence partition function calculations and a similarity score for the alignment. LocARNA runs quickly, as only significant base pairs are considered, reducing the order of the algorithm to $O(N^2(N^2+M^2))$, for two sequences of length N and where M is the number of significant base pairs, also on the order of N. It does sacrifice some accuracy compared to FOLDALIGN and Dynalign, however (96).

Also in the paradigm that simultaneously folds and aligns sequences is the PARTS (probabilistic alignment for RNA joinT secondary structure prediction) algorithm (96). PARTS is a partition function for conserved structures for two sequences. Like LocARNA, instead of using free energy change for structures, it uses pair probabilities calculated using single sequences. PARTS uses a "matched helical region" formalism to align conserved structures instead of a more rigid definition of conservation used in above algorithms, and a probabilistic approach to identifying the most probable structures. For RNAs with a mixture of conservation and diversity, such as the RNase P family of RNAs, PARTS offers particular advantages over algorithms with more rigid constraints on alignments (96). PARTS has also been extended to allow stochastic sampling of structures, which may also be important for sequences with more than one structure, such as riboswitches (97).

While directly applying the Sankoff algorithm to more than two sequences is generally infeasible in terms of computational cost, other ways of applying it to an arbitrary number of sequences at significantly reduced complexity have been implemented. FOLDALIGNM uses pair frequencies observed in a set of all pairwise FOLDALIGN calculations of the sequences (98). A descendant of Dynalign is Multilign, which uses successive pairwise Dynalign calculations to identify base pairs conserved in all sequences under consideration (87). Multilign is notable because it scales linearly with the number of sequences. LocARNA was also originally capable of working on multiple sequences (95). To generalize to multiple sequences, mLocARNA uses the results of all pairwise alignments from LocARNA calculations to compute the multiple-sequence alignment. The probability of a pair in aligned columns of successive alignments is the square root of the product of the pairing probabilities from the two alignments. A distinct approach that simultaneously aligns and folds an arbitrary number of sequences is RAF (99). RAF operates on two sequences at a time, and then it aligns alignments in subsequent iterations instead of sequences. In benchmarks, FOLDALIGNM and mLocARNA tended to perform better on shorter sequences, and RAF and Multilign tended to perform better on longer ones, and they all tended to outperform single- and double-sequence methods for most kinds of RNAs (87). Most of these algorithms were of reasonable computational cost on modern hardware.

5.2. Algorithms That First Align, Then Fold

The second paradigm for the structure prediction of more than one sequence is to first align the sequences and then fold them. This paradigm is exemplified by RNAalifold (100). RNAalifold identifies the minimum free energy consensus structure that can be formed by a set of input aligned sequences. The input alignment often comes from a sequence alignment algorithm, although manually curated alignments are supported and can offer increased accuracy. RNAalifold is fast and accurate for well-aligned sequences. Its accuracy suffers when the input alignment is poor, as may be the case when the pairwise sequence identity is below 60%. Automated sequence alignment algorithms struggle in such cases (101). An alignment of low-identity sequences performed, or at least augmented, by a skilled researcher may still offer good results.

Also in the “align, then fold” paradigm is the CentroidFold algorithm (102). Rather than looking for the minimum free energy consensus structure, it computes the centroid structure in a manner analogous to Sfold (45). The centroid structure represents the central structure of the largest cluster of similar structures that was generated by stochastically sampling the ensemble of homologous structures. The probabilities necessary to determine the centroid consensus structure be determined using the nearest neighbor model as determined by experimental measurements (34) or by learning from the database of sequences with known structure (36).

Most recently in the “align, then fold” paradigm, the TurboFold algorithm takes an arbitrary number of sequences and then computes their pairwise probabilistic alignments and their single-sequence base pair probabilities (103). The alignments are used to map putative structures between sequences. The single-sequence base pair probabilities for a given sequence in the set is referred to as its “intrinsic information.” The combined probabilistic alignments and the base pair probabilities for all the other sequences is referred to as the "extrinsic information." The extrinsic information is used to revise the base pair probabilities for each sequence. Subsequently, the revised pair probabilities are used to recompute the extrinsic information. Multiple iterations refine and improve the predicted pair probabilities for each sequence. After the desired number of iterations, structures are assembled using a maximum expected accuracy algorithm. TurboFold usually outperforms other multiple-sequence prediction algorithms in sensitivity on random assortments of homologous RNA sequences that typically have pairwise identities under 60%, and it is usually comparable in PPV (103). While it may be more computationally expensive than most of the align and then fold algorithms discussed above, it only takes on the order of minutes for up to ten RNA sequences of typical lengths. One important feature of TurboFold compared to most of the algorithms discussed above is that it does not enforce a common structure. Variable elements of homologous sequences, such as the variable stem in tRNAs, may still be predicted correctly, making TurboFold a compelling choice for structure prediction when multiple sequences of divergent identities and unknown alignments are available.

5.3. An Algorithm That Folds, Then Aligns

The third paradigm in multiple-sequence structure prediction is to "fold, then align." This approach is taken by RNAshapes (104). This algorithm enumerates the abstract "shape" space available to each sequence independently and computes the probabilities of each shape, and then identifies the thermodynamically optimal structure that has the common shape. Abstract shapes encode features of an RNA structure instead of the complete pairing information. There are many fewer low free energy shapes than structures, and this fact makes the approach feasible. RNAshapes is fast; it is approximately linear in time after the single-sequence structure calculations. It offers accuracy comparable to multiple-sequence methods above. It does not provide sequences alignments, but those can be built subsequently from the conserved structure using RNAforester (105).

5.4. Which Algorithm to Choose

Many multiple-sequence methods have been presented. The choice of which to use depends on many factors. When more than two sequences are available, using a method that can predict the conserved structure for all the sequences together is the best choice. Accuracy tends to improve for all of the algorithms described above with an increasing number of sequences, although there can be diminishing returns, and there is increasing computational expense (87, 103). When pairwise sequence identity is high (roughly above 70%) and therefore a sequence alignment algorithm (or human) will do a good job of aligning the sequences, RNAalifold is a fast and accurate algorithm to use (100). The speed of RNAalifold also makes it an attractive algorithm to use when the RNA sequences in question are very long, i.e., over 1,500 nucleotides in length. When sequence identity is low, however, an algorithm that folds and aligns sequences simultaneously, such as Multilign (87) or RAF (99), may often be a better choice because they can construct superior alignments based on structure than a sequence alignment algorithm. When the structural homology of the sequences in question may be lower, TurboFold's lack of rigidly enforced structural homology constraints enable the accurate prediction of structures that have many regions in common but that also have variable regions (103). TurboFold also computes base pair probabilities, which, as discussed above for single sequences, can highlight regions of a resulting structure that are more likely to be correct than others.

Ultimately, using multiple algorithms based on different paradigms is the best practice. Ideally, they will all give the same answers, but when they differ, the predictions represent important alternative hypotheses about the structure, as no algorithm can correctly identify all the base pairs in all the sequences of all the classes of RNA all the time. Investigators are encouraged to critically examine the varying results, and to use them to seed a manual comparative sequence analysis, which will offer the best results.

6. RNA Tertiary Structure Prediction

Tertiary structure prediction is a developing frontier in RNA structure prediction. While the critical assessment of techniques for protein structure prediction (CASP) competition attracts hundreds of participating automated web servers and human expert groups that construct tens of thousands of models for evaluation (106), the field attempting to model RNAs in similar ways is smaller but growing.

The original approach for RNA tertiary structure prediction was comparative sequence analysis, just like secondary structure prediction. Important tertiary contacts tend to be conserved through evolution similarly to the conservation of base pairs, and this was first used to model tRNA structure (17). Comparative sequence analysis was also applied to find higher order structures in the small and large subunit rRNAs with high accuracy (20, 107–110). In addition to identifying higher order contacts, comparative sequence analysis has been used as the basis for constructing a complete 3D model of the group I intron with high accuracy (111), although it required a large amount of human insight.

Recent work demonstrated success at automating the construction of all-atom models from restraints derived from comparative sequence analysis (112). Restraints from biochemical data are also supported. This approach has been applied to RNA sequences as large and as complex as the P4–P6 domain of the group I intron, although the larger the RNA, the more restraints are necessary for an accurate prediction. Also, the computational cost is high compared to secondary structure prediction methods, but this method is still approachable for modest modern computer clusters.

Other approaches to tertiary structure prediction include the constraint satisfaction approach of the MC-Sym/MC-Fold pipeline (113, 114). MC-Fold predicts base pairs for sequences including both canonical and non-canonical pairs using a scoring function based on nucleotide-based motifs with scores learned from the database of experimentally determined tertiary structures. The program is successful at predicting motifs such as non-canonical base pairs, but it is limited to sequences of less than about 100 nucleotides. MC-Sym constructs a 3D model from either an MC-Fold prediction or a known secondary structure. It is especially useful when combined with experimental data to constrain the set of models. For example, models were built for the hairpin ribozyme using the secondary structure and cross-linking data (115).

Successful prediction of the structures of small RNAs has also been shown using molecular mechanics potential energy functions (116). Like the MC pipeline, the Rosetta energy function can be used on smaller RNAs to identify structural motifs such as non-canonical pairs, base triples, and pseudoknots. iFoldRNA is a web server for the 3D structure prediction of RNAs under 50 nucleotides that also uses molecular mechanics force fields (117).

Reduced representations have also been used to model the overall folds of RNAs. Methods using this strategy include NAST (118), DMD (119), YAMMP (120), and YUP (121). These programs sacrifice atomic resolution, but can offer insight into the overall fold of the molecule. It is possible to reconstruct an atomic-resolution model from a coarse-grained representation, but this remains an ongoing research problem (117, 122).

7. Summary and Guidance

As emphasized at the beginning of this chapter, a manually curated comparative sequence analysis remains the gold standard for RNA secondary structure prediction, and it may even offer insight into tertiary structure. Comparative sequence analysis is time-consuming and it requires both a moderate to large number of sequences with at least some divergence in the pairwise sequence identity and considerable human insight. Despite the difficulty, no computational algorithm currently offers comparable quality. For accuracy, there is no currently available substitute.

It is understood, however, that an acceptable sample of sequences is not always available. In such cases the computer algorithms

Table 1
RNA secondary structure prediction packages

Program	Description	URL	References
RNAstructure	Includes free energy minimization, partition function calculations, stochastic sampling, maximum expected accuracy structure prediction, ProbKnot, Dynalign, PARTS, Multilign, and TurboFold	http://rna.urmc.rochester.edu/RNAstructure.html	(28)
Vienna RNA Package	Includes free energy minimization, partition function calculations, stochastic sampling, maximum expected accuracy structure prediction, LocARNA, and RNAalifold	http://www.tbi.univie.ac.at/	(130, 131)
SFold	Stochastic sampling of RNA structures.	http://sfold.wadsworth.org/cgi-bin/index.pl	(132)
Foldalign/FoldalignM	Simultaneous alignment and folding of multiple RNA sequences.	http://foldalign.ku.dk/software/index.html	(98)
CentroidFold	Computes a centroid structure from an alignment of multiple sequences	http://www.ncrna.org/centroidfold/	(102)
RAF	Simultaneous alignment and folding of multiple RNA sequences.	http://contra.stanford.edu/contrafold/	(99)
RNAshapes	Folding of multiple sequences, followed by determination of conserved structure	http://bibiserv.techfak.uni-bielefeld.de/rnashapes/	(104)
MC-FOLD	Prediction of secondary structure and non-canonical pairs	http://www.major.iric.ca/MC-Pipeline/	(114)

This table summarizes the location of the programs mentioned in this chapter. A more complete list of secondary structure prediction programs is available on Wikepedia: http://en.wikipedia.org/wiki/List_of_RNA_structure_prediction_software

that predict homologous structures from multiple homologous sequences are the next-best available methods. As stated, different algorithms have their strengths and weaknesses with respect to the length, sequence identity, and estimated structural homology of the RNA sequences in question. While using one algorithm that has strengths corresponding to the RNA sequences in question is good, it is much better to use multiple appropriate algorithms to generate competing hypotheses about the structures of the RNAs in question. Table 1 lists the programs presented in this chapter, along with web links to their location. Guiding the structure prediction with the common features of predictions from multiple algorithms and resolving the differences with comparative sequence analysis is an excellent approach to structure prediction when comparative sequence analysis is impractical.

If only a single sequence is available, then there are multiple approaches that are superior to considering only the predicted minimum free energy structure, and many of these approaches can be combined. First, given the lower accuracy of single-sequence prediction compared to multiple-sequence methods or especially comparative sequence analysis, it is important to predict suboptimal structures. These suboptimal structures represent important alternative hypotheses, which may contain structural features not seen in the minimum free energy structure or may be more accurate than the minimum free energy structure. Second, the color annotation of structures according to their predicted base pair probabilities is critical to estimating the accuracy of a predicted structure. Base pairs predicted to be highly probable are highly accurate, and base pairs predicted to have low probability are suspect (29). Third, maximum expected accuracy structures offer a modest improvement over minimum free energy structures.

Pseudoknots are a continuing challenge for RNA secondary structure prediction algorithms. The accuracy with which current algorithms predict pseudoknots remains low. If a researcher suspects that there may be a pseudoknot in the RNA of interest, then it is most important to get an accurate secondary structure of the nested base pairs using the best practices described above. Then, the various algorithms capable of predicting pseudoknots may provide testable hypotheses regarding the various possible pseudoknots. Comparative sequence analysis is the only method that can conclusively predict the existence of pseudoknots with accuracy.

This chapter focused on secondary structure prediction methods that rely on folding thermodynamics. The thermodynamic-based programs are still probably the most popular programs for secondary structure prediction, but other important approaches exist. Comprehensive reviews of methods that predict a structure common to two or more sequences are available (123, 124). Other review articles provide information about alternative approaches for secondary structure prediction that use, for example, kinetics or stochastic context-free grammars (125, 126).

Secondary structure prediction is commonly used, but there is also significant interest in predicting tertiary structure. Tertiary structure prediction requires expert interpretation and still requires experimental mapping data or significant human insight to build accurate models. Recent successes in this area, however, suggest that tertiary structure prediction may become routine and accurate in the next decade or two.

Acknowledgments

The authors thank Joseph E. Wedekind for helpful comments. This contribution was supported by the National Institutes of Health grant R01GM076485 to D.H.M.

References

1. Crick F (1970) Central dogma of molecular biology. Nature 227:561–563
2. Guerrier-Takada C, Gardiner K, Marsh T, Pace N, Altman S (1983) The RNA moiety of ribonuclease P is the catalytic subunit of the enzyme. Cell 35:849–857
3. Cech TR, Zaug AJ, Grabowski PJ (1981) In vitro splicing of the ribosomal RNA precursor of *Tetrahymena*: involvement of a guanosine nucleotide in the excision of the intervening sequence. Cell 27:487–496
4. Greider CW, Blackburn EH (1985) Identification of a specific telomere terminal transferase activity in *Tetrahymena* extracts. Cell 43:405–413
5. Fire A, Xu S, Montgomery MK, Kostas SA, Driver SE, Mello CC (1998) Potent and specific genetic interference by double-stranded RNA in *Caenorhabditis elegans*. Nature 391:806–811
6. Ban N, Nissen P, Hansen J, Moore PB, Steitz TA (2000) The complete atomic structure of the large ribosomal subunit at 2.4 Å resolution. Science 289:905–920
7. Schluenzen F, Tocilj A, Zarivach R, Harms J, Gluehmann M, Janell D, Bashan A, Bartels H, Agmon I, Franceschi F et al (2000) Structure of functionally activated small ribosomal subunit at 3.3 Å resolution. Cell 102: 615–623
8. Wimberly BT, Brodersen DE, Clemons WM Jr, Morgan-Warren RJ, Carter AP, Vonrhein C, Hartsch T, Ramakrishnan V (2000) Structure of the 30S ribosomal subunit. Nature 407:327–339
9. Gottesman S, Storz G (2010) Bacterial small RNA regulators: versatile roles and rapidly evolving variations. Cold Spring Harb Perspect Biol. doi:10.1101/cshperspect.a003798
10. Montange RK, Batey RT (2008) Riboswitches: emerging themes in RNA structure and function. Annu Rev Biophys 37:117–133
11. Serganov A, Patel DJ (2007) Ribozymes, riboswitches and beyond: regulation of gene expression without proteins. Nat Rev Genet 8:776–790
12. Marraffini LA, Sontheimer EJ (2010) CRISPR interference: RNA-directed adaptive immunity in bacteria and archaea. Nat Rev Genet 11:181–190
13. Onoa B, Tinoco I Jr (2004) RNA folding and unfolding. Curr Opin Struct Biol 14:374–379
14. Latham JA, Cech TR (1989) Defining the inside and outside of a catalytic RNA molecule. Science 245:276–282
15. Flor PJ, Flanegan JB, Cech TR (1989) A conserved base pair within helix P4 of the *Tetrahymena* ribozyme helps to form the tertiary structure required for self-splicing. EMBO J 8:3391–3399
16. Kim YK, Furic L, Parisien M, Major F, DesGroseillers L, Maquat LE (2007) Staufen1 regulates diverse classes of mammalian transcripts. EMBO J 26:2670–2681
17. Levitt M (1969) Detailed molecular model for transfer ribonucleic acid. Nature 224:759–763
18. Madison JT, Everett GA, Kung H (1966) Nucleotide sequence of a yeast tyrosine transfer RNA. Science 153:531–534
19. Robertus JD, Ladner JE, Finch JT, Rhodes D, Brown RS, Clark BFC, Klug A (1974) Structure of yeast phenylalanine tRNA at 3 Å resolution. Nature 250:546–551
20. Gutell RR, Lee JC, Cannone JJ (2002) The accuracy of ribosomal RNA comparative structure models. Curr Opin Struct Biol 12: 301–310
21. Gardner PP, Daub J, Tate J, Moore BL, Osuch IH, Griffiths-Jones S, Finn RD, Nawrocki EP, Kolbe DL, Eddy SR et al (2010) Rfam: Wikipedia, clans and the "decimal" release. Nucleic Acids Res 39:D141–D145
22. Mathews DH, Disney MD, Childs JL, Schroeder SJ, Zuker M, Turner DH (2004) Incorporating chemical modification constraints into a dynamic programming algorithm for prediction of RNA secondary structure. Proc Natl Acad Sci U S A 101: 7287–7292
23. Mathews DH, Sabina J, Zuker M, Turner DH (1999) Expanded sequence dependence of thermodynamic parameters provides improved prediction of RNA secondary structure. J Mol Biol 288:911–940
24. Mathews DH, Turner DH (2002) Experimentally derived nearest neighbor parameters for the stability of RNA three- and four-way multibranch loops. Biochemistry 41:869–880
25. Turner DH, Mathews DH (2010) NNDB: the nearest neighbor parameter database for predicting stability of nucleic acid secondary structure. Nucleic Acids Res 38:D280–D282
26. Mathews DH, Turner DH (2006) Prediction of RNA secondary structure by free energy minimization. Curr Opin Struct Biol 16: 270–278
27. Eddy SR (2004) How do RNA folding algorithms work? Nat Biotechnol 22:1457–1458

28. Reuter J, Mathews D (2010) RNAstructure: software for RNA secondary structure prediction and analysis. BMC Bioinformatics 11:129
29. Mathews DH (2004) Using an RNA secondary structure partition function to determine confidence in base pairs predicted by free energy minimization. RNA 10:1178–1190
30. Bellaousov S, Mathews DH (2010) ProbKnot: fast prediction of RNA secondary structure including pseudoknots. RNA 16:1870–1880
31. Kierzek R, Burkard M, Turner D (1999) Thermodynamics of single mismatches in RNA duplexes. Biochemistry 38: 14214–14223
32. Longfellow CE, Kierzek R, Turner DH (1990) Thermodynamic and spectroscopic study of bulge loops in oligoribonucleotides. Biochemistry 29:278–285
33. Theimer C, Wang Y, Hoffman D, Krisch H, Giedroc D (1998) Non-nearest neighbor effects on the thermodynamics of unfolding of a model mRNA pseudoknot. J Mol Biol 279:545–564
34. McCaskill JS (1990) The equilibrium partition function and base pair probabilities for RNA secondary structure. Biopolymers 29: 1105–1119
35. Hofacker IL, Stadler PF (2004) Computational science—ICCS 2004. In: Bubak M, van Albada GD, Sloot PMA, Dongarra J (eds) Lecture notes in computer science, vol 3039, Kraków, pp 728–735
36. Do CB, Woods DA, Batzoglou S (2006) CONTRAfold: RNA secondary structure prediction without physics-based models. Bioinformatics 22:e90–e98
37. Knudsen B, Hein J (2003) Pfold: RNA secondary structure prediction using stochastic context-free grammars. Nucleic Acids Res 31: 3423–3428
38. Lu ZJ, Gloor JW, Mathews DH (2009) Improved RNA secondary structure prediction by maximizing expected pair accuracy. RNA 15:1805–1813
39. Stoddard CD, Montange RK, Hennelly SP, Rambo RP, Sanbonmatsu KY, Batey RT (2010) Free state conformational sampling of the SAM-I riboswitch aptamer domain. Structure 18:787–797
40. Mathews DH (2006) Revolutions in RNA secondary structure prediction. J Mol Biol 359:526–532
41. Steger G, Hofmann H, Fortsch J, Gross HJ, Randles JW, Sanger HL, Riesner D (1984) Conformational transitions in viroids and virusoids: Comparison of results from energy minimization algorithm and from experimental data. J Biomol Struct Dyn 2:543–571
42. Zuker M (1989) On finding all suboptimal foldings of an RNA molecule. Science 244: 48–52
43. Wuchty S, Fontana W, Hofacker IL, Schuster P (1999) Complete suboptimal folding of RNA and the stability of secondary structures. Biopolymers 49:145–165
44. Morgan SR, Higgs PG (1998) Barrier heights between ground states in a model of RNA secondary structure. J Phys A Math Gen 31: 3153
45. Ding Y, Lawrence C (2001) Statistical prediction of single-stranded regions in RNA secondary structure and application to predicting effective antisense target sites and beyond. Nucleic Acids Res 29:1034–1046
46. Ding Y, Lawrence CE (2003) A statistical sampling algorithm for RNA secondary structure prediction. Nucleic Acids Res 31:7280–7301
47. Ding Y, Chan CY, Lawrence CE (2005) RNA secondary structure prediction by centroids in a Boltzmann weighted ensemble. RNA 11: 1157–1166
48. Lyngsø R, Pederson C (2000) RNA pseudoknot prediction in energy-based models. J Comput Biol 7:409–427
49. Aalberts DP, Hodas NO (2005) Asymmetry in RNA pseudoknots: observation and theory. Nucleic Acids Res 33:2210–2214
50. Cao S, Chen SJ (2006) Predicting RNA pseudoknot folding thermodynamics. Nucleic Acids Res 34:2634–2652
51. Dirks RM, Pierce NA (2003) A partition function algorithm for nucleic acid secondary structure including pseudoknots. J Comput Chem 24:1664–1677
52. Adams PL, Stahley MR, Kosek AB, Wang J, Strobel SA (2004) Crystal structure of a self-splicing group I intron with both exons. Nature 430:45–50
53. Liu B, Mathews DH, Turner DH (2010) RNA pseudoknots: folding and finding. F1000 Biol Rep 2:8
54. Rivas E, Eddy SR (1999) A dynamic programming algorithm for RNA structure prediction including pseudoknots. J Mol Biol 285:2053–2068
55. Uemura Y, Hasegawa A, Kobayashi S, Yokomori T (1999) Tree adjoining grammars for RNA structure prediction. Theor Comput Sci 210:277–303
56. Akutsu T (2000) Dynamic programming algorithms for RNA secondary structure prediction with pseudoknots. Discrete Appl Math 104:45–62

57. Reeder J, Giegerich R (2004) Design, implementation and evaluation of a practical pseudoknot folding algorithm based on thermodynamics. BMC Bioinformatics 5:104
58. Condon A, Davy B, Rastegari B, Zhao S, Tarrant F (2004) Classifying RNA pseudoknotted structures. Theor Comput Sci 320: 35–50
59. Ruan J, Stormo GD, Zhang W (2004) An iterated loop matching approach to the prediction of RNA secondary structures with pseudoknots. Bioinformatics 20:58–66
60. Ren J, Rastegari B, Condon A, Hoos HH (2005) HotKnots: heuristic prediction of RNA secondary structures including pseudoknots. RNA 11:1494–1504
61. Jabbari H, Condon A, Zhao S (2008) Novel and efficient RNA secondary structure prediction using hierarchical folding. J Comput Biol 15:139–163
62. Abrahams JP, van den Berg M, van Batenburg E, Pleij C (1990) Prediction of RNA secondary structure, including pseudoknotting, by computer simulation. Nucleic Acids Res 18:3035–3044
63. Gultyaev AP, van Batenburg FHD, Pleij CWA (1995) The computer simulation of RNA folding pathways using a genetic algorithm. J Mol Biol 250:37–51
64. Isambert H, Siggia ED (2000) Modeling RNA folding paths with pseudoknots: application to hepatitis delta virus ribozyme. Proc Natl Acad Sci U S A 97:6515–6520
65. Dawson WK, Fujiwara K, Kawai G (2007) Prediction of RNA pseudoknots using heuristic modeling with mapping and sequential folding. PLoS One 2:e905
66. Meyer IM, Miklos I (2007) SimulFold: simultaneously inferring RNA structures including pseudoknots, alignments, and trees using a Bayesian MCMC framework. PLoS Comput Biol 3:e149
67. Lockard RE, Kumar A (1981) Mapping tRNA structure in solution using double-strand-specific ribonuclease V1 from cobra venom. Nucleic Acids Res 9:5125–5140
68. Lowman HB, Draper DE (1986) On the recognition of helical RNA by cobra venom V1 nuclease. J Biol Chem 261:5396–5403
69. Auron PE, Weber LD, Rich A (1982) Comparison of transfer ribonucleic acid structures using cobra venom and S1 endonucleases. Biochemistry 21:4700–4706
70. Speek M, Lind A (1982) Structural analyses of *E. coli* 5S RNA fragments, their associates and complexes with proteins L18 and L25. Nucleic Acids Res 10:947–965
71. Kertesz M, Wan Y, Mazor E, Rinn JL, Nutter RC, Chang HY, Segal E (2010) Genome-wide measurement of RNA secondary structure in yeast. Nature 467:103–107
72. Underwood JG, Uzilov AV, Katzman S, Onodera CS, Mainzer JE, Mathews DH, Lowe TM, Salama SR, Haussler D (2010) FragSeq: transcriptome-wide RNA structure probing using high-throughput sequencing. Nat Methods 7:995–1001
73. Knapp G (1989) Enzymatic approaches to probing RNA secondary and tertiary structure. Methods Enzymol 180:192–212
74. Miura K, Tsuda S, Ueda T, Harada F, Kato N (1983) Chemical modification of guanine residues of mouse 5S ribosomal RNA with kethoxal. Biochim Biophys Acta 739: 281–285
75. Inoue T, Cech TR (1985) Secondary structure of the circular form of the *Tetrahymena* rRNA intervening sequence: a technique for RNA structure analysis using chemical probes and reverse transcriptase. Proc Natl Acad Sci U S A 82:648–652
76. Rocca-Serra P, Bellaousov S, Birmingham A, Chen C, Cordero P, Das R, Davis-Neulander L, Duncan CD, Halvorsen M, Knight R et al (2011) Sharing and archiving nucleic acid structure mapping data. RNA 17:1204–1212
77. Ehresmann C, Baudin F, Mougel M, Romby P, Ebel J, Ehresmann B (1987) Probing the structure of RNAs in solution. Nucleic Acids Res 15:9109–9128
78. Fritz JJ, Lewin A, Hauswirth W, Agarwal A, Grant M, Shaw L (2002) Development of hammerhead ribozymes to modulate endogenous gene expression for functional studies. Methods 28:276–285
79. Karaduman R, Fabrizio P, Hartmuth K, Urlaub H, Lührmann R (2006) RNA structure and RNA-protein interactions in purified yeast U6 snRNPs. J Mol Biol 356:1248–1262
80. Merino EJ, Wilkinson KA, Coughlan JL, Weeks KM (2005) RNA structure analysis at single nucleotide resolution by selective 2′-hydroxyl acylation and primer extension (SHAPE). J Am Chem Soc 127:4223–4231
81. Mortimer SA, Weeks KM (2007) A fast-acting reagent for accurate analysis of RNA secondary and tertiary structure by SHAPE chemistry. J Am Chem Soc 129:4144–4145
82. Wilkinson KA, Gorelick RJ, Vasa SM, Guex N, Rein A, Mathews DH, Giddings MC, Weeks KM (2008) High-throughput SHAPE analysis reveals structures in HIV-1 genomic RNA strongly conserved across distinct biological states. PLoS Biol 6:e96

83. Watts JM, Dang KK, Gorelick RJ, Leonard CW, Bess JW Jr, Swanstrom R, Burch CL, Weeks KM (2009) Architecture and secondary structure of an entire HIV-1 RNA genome. Nature 460:711–716
84. Lucks JB, Mortimer SA, Trapnell C, Luo S, Aviran S, Schroth GP, Pachter L, Doudna JA, Arkin AP (2011) Multiplexed RNA structure characterization with selective 2′-hydroxyl acylation analyzed by primer extension sequencing (SHAPE-Seq). Proc Natl Acad Sci U S A 108:11063–11068
85. Deigan KE, Li TW, Mathews DH, Weeks KM (2009) Accurate SHAPE-directed RNA structure determination. Proc Natl Acad Sci U S A 106:97–102
86. Low JT, Weeks KM (2010) SHAPE-directed RNA secondary structure prediction. Methods 52:150–158
87. Xu Z, Mathews DH (2011) Multilign: an algorithm to predict secondary structures conserved in multiple RNA sequences. Bioinformatics 27:626–632
88. Sankoff D (1985) Simultaneous solution of the RNA folding, alignment and protosequence problems. SIAM J Appl Math 45:810–825
89. Gorodkin J, Heyer LJ, Stormo GD (1997) Finding the most significant common sequence and structure in a set of RNA sequences. Nucleic Acids Res 25:3724–3732
90. Havgaard JH, Torarinsson E, Gorodkin J (2007) Fast pairwise structural RNA alignments by pruning of the dynamtical programming matrix. PLoS Comput Biol 3: e193
91. Mathews DH, Turner DH (2002) Dynalign: an algorithm for finding the secondary structure common to two RNA sequences. J Mol Biol 317:191–203
92. Uzilov AV, Keegan JM, Mathews DH (2006) Detection of non-coding RNAs on the basis of predicted secondary structure formation free energy change. BMC Bioinformatics 7:173
93. Harmanci AO, Sharma G, Mathews DH (2007) Efficient pairwise RNA structure prediction using probabilistic alignment constraints in Dynalign. BMC Bioinformatics 8:130
94. Mathews DH (2005) Predicting a set of minimal free energy RNA secondary structures common to two sequences. Bioinformatics 21:2246–2253
95. Will S, Reiche K, Hofacker IL, Stadler PF, Backofen R (2007) Inferring noncoding RNA families and classes by means of genome-scale structure-based clustering. PLoS Comput Biol 3:e65
96. Harmanci AO, Sharma G, Mathews DH (2008) PARTS: probabilistic alignment for RNA joinT secondary structure prediction. Nucleic Acids Res 36:2406–2417
97. Harmanci AO, Sharma G, Mathews DH (2009) Stochastic sampling of the RNA structural alignment space. Nucleic Acids Res 37:4063–4075
98. Torarinsson E, Havgaard JH, Gorodkin J (2007) Multiple structural alignment and clustering of RNA sequences. Bioinformatics 23:926–932
99. Do CB, Foo CS, Batzoglou S (2008) A max-margin model for efficient simultaneous alignment and folding of RNA sequences. Bioinformatics 24:i68–i76
100. Bernhart SH, Hofacker IL, Will S, Gruber AR, Stadler PF (2008) RNAalifold: improved consensus structure prediction for RNA alignments. BMC Bioinformatics 9:474
101. Gardner PP, Wilm A, Washietl S (2005) A benchmark of multiple sequence alignment programs upon structural RNAs. Nucleic Acids Res 33:2433–2439
102. Hamada M, Kiryu H, Sato K, Mituyama T, Asai K (2009) Prediction of RNA secondary structure using generalized centroid estimators. Bioinformatics 25:465–473
103. Harmanci AO, Sharma G, Mathews DH (2011) TurboFold: iterative probabilistic estimation of secondary structures for multiple RNA sequences. BMC Bioinformatics 12:108
104. Steffen P, Voss B, Rehmsmeier M, Reeder J, Giegerich R (2006) RNAshapes: an integrated RNA analysis package based on abstract shapes. Bioinformatics 22:500–503
105. Höchsmann M, Voss B, Giegerich R (2004) Pure multiple RNA secondary structure alignments: a progressive profile approach. IEEE Trans Comput Biol Bioinform 1:1–10
106. Moult J, Fidelis K, Kryshtafovych A, Rost B, Tramontano A (2009) Critical assessment of methods of protein structure prediction—round VIII. Proteins 77:1–4
107. Gutell RR, Noller HF, Woese CR (1986) Higher order structure in ribosomal RNA. EMBO J 5:1111–1113
108. Cannone JJ, Subramanian S, Schnare MN, Collett JR, D'Souza LM, Du Y, Feng B, Lin N, Madabusi LV, Muller KM et al (2002) The comparative RNA web (CRW) site: an online database of comparative sequence and structure information for ribosomal, intron, and other RNAs. BMC Bioinformatics 3:2
109. Gutell RR, Gray MW, Schnare MN (1993) A compilation of large subunit (23S- and 23S-like) ribosomal RNA structures. Nucleic Acids Res 21:3055–3074

110. Gutell RR (1994) Collection of small subunit (16S- and 16S-like) ribosomal RNA structures. Nucleic Acids Res 22:3502–3507
111. Michel F, Westhof E (1990) Modeling of the three-dimensional architecture of group I catalytic introns based on comparative sequence analysis. J Mol Biol 216:585–610
112. Seetin MG, Mathews DH (2011) Automated RNA tertiary structure prediction from secondary structure and low-resolution restraints. J Comput Chem 32:2232–2244
113. Major F, Turcotte M, Gautheret D, Lapalme G, Fillion E, Cedergren R (1991) The combination of symbolic and numerical computation for three-dimensional modeling of RNA. Science 253:1255–1260
114. Parisien M, Major F (2008) The MC-Fold and MC-Sym pipeline infers RNA structure from sequence data. Nature 452:51–55
115. Pinard R, Lambert D, Walter NG, Heckman JE, Major F, Burke JM (1999) Structural basis for the guanosine requirement of the hairpin ribozyme. Biochemistry 38:16035–16039
116. Das R, Baker D (2007) Automated de novo prediction of native-like RNA tertiary structures. Proc Natl Acad Sci U S A 104:14664–14669
117. Sharma S, Ding F, Dokholyan NV (2008) iFoldRNA: three-dimensional RNA structure prediction and folding. Bioinformatics 24: 1951–1952
118. Jonikas MA, Radmer RJ, Laederach A, Das R, Pearlman S, Herschlag D, Altman RB (2009) Coarse-grained modeling of large RNA molecules with knowledge-based potentials and structural filters. RNA 15:189–199
119. Ding F, Sharma S, Chalasani P, Demidov VV, Broude NE, Dokholyan NV (2008) Ab initio RNA folding by discrete molecular dynamics: from structure prediction to folding mechanisms. RNA 14:1164–1173
120. Malhotra A, Tan RK, Harvey SC (1994) Modeling large RNAs and ribonucleoprotein particles using molecular mechanics techniques. Biophys J 66:1777–1795
121. Tan RKZ, Petrov AS, Harvey SC (2006) YUP: a molecular simulation program for coarse-grained and multiscaled models. J Chem Theory Comput 2:529–540
122. Jonikas MA, Radmer RJ, Altman RB (2009) Knowledge-based instantiation of full atomic detail into coarse grain RNA 3D structural models. Bioinformatics 25:3259–3266
123. Bernhart SH, Hofacker IL (2009) From consensus structure prediction to RNA gene finding. Brief Funct Genomic Proteomic 8:461–471
124. Mathews DH, Moss WN, Turner DH (2010) Folding and finding RNA secondary structure. Cold Spring Harb Perspect Biol 2: a003665
125. Shapiro BA, Yingling YG, Kasprzak W, Bindewald E (2007) Bridging the gap in RNA structure prediction. Curr Opin Struct Biol 17:157–165
126. Jossinet F, Ludwig TE, Westhof E (2007) RNA structure: bioinformatic analysis. Curr Opin Microbiol 10:279–285
127. Szymanski M, Barciszewska MZ, Erdmann VA, Barciszewski J (2002) 5S ribosomal RNA database. Nucleic Acids Res 30:176–178
128. Crick FH (1958) On protein synthesis. Symp Soc Exp Biol 12:138–163
129. Edgar RC (2004) MUSCLE: multiple sequence alignment with high accuracy and high throughput. Nucleic Acids Res 32: 1792–1797
130. Hofacker IL (2003) Vienna RNA secondary structure server. Nucleic Acids Res 31:3429–3431
131. Hofacker IL, Fontana W, Stadler PF, Bonhoeffer LS, Tacker M, Schuster P (1994) Fast folding and comparison of RNA secondary structures. Monatsh Chem 125: 167–168
132. Ding Y, Chan CY, Lawrence CE (2004) Sfold web server for statistical folding and rational design of nucleic acids. Nucleic Acids Res 32:W135–W141

Chapter 9

Crystallization of RNA–Protein Complexes: From Synthesis and Purification of Individual Components to Crystals

Anna Perederina and Andrey S. Krasilnikov

Abstract

A broad range of biological processes relies on complexes between RNA and proteins. Crystallization of RNA–protein complexes can yield invaluable information on structural organizations of key elements of cellular machinery. However, crystallization of RNA–protein complexes is often challenging and requires special approaches. Here we review the purification of RNA, RNA-binding proteins, and the formation and crystallization of RNA–protein complexes, using the crystallization of the P3 RNA domain of ribonuclease MRP, a multicomponent ribonucleoprotein complex involved in the metabolism of various RNA molecules, as an example. The RNA–protein complex was formed using gel-purified RNA, produced by run-off transcription with T7 RNA polymerase in vitro, and proteins that were overexpressed in *Escherichia coli* and purified to be RNase-free. The complex was crystallized using a sitting drop setup; initial screening for suitable crystallization conditions was performed using a sparse matrix approach.

Key words: RNA–protein complexes, In vitro transcription, RNA purification, Protein purification, Crystallization

1. Introduction

Complexes between proteins and RNA play key roles in a wide variety of biological processes such as transcription, translation, RNA metabolism (including splicing), telomere maintenance, and many others. Some complexes are transient (e.g., complexes between diverse RNA-binding or RNA processing proteins and mRNAs), while others form stable functional units where RNA and protein act together. A special place among RNA–protein complexes is occupied by catalytic ribonucleoprotein complexes where RNA participates in or is solely responsible for catalysis (such as the ribosome (1), the spliceosome (2), ribonucleases (RNases) P and MRP (3), and others). Understanding their structural organizations is crucial for the understanding of a broad range of biological

Kenneth C. Keiler (ed.), *Bacterial Regulatory RNA: Methods and Protocols*, Methods in Molecular Biology, vol. 905, DOI 10.1007/978-1-61779-949-5_9,

processes, and X-ray crystallography is an indispensable tool in structural studies of RNA–protein complexes. However, crystallization of RNA–protein complexes is usually difficult and presents multiple, often unique challenges, especially in the case of large complexes. Here we briefly review approaches to the crystallization of RNA–protein complexes using the crystallization of the P3 RNA domain of ribonucleoprotein enzyme RNase MRP (4) in a complex with protein components of yeast RNases P/MRP (5) Pop6 and Pop7. The complex of Pop6 and Pop7 with the P3 RNA domain plays important structural roles in the enzymes of the RNase P/MRP family (6, 7). Crystallization of this complex (4, 6) provided the first view of interactions between RNA and proteins in large multicomponent enzymes of the RNase P/MRP family (5), and may serve as a useful example for the crystallization of both bacterial and eukaryotic RNA–protein complexes. The RNA component of the complex was produced by in vitro transcription and purified using denaturing gel electrophoresis; the protein components were overexpressed in *Escherichia coli* and purified using ion exchange and size exclusion chromatography. The RNA–protein complex was formed by incubation of RNA with proteins and crystallized using a sitting drop setup.

2. Materials

2.1. RNA Synthesis and Purification

All reagents must be RNase-free unless stated otherwise (see Notes 1 and 2).

1. Dry or water baths capable of maintaining 37°C and 85–90°C.
2. 0.5 M ethylenediaminetetraacetic acid (EDTA). Adjust pH to 7.8 with NaOH. Store at 4°C.
3. 3 M sodium acetate, pH 7.0. Store at 4°C.
4. Phenol/chloroform mix: combine 1 volume of Tris–HCl (pH 8.0) buffered phenol with an equal volume of a mix of chloroform with isopropyl alcohol (24:1 v:v); shake well. Store in a dark bottle at 4°C.
5. Ethanol and 70% (v:v) ethanol solution (use RNase-free water). Store at room temperature.
6. Oligonucleotide annealing buffer: 10 mM Tris–HCl (pH 8.0), 10 mM NaCl. Use RNase-free reagents. Store at 4°C.
7. Diethylpyrocarbonate (DEPC). Store in a dark bottle at 4°C.
8. 5× transcription buffer: 200 mM Tris–HCl (pH 8.0), 5 mM spermidine, 250 μg/mL acetylated BSA. Store at –20°C. Add DTT to 100 mM immediately prior to use.

9. 1 M $MgCl_2$. Store at −20°C.
10. 50 mM solutions of GTP, ATP, UTP, and CTP. Important: adjust pH to 8.0 with 1 M Tris base. Store at −20°C.
11. T7 RNase polymerase (5–10 mg/mL in 50% (v:v) glycerol). Store at −20°C.
12. Owl Nucleic Acid Sequencing System (Thermo Fisher) or a similar sequencing gel system with corresponding glass plates and a high-voltage (maximum 2,000 V or more) power supply.
13. Two glass plates (48 × 35 × 0.5 cm and 45 × 35 × 0.5 cm), two spacers (48 × 1.2 × 0.2 cm), and a single-well comb (6 × 27 × 0.2 cm) (see Note 3).
14. Portable UV lamp (254 nm).
15. Savant DNA 120 speedvac concentrator (Thermo Fisher) or a similar concentrator.
16. Beckman Avanti JE centrifuge (Beckman) or a similar centrifuge with a set of rotors.
17. 10× concentrated solution of Tris-Borate-EDTA (TBE) buffer: 890 mM Tris base, 890 mM boric acid, 20 mM EDTA, pH 8.4. TBE buffer used to make preparative polyacrylamide gels and gel loading buffer (below) must be RNase-free; the buffer used to run electrophoresis (i.e., added to the electrophoresis chambers) does not have to be RNase-free. Store at room temperature.
18. Gel loading buffer: dissolve urea (to 8 M) in 1× TBE buffer, add a small amount of bromophenol blue and xylene cyanol dyes (just enough to make dyes clearly visible when run on a gel). Store at 4°C; re-solubilize urea before use if necessary.
19. Components of denaturing polyacrylamide gels: 40% acrylamide: bisacrylamide (19:1) solution, tetramethylethylenediamine (TEMED), 10% (w:v) ammonium persulfate solution, urea powder. Prepare the gel mix immediately before use.
20. Purified plasmid DNA encoding the RNA component of the complex (8), see Note 4.
21. RNA elution buffer: 50 mM potassium acetate (pH 7.0), 200 mM KCl, 700 mM NaCl.
22. Crash & soak buffer: 50 mM potassium acetate (pH 7.0), 200 mM KCl.

2.2. Purification of the Recombinant Pop6/Pop7 Protein Complex

All reagents must be RNase-free (see Note 1).

1. Sonic Dismembrator Model 500 sonicator (Fisher Scientific) or a similar sonicator.
2. Beckman Avanti JE centrifuge (Beckman) or a similar centrifuge with a set of rotors.

3. Biologic LP (Bio-Rad) chromatograph or a similar low pressure chromatographic system.
4. AKTA FPLC chromatograph (GE Healthcare) or a similar chromatographic system.
5. Equipment to run denaturing SDS-polyacrylamide gels: Mini Protean 3 gel system (Bio-Rad) or a similar system with a suitable power supply, and a set of glass plates, spacers, and combs.
6. Chromatographic columns: SP-sepharose; Hi-Trap Heparin HP and Superdex 75 gel-filtration columns (Amersham). All chromatography columns must be cleaned with 0.5–1 M NaOH, followed by a wash with RNase-free water; follow manufacturer recommendations on column wash.
7. Amicon Ultra-15 concentrators (10,000 MWCO) (Millipore).
8. *E. coli* cell line Rosetta DE3 co-expressing Pop6 and Pop7: see refs. (8, 9). Store *E. coli* cells at –70°C.
9. 1 M NaOH solution. Store at room temperature.
10. Lysis Buffer: 50 mM Tris–HCl (pH 7.5), 300 mM NaCl, 2 mM DTT, 1 mM EDTA, 1 mM PMSF, 1 mg/mL lysozyme (add last). Prepare immediately before use.
11. 50% polyethylenimine solution (Fluka). Prepare a 10% stock solution, adjust pH to ~7.5 with HCl. Store at 4°C.
12. Ammonium sulfate, powder.
13. 5 M NaCl solution. Store at 4°C.
14. Buffer A: 50 mM MES-NaOH (pH 6.0), 50 mM KCl, 5 mM NaCl, 2 mM DTT, and 0.1 mM PMSF. Prepare before use.
15. Buffer B: 10 mM Tris–HCl (pH 7.8), 100 mM NaCl, 50 mM KCl, 2 mM DTT, 0.1 mM EDTA, 0.1 mM PMSF. Prepare before use.

2.3. RNA Annealing and Complex Formation: Pop6/Pop7-RNA Complex Formation

All reagents must be RNase-free (see Note 1) unless stated otherwise.

1. Equipment to run native polyacrylamide gels at 4°C: a chromatography cabinet (4°C) or a cold room, Mini Protean 3 gel system (Bio-Rad) or a similar system with a suitable power supply, and a set of glass plates, spacers, and combs.
2. Three dry baths capable of maintaining 85, 50, and 28°C.
3. 1 M Tris–HCl (pH 7.4) buffer. Store at 4°C.
4. 1 M $MgCl_2$ solution. Store at 4°C.
5. 2 M KCl solution. Store at 4°C.
6. 10× Tris-borate (TB) buffer as a 10× concentrated solution: 890 mM Tris base, 890 mM boric acid (pH 8.4). Store at room temperature.

7. 1 M Magnesium acetate solution. Store at 4°C.
8. 100% Glycerol. Store at room temperature.
9. Toluidine Blue staining solution: 0.25% toluidine blue, 10% ethanol, 5% acetic acid. This solution does not have to be RNase-free. Store at room temperature.
10. RNA dilution buffer: 10 mM Tris–HCl (pH 7.4), 5 mM $MgCl_2$, 200 mM KCl.

2.4. Crystallization of RNA–Protein Complexes

1. Zeiss Discovery V12 stereo microscope (Zeiss) or a similar stereo microscope.
2. Type 815 (VWR) or similar heated/refrigerated incubators for crystallization.
3. Sparse matrix screening kits (commercially available from Hampton Research, Sigma, or home-made).
4. 24-Well Cryschem crystallization plates (Hampton Research) or similar crystallization plates.

3. Methods

3.1. Preparation of Plasmid DNA for In Vitro Transcription

If oligonucleotides are to be used as the template for in vitro transcription (see Note 5), start at Subheading 3.2.

1. Linearize the plasmid DNA (see Note 6): digest it with the selected restriction endonuclease in the buffer recommended by the manufacturer. As a general rule, when a given amount of DNA is being digested, higher volumes of reaction result in better efficiency of cleavage but can make subsequent steps of template preparation more cumbersome; normally, digest 100 μg of plasmid DNA in 1 mL of the reaction mix. Use 0.3–0.5 units of the restriction endonuclease per 1 μg of plasmid DNA and digest for 3 h.
2. After 3 h, take an aliquot containing 0.1–0.5 μg of plasmid, add EDTA to 10 mM (to the aliquot only) and analyze the aliquot for the completion of digestion using an ethidium bromide-stained agarose gel. If a single restriction site is expected, compare the mobilities of the digested sample and the initial plasmid on a 0.7% agarose gel. Linearized DNA will have a lower mobility than the uncut, supercoiled DNA. If multiple restriction sites are expected, adjust the percentage of the gel according to the sizes of the expected fragments. Keep in mind that the efficiency of cleavage may differ for different sites on the same plasmid.
3. If the aliquot analysis demonstrates complete digestion, stop the reaction by the addition of EDTA to 10 mM; otherwise

continue digestion for 2 more hours and check for the completeness of the digestion again. To speed up the digestion, more enzyme can be used (up to 5 units per microgram of plasmid DNA), but the cost of using hundreds of units of enzyme should be considered. Upon the addition of EDTA, digested DNA can be stored at −20°C indefinitely.

4. To prepare linearized DNA for in vitro transcription with T7 RNA polymerase, transfer the digested DNA into a 50-mL centrifuge tube and add sodium acetate (pH 7.0) to a final concentration of 0.6 M.
5. Purify the template DNA by three sequential extractions with phenol/chloroform followed by ethanol precipitation. For each extraction, add an equal volume of phenol/chloroform/isoamyl alcohol mix (25:24:1), vortex for 30 s, and centrifuge for 5 min at 14,000 × *g* (or 10 min at 5,000 × *g*) at room temperature (see Note 7). Collect the aqueous (typically, the upper) fraction, transfer it into a fresh centrifuge tube, and repeat the procedure two more times, carefully avoiding the organic phase and interphase carryover.
6. After phenol/chloroform extraction, add 3 volumes of cold (−20°C) ethanol, mix, and keep at −20°C for at least 1 h. It may be convenient to leave the sample overnight (if necessary, the sample can be stored at −20°C indefinitely).
7. Pellet DNA by centrifugation (30 min at 7,000 × *g* at 4°C) and carefully remove supernatant.
8. Add 20 mL of cold 70% ethanol, vortex, pellet DNA by centrifugation (5 min at 7,000 × *g* at 4°C), and carefully remove the ethanol; repeat this ethanol wash two more times.
9. After the final wash, remove as much ethanol as possible (usually it requires additional short centrifugation runs to collect liquid accumulated on the walls of the tube), and dry the sample in a speedvac until it appears completely dry.
10. Dissolve DNA in 1–5 mL of RNase-free water (see Notes 1 and 2), and store at −20°C.

3.2. Preparation of Oligonucleotide Templates for In Vitro Transcription

Oligonucleotides can be used in run-off transcription with T7 RNA polymerase as an alternative to the linearized plasmid DNA as described in Subheading 3.1 (see Note 5).

1. Synthesize or order the required deoxyribo-oligonucleotides. One of the required oligonucleotides is complementary to the T7 RNA polymerase promoter region (−17 to +1) of the template oligonucleotide and should have the following sequence: 5′-TAATACGACTCACTATA<u>G</u>-3′. This oligonucleotide can be combined with practically any template oligonucleotide (below); the underlined <u>G</u> corresponds to the first

(5′-terminal) nucleotide of the transcript. The second (template) oligonucleotide should contain the sequence of the desired RNA (as a reverse-complement) plus the T7 RNA polymerase promoter region. The template oligonucleotide should have the following sequence: 5′-(reverse-complement of the desired RNA sequence, excluding the 5′-end guanine)-CTATAGTGAGTCGTATTA-3′. The underlined C corresponds to the first (5′-terminal) nucleotide of the transcript (which must be a G); the 5′-end of the template nucleotide corresponds to the 3′-end of the desired RNA.

2. (Optional). Purify the oligonucleotides on a denaturing polyacrylamide gel to remove the shorter products of aborted synthesis. This step may be necessary for long template oligonucleotides when a substantial fraction of oligonucleotides are in fact products of aborted synthesis. When gel purification is deemed beneficial, this step can be performed using a procedure described in Subheading 3.4.
3. Anneal oligonucleotides for transcription. Mix the template oligonucleotide (final concentration 7–10 μM) with an equal molar amount of the oligonucleotide complementary to the T7 promoter region in the oligonucleotide annealing buffer (see Note 8).
4. Keep the mix 2 min at 90°C, then cool it for 10 min on ice. Annealed oligonucleotides can be stored at −20°C.

3.3. In Vitro Transcription

The optimal conditions for in vitro transcription may depend on the type, length, and sequence of the template. Therefore, it is recommended to perform several small-scale (~100 μL) optimization runs before attempting large-scale transcription. A good starting point for optimization is described in step 1. Concentrations of the template, T7 RNA polymerase, ribonucleotide triphosphates, and $MgCl_2$ (see Note 9) can be adjusted to maximize the yield of the RNA of interest.

1. Mix 20 μL of 5× transcription buffer with the required volume of water (so the final volume of the reaction will be 100 μL), ribonucleotide triphosphates (GTP, ATP, UTP, and CTP (pH 8.0), the final concentration 4 mM of each), DNA template (final concentration 0.15 μM if oligonucleotide template is used, or 60 μg/mL if linearized plasmid DNA template is used), T7 RNA polymerase to a final concentration of 60 μg/mL (add the enzyme last). Assemble the reaction at room temperature; do not put it on ice.
2. Run the test reactions at 37°C for 2–4 h; the appearance of a white precipitate within an hour of the start of the reaction is normal.

3. Stop the reactions with an equal volume of a buffer containing 8 M urea and analyze the results using a 5–15% denaturing (8 M urea) polyacrylamide gel.
4. To analyze the products of transcription on a denaturing (8 M urea) polyacrylamide gel, prepare a small 15 × 15 cm gel and use it to analyze the RNA following an appropriately scaled-down version of the procedures described in Subheading 3.4, steps 12–17. The percentage of the gel depends on the length of the expected product—for 40–50 nt-long RNAs use 15% gels and for RNAs that are several hundred nucleotides long, 5% gels should be used. A 15 × 15 cm denaturing gel should be run at a constant 12–15 W power; the temperature of the glass plates should be about 60°C.
5. Use the optimized conditions identified in steps 1–4 to set up a large-scale (10–15 mL) in vitro transcription reaction (see Notes 1 and 2).
6. Run the reaction for 4 h at 37°C.
7. Pellet the inorganic precipitate by centrifugation at 10,000 × *g* for 5 min at room temperature.
8. Transfer the supernatant to a fresh 50-mL centrifuge tube, and perform a single round of phenol/chloroform extraction followed by ethanol precipitation as described in Subheading 3.1, steps 5–7. After this stage, the sample can be safely stored at –20°C indefinitely.

3.4. RNA Purification Using Denaturing Polyacrylamide Gels

Make a preparative gel for RNA purification (see Note 10). For large-scale RNA purification we use the Thermo Scientific Owl Nucleic Acid Sequencing System.

1. Pellet RNA from Subheading 3.3, step 8 by centrifugation at 7,000 × *g* for 30 min at 4°C.
2. Dissolve the pellet in ~1 mL of gel loading buffer (see Note 11).
3. Wash the glass plates, spacers, and comb carefully (we use DRNAse-Free or similar decontaminant), rinse with RNase-free water, and dry.
4. Position the spacers between the glass plates along their sides, align the bottom sides of the glass plates, and tape the sides of the glass plates using gel sealing tape, leaving the upper side of the assembly open; the goal is to create a water-tight assembly (see Note 12).
5. Clamp the sides with large binder clips (three clips on each side; do not clamp the top and bottom sides of the assembly).
6. To purify ~50 nt-long RNA, prepare 300 mL of 15% acrylamide with 8 M of urea and 1× TBE (RNase-free); for longer RNAs a lower percentage gel will give better resolution.

7. Filter the solution through a 0.22-μm nitrocellulose filter, add 1 mL of a 10% solution of ammonium persulfate, mix well (avoid air bubbles), add 100 μL of TEMED, and mix again.
8. Apply the solution between the glass plates using a 60-mL syringe; keep the glass plates assembly at a slight angle; avoid air bubbles.
9. When the acrylamide solution reaches about 2 cm from the open upper side, put the assembly on a flat horizontal surface.
10. Insert the comb, clamp it with four large binder clips (make sure that the clips fit over the comb before you start!), and allow the gel to polymerize (typically, about 1 h) (see Note 13).
11. Remove all clips, the tape, and the comb, and rinse the well with distilled water.
12. Set up the electrophoresis system. The running buffer (1× TBE) and the electrophoresis system itself do not necessarily have to be RNase-free. Preheat the gel by running electrophoresis at a constant power of 60–70 W for approximately 1 h; the temperature of the glass plates should reach about 60°C. Turn off power, rinse the well with the running buffer, load RNA (see Note 14), restore power, and run the gel for as long as it takes the fragment of interest to cross about ¾ of the gel (see Note 15) (typically, about 10–15 h).
13. Remove the glass plate assembly from the electrophoresis system. Let the plates cool to room temperature. Do not try to cool them with running water: they may crack.
14. Pull out the spacers and remove all traces of the running buffer.
15. Carefully remove one of the glass plates, leaving the gel on the other one. Carefully remove any remaining buffer, avoiding any contact between the buffer and the area of the gel containing the RNA of interest (the running buffer is not RNase-free!).
16. Put clear plastic wrap over the gel, flip the glass plate so the gel and plastic wrap end up on the plate's bottom side, and carefully pull the gel (together with plastic wrap) away from the glass plate. Cover the gel with the second piece of plastic wrap.
17. Put the gel on a piece of white paper and use a portable UV lamp (254 nm) to illuminate it (see Note 16). You should see clear shadows corresponding to the RNA's position in the gel. Do not expose RNA to UV light for longer than necessary; keep the UV lamp as far from the gel as possible (UV light crosslinks RNA, and this undesirable effect should be minimized). Fluorescent phosphorous plates can be used instead of white paper to increase the sensitivity of RNA detection.
18. Use a marker to contour the area of interest on one of the pieces of plastic wrap covering the gel.

19. Flip the gel (so the contour ends up on the bottom piece of plastic wrap), remove the top piece of plastic wrap and cut the area of interest out with a clean (RNase-free) razor blade (see Note 17).
20. Put the piece of the gel containing the RNA fragment of interest in a centrifuge tube, and break the gel into small pieces using a sealed sterile 5–10-mL pipette.
21. Add approximately 1 mL of RNA elution buffer and 1 mL of phenol (not phenol-chloroform!) per milliliter of the cut gel fragment's volume.
22. Incubate the gel fragment in the RNA elution buffer/phenol mix at room temperature for 4–5 h with vigorous agitation.
23. Spin the centrifuge tube with the gel at 17,000 × *g* for 5 min at room temperature, and transfer the upper (aqueous) phase to a new 50-mL centrifuge tube.
24. Add 3 volumes of cold (–20°C) ethanol and keep at –20°C for at least 2 h or overnight.
25. Pellet the RNA at 17,000 × *g* for 30 min at 4°C.
26. Dissolve the RNA pellet in crash & soak buffer (~300 μL of the buffer per milliliter of crashed gel, step 20), and precipitate it with ethanol again.
27. Pellet the RNA (17,000 × *g* for 30 min at 4°C), wash it three times with cold 70% ethanol as described in Subheading 3.1, steps 6–9.
28. Dry the RNA in a Speedvac and dissolve it in RNase-free water to the desired final concentration, typically 5–10 mg/mL. Store the sample at –70°C.

3.5. Purification of the Recombinant RNA-Binding Pop6/Pop7 Protein Complex

All purification procedures (see Note 18) should be carried out at 4°C unless specified otherwise. Aliquots of the sample should be taken after each purification step: analyze the results of each step using denaturing SDS-polyacrylamide gels before proceeding to the next step.

1. Harvest *E. coli* cells (8, 9) by centrifugation at 4,000 × *g* for 10 min. Harvested cells can be frozen and stored at –70°C or used immediately.
2. Resuspend cells in 5–10 mL of Lysis Buffer.
3. Incubate cell suspension for 1 h on ice, agitating it gently every 10 min. Due to the release of genomic DNA, the cell suspension will become very viscous.
4. After the 1 h incubation, disrupt genomic DNA and any remaining cells by sonication until the viscosity of the suspension is significantly reduced. During sonication, keep the sample on ice, avoid overheating the sample by sonicating in short bursts followed by prolonged cool-down periods. When using

Fisher Scientific Sonic Dismembrator Model 500 sonicator, perform 6 sonication bursts (7 s each, followed by 1 min cool-down period).

5. Clear the cell lysate by centrifugation at 17,000 × *g* for 30 min.
6. Treat the clarified lysate with 0.1% (v:v) of polyethylenimine (see Note 19) for 10 min with light agitation.
7. Add 0.48 g of ammonium sulfate powder per milliliter of the protein solution, dissolve it, and let the protein precipitate overnight (see Note 20).
8. Collect the protein precipitate by centrifugation at 17,000 × *g* for 30 min, and dissolve it in Buffer A.
9. Clear the protein solution by centrifugation at 17,000 × *g* for 30 min.
10. Equilibrate a 5 mL SP-sepharose column (Amersham) with Buffer A supplemented with 100 mM NaCl.
11. Load the protein sample on the SP-sepharose column. Wash the column with Buffer A supplemented with 100 mM NaCl. Elute the Pop6/Pop7 protein complex with a linear gradient of NaCl (0.1–0.5 M in Buffer A; the length of the gradient should be 100 mL) at a 1 mL/min flow rate. Collect 0.5 mL fractions.
12. Combine the major protein-containing fractions and dilute them threefold with Buffer A.
13. Equilibrate a 1 mL Hi-Trap Heparin HP column (Amersham) with Buffer A supplemented with 100 mM NaCl.
14. Load the Pop6/Pop7 complex on the Heparin HP column, wash the column with Buffer A supplemented with 100 mM NaCl, and then elute the protein complex using a linear gradient of salt (0.1–0.5 M NaCl in Buffer A; the length of the gradient should be 100 mL with a flow rate of 1 mL/min; collect 1.1 mL fractions).
15. Combine fractions of interest and concentrate them to ~10–20 mg/mL using an Amicon Ultra-15 concentrator (10,000 Da MWCO, Millipore).
16. Equilibrate a Superdex 75 gel-filtration column (Amersham) with Buffer B.
17. Load the protein sample on the column and run it through the column in Buffer B. Collect 0.5 mL fractions.
18. Combine fractions of interest and concentrate them to 20 mg/mL using an Amicon Ultra-15 concentrator (10,000 Da MWCO, Millipore).

This procedure should yield about 8 mg of RNase-free Pop6/Pop7 complex per liter of culture. The protein solution can be stored at 4°C for several months.

3.6. Pop6/Pop7-RNA Complex Formation

1. Refold the P3 domain RNA (6 mg/mL) by incubation at 85°C for 2 min in a buffer containing 10 mM Tris–HCl (pH 7.4), followed by 5 min cooling to room temperature in a styrofoam rack.
2. Add $MgCl_2$ to 5 mM and incubated the sample at 50°C for 10 min.
3. Cool the sample to room temperature in a styrofoam rack, and add KCl to 200 mM (see Note 21).
4. Adjust RNA concentration to 5 mg/mL using RNA dilution buffer.
5. To form the Pop6/Pop7-RNA complex, mix 13 μL of RNA solution with 7 μL of the Pop6/Pop7 heterodimer solution at room temperature, and incubate the mix at 28°C for 15 min (see Note 22). Keep at room temperature and use for crystallization immediately.
6. To monitor the formation of RNA–protein complexes, use gel mobility shift assays. Use 5–10% native polyacrylamide gels (the gel percentage depends on the length of the RNA; for ~50-nt-long RNA such as the P3 RNA domain use 10% gels; see Note 23).
7. Add glycerol (to 10–15% (v:v)) to aliquots (3 μg of RNA per aliquot) of the samples.
8. Load the aliquots on the gel.
9. Run the gel at 4°C at low power to avoid overheating (see Note 24).
10. Stain the gel with Toluidine Blue dye (see Note 25). The stained gel should look similar to the gel shown in Fig. 1. If necessary, the gel can be stained with the standard Coomassie Blue stain to visualize proteins.

3.7. Crystallization of the P3 RNA Domain of Yeast RNase MRP in a Complex with Proteins Pop6 and Pop7

1. Prepare the RNA–protein complex as described in Subheadings 3.1–3.6 (see Note 26).
2. Put 0.5 mL of the crystallization solution in each of the bottom (large) wells of a crystallization plate. For initial screening, sparse matrix crystallization screens give good results (see Note 27).
3. Put 0.5–2 μL of the solution of the RNA–protein complex in the middle of the small (elevated) well and mix with an equal volume of the solution from the bottom well. For initial screening or optimization runs, the ratio between the sample and the crystallization solution is often varied.
4. Seal the wells with clear tape (see Note 28). We obtain tape from Hampton Research; cheaper versions of similarly looking tape often degrade when exposed to the vapors in crystallization wells.

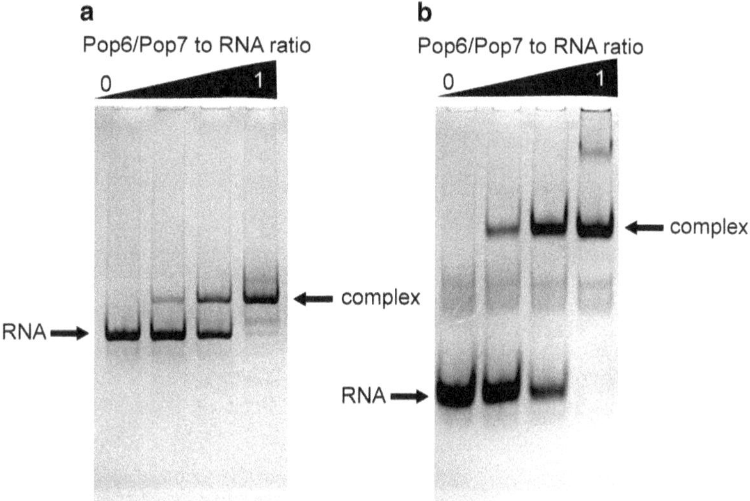

Fig. 1. Gel mobility shift assays for Pop6/Pop7-RNA complexes (8). (**a**) The binding of the Pop6/Pop7 heterodimer to the 340-nt-long full-length yeast RNase MRP RNA, analyzed on a Toluidine Blue-stained 5% native polyacrylamide gel. (**b**) The binding of the Pop6/Pop7 heterodimer to the 50-nt-long P3 domain RNA from yeast RNase MRP RNA, analyzed on a Toluidine Blue-stained 10% native polyacrylamide gel. The protein to RNA ratio increases from *left* to *right* as indicated; the positions of free RNA and RNA–protein complexes are shown by *arrows*.

5. Put the crystallization tray in an incubator set at 4–28°C (see Note 29); the exact temperature should be optimized for each of the crystallization conditions and for each of the RNA–protein complexes.
6. Check for crystals (see Note 30) the next day, in 3 days, in a week, and then every 2 weeks thereafter. Crystals may appear after several hours, after several months, or—most likely—never. It is important to check for crystals on a regular basis as often they disappear when kept in the well too long.
7. If small crystals are detected, measure them, allow them to grow for a week (less if crystals appeared after just a day or two, and more if it took months for them to appear), and then check their dimensions again. Harvest crystals that are deemed suitable for diffraction testing.
8. To optimize crystallization conditions, vary the ratio between the volumes of the sample and that of the reservoir solution in the drop, the concentration of the complex in the drop, the concentration of the precipitant or salts in the reservoir solution, the buffer, the pH, various additives, etc. (Fig. 2).

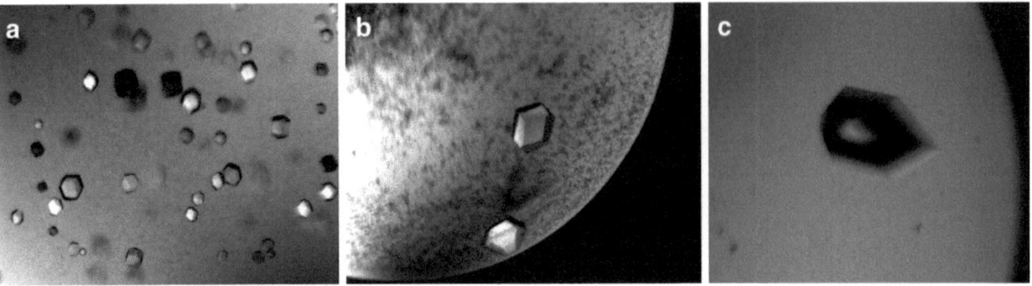

Fig. 2. Crystallization of the P3 RNA domain of yeast RNase MRP in a complex with proteins Pop6/Pop7 (4, 6). (**a**) Initial crystallization screening. Crystals appeared in 5 days in reagent #39 of Crystal Screen 1 (Hampton Research) at 20°C. (**b**) Crystals were improved by the addition of 2 mM $ZnCl_2$. Crystals grew to ~$0.1 \times 0.1 \times 0.1$ mm^3. (**c**) Crystals were further improved by the addition of 2 mM $ZnCl_2$ and 5% (w:v) D-trehalose. Crystals grew to ~$0.65 \times 0.65 \times 0.65$ mm^3.

4. Notes

1. The best way to prepare RNase-free solutions is to start with RNase-free initial components and use preparation techniques that exclude any possible contamination with RNases. Use disposable certified RNase-free plasticware; glassware should be carefully washed and baked at 180°C overnight. Deionized water that was autoclaved and filtered through a 0.22–0.45-μm cellulose filter is RNase-free, provided that the water purification system is well maintained. If the initial components for a solution are not available in an RNase-free form, prepare the solution, add 1% (v:v) of DEPC, incubate overnight at 30–37°C with agitation, and then autoclave and filter the solution. Keep in mind that treatment with DEPC will result in the presence of residual ethanol and ethers in your solution; this may negatively affect activities of some enzymes. Additionally, DEPC treatment cannot be used with reagents that do not tolerate autoclaving (such as MES) or contain chemical groups that interact with DEPC (such as Tris).

   When adjusting the pH of an RNase-free solution, avoid inserting a pH meter electrode into the solution. Instead, take small aliquots, use them to measure pH, and then discard them; paper pH-indicator strips are a practical solution for the initial stages of pH adjustments.

   Store RNase-free solutions at a low temperature or frozen if possible, and keep them sterile: bacterial growth will result in contamination with RNases.

   It is advisable to test the solution as well as the RNA and protein preparations for contamination with RNases. Remember that for a successful crystallization of an RNA or RNA–protein complex, the RNA should not noticeably degrade over the course of weeks at the crystallization temperature.

2. The presence of residual products of DEPC decay noticeably reduces the efficiency of transcription with T7 RNA polymerase; the amount of DEPC used in the preparation of reagents for transcription should be minimized.
3. Spacers and combs of this thickness are not easily available commercially; however they can easily be cut from a 2-mm thick piece of Teflon.
4. When plasmid DNA is used as the template, required amounts of DNA (typically, hundreds of micrograms) can be obtained from 0.2 to 1 L of a saturated culture of *E. coli* cells (usually DH5a) carrying the plasmid of interest by purification of the plasmid in CsCl gradient (a time-consuming, but easily scalable method) or, alternatively, using any of the commercially available DNA purification kits capable of yielding the required quantity of DNA.
5. In vitro run-off transcription with T7 RNA polymerase (10) is commonly used to prepare large (milligram) quantities of RNA required for crystallization experiments. For efficient transcription, the promoter region on the DNA template should be double-stranded and the first (5′-terminal) nucleotide in the product RNA should be a guanine (10). Run-off transcription with T7 RNA polymerase usually results in product RNA that is somewhat heterogeneous at its 3′ (and sometimes 5′) end. Since homogeneous samples are crucial for the success of crystallization attempts, special measures (such as product RNA purification on nucleotide-resolution denaturing polyacrylamide gels, incorporation of self-cleaving ribozymes into the transcript (11), or the use of modified nucleotides in the template (12)) are usually taken. As a DNA template for transcription, one can use either (a) a single-stranded template oligonucleotide annealed to a 18-nucleotide-long (or longer) oligonucleotide complementary to the promoter region (nucleotides −17 to +1) (13), or (b) a purified PCR product containing the sequence of interest and the promoter or, most commonly, (c) a linearized plasmid. The major advantage of the use of the single-stranded template oligonucleotide (a) is that it allows for a considerable shortening of the time between the design of a specific template and the actual production of RNA (as no cloning is necessary). Due to limitations on the length of chemically synthesized oligonucleotides (which must include, in addition to the sequence of interest, the promoter region), this approach cannot be used for the synthesis of long RNAs. The use of purified PCR product as the template for run-off transcription (b) practically eliminates the restriction on the length of the RNA product, while allowing the incorporation of modified nucleotides into terminal regions of the template to reduce the heterogeneity of the product RNA (12).

However, linearized plasmid DNA remains the most commonly used type of template for run-off transcription with T7 RNA polymerase when large quantities of RNA are needed.

6. The plasmid used as the template in run-off transcription with T7 RNA polymerase must contain a restriction site incorporated at the 3′-end of the desired RNA sequence. The best results are obtained when the restriction enzyme used for DNA linearization produces either blunt ends or ends with 5′-overhangs. In many cases the sequence requirements imposed by the necessity to have a 3′-end sequence recognizable by a suitable restriction endonuclease are not acceptable; in such cases restriction endonucleases that cleave DNA at some distance from the recognition site (such as *Fok*I, *Bsa*I, or *Bsm*AI) are commonly used. The cost of the enzyme considered for the linearization of large quantities of plasmid DNA is also often taken into account.
7. High concentrations of salt will increase the density of the solution. When the density of the solution exceeds that of the phenol/chloroform mix, the aqueous phase will end up at the bottom of the tube after centrifugation. When performing phenol or phenol/chloroform extraction from solutions with considerable concentrations of salt, be sure to correctly identify the aqueous phase. If the densities of the aqueous and the organic phases are too close for efficient phase separation, slightly dilute the solution with water and repeat the phenol extraction.
8. For some templates, better results are obtained when the shorter oligonucleotide is added in a 2–3 M excess over the template oligonucleotide.
9. Nucleotide triphosphates bind magnesium, thus the concentration of $MgCl_2$ should exceed the combined concentration of NTP. A 4 mM excess of $MgCl_2$ (e.g., 20 mM $MgCl_2$ at 4 mM of each NTP) is a good starting point.
10. The goal of the gel purification of the product of in vitro transcription is to eliminate unincorporated nucleotide triphosphates, products of aborted transcription, and any other undesired RNAs, as well as template DNA. Importantly, gel purification typically serves to produce RNA with homogeneous ends. The latter requires a nucleotide-level resolution of the gel. The nucleotide-level resolution on a preparative gel can easily be achieved for relatively short RNAs, but care should be taken to avoid overloading the gel.
11. It may take up to half an hour to dissolve the RNA. Make sure the RNA is completely dissolved before proceeding with the next step.

12. Siliconizing glass plates will make gel preparation and subsequent manipulations easier; however, be careful not to over-siliconize them as the gel may start moving between the plates during electrophoresis.
13. One can use a hair drier or a table lamp to speed up polymerization.
14. Load the sample very slowly; avoid mixing it with the running buffer. Use long pipette tips or a syringe with a suitable needle to load the sample at the very bottom of the well.
15. The mobility of the loading dye can be used to estimate the progress of electrophoresis. On a 5% denaturing polyacrylamide gel, Bromophenol Blue/Xylene Cyanol dyes run approximately as ~35/130 nt-long RNAs, respectively, on a 8% gel—as ~20/75 nt-long RNAs, and on a 10% gel—as ~12/55 nt-long RNAs.
16. UV light is harmful. Protect your eyes and always wear suitable protective goggles or a face shield when working with UV light.
17. Often several bands (corresponding to heterogeneous RNA ends) will appear. It is advisable to extract all major bands separately and use them all in crystallization trials (separately).
18. In most cases, structural studies of RNA–protein complexes involve purification (typically, from *E. coli*) of recombinant proteins of interest. Proteins can be purified using their individual properties by hydrophobic, ion-exchange, gel-filtration, or affinity chromatography. Often, affinity purification tags are fused to proteins of interest to facilitate their purification. This approach allows for rather straightforward protein purification; however, the possible influence of the affinity tag on the protein's interactions with RNA, as well as its potential negative effect on the crystallization of the RNA–protein complex of interest, should be considered. It is generally a good idea to design a protein construct that allows for the removal of the purification tag (usually, with a site-specific protease). While affinity chromatography allows one-step protein purification that is suitable for many applications, crystallization of RNA–protein complexes requires highly pure RNase-free protein preparations which typically require multiple purification steps even when an affinity tag is used. Gel-filtration is usually used as a final purification step to remove minor contaminants.

    Becausee RNA-binding proteins are typically basic, they can usually be purified by cation-exchange chromatography using carboxymethyl (CM), sulfopropyl (SP), or heparin resins. Purification of RNA-binding proteins (especially recombinant ones) is often complicated by nonspecific binding to nucleic acids in cell lysate. In this case, the protein will not be able to bind cation-exchange resins (and will not be suitable for crystallization in general) unless the nonspecifically bound nucleic acids are removed. The removal of the nonspecifically bound nucleic acids while preserving the native fold of the protein is often a key step in the preparation of a recombinant protein sample suitable for crystallization as a part of a complex with RNA.

Denaturing the protein of interest allows for a straightforward removal of any bound nucleic acids, but since sufficiently complete refolding is typically difficult (or most often impossible for larger proteins) to achieve, we avoid this approach in our structural studies of RNA-binding proteins. As alternative approaches we typically try to prevent the nonspecific binding of nucleic acids by (a) increasing the concentration of salt in the buffer used for cell lysis (up to 1.5 M of NaCl can be used, but see Note 31), (b) using hydrophobic interaction chromatography (in a high concentration of ammonium sulfate (see Note 32)), or (c) using the extraction of nucleic acids by polyethyleneimine followed by ammonium sulfate precipitation (see Note 33).

19. The concentration of polyethylenimine should be optimized for each protein. We usually try three concentrations (0.1, 0.5, and 1%) to find the one that yields a heavy precipitate (mostly nucleic acids) without precipitating the protein.
20. After longer precipitation some proteins tend to stay insoluble.
21. The refolding of RNA prior to the formation of a complex with proteins is very important, especially for large RNAs. Typically, RNA refolding involves thermal denaturation at 70–95°C, followed by cooling down at a specific rate (which may vary from practically instant cooling achieved by putting the sample on ice to slow overnight cooling achieved by putting the sample in a large vessel with hot water that is allowed to cool down). The rate of cooling is often crucial to proper RNA folding and should be adjusted experimentally for each RNA of interest. The composition of the buffer used to refold the RNA (especially the presence and concentration of magnesium) may have dramatic effects on RNA refolding. The ability of magnesium to cause degradation of RNA at elevated temperatures has to be taken into account; a common way to reduce magnesium-induced degradation of RNA is to perform the high-temperature denaturation of RNA in the absence of magnesium and add it at a later step.

22. It is typical to see light white precipitate when an RNA-binding protein is added to RNA. Usually this precipitate quickly disappears when the solution is pipetted. If the precipitate does not disappear, add KCl to 200–300 mM; if salt does not help, try lowering the RNA and protein concentrations.
23. If the RNA–protein complex of interest is formed in the presence of magnesium, use a Tris-Borate buffer (a standard 1× TBE buffer, but without EDTA) supplemented with magnesium acetate (typically, 0.5 mM of magnesium acetate). Avoid $MgCl_2$ in electrophoresis: toxic chlorine gas will form. It is recommended that you occasionally transfer the buffer between the upper and the lower chambers. If no magnesium was used in the formation of the RNA–protein complex, the standard 1× TBE buffer can be used.
24. Keep in mind that the addition of metal ions to the gel and the running buffer will dramatically increase their conductivity, requiring reduced voltage. Normally, ~8 × 10 cm native gels are run at a constant power of 1–2 W.
25. In our experience, staining with Toluidine Blue dye gives better results than staining with ethidium bromide.
26. The availability of structurally homogeneous samples of the RNA–protein complex of interest is absolutely crucial for successful crystallization. Accordingly, the fact that an RNA–protein complex consists of several components and usually has to be reconstituted presents additional, often considerable challenges: each individual component has to be structurally homogeneous (which means the absence of alternative three-dimensional folds—a requirement that is more stringent than a simple absence of contamination as observed under denaturing conditions), additionally, the reconstitution of the complex must be complete and result in a homogeneous sample.

    Crystals containing large RNA molecules with open helical stems (that is, RNA where both the 5′- and the 3′-end belong to the same blunt-end double helix) are very often packed in such a way that two RNA molecules form a pseudo-continuous helix with their blunt ends pressed against each other. To facilitate such packing, it is important to ensure that both 5′ and 3′ ends of the RNA are homogeneous.
27. To improve the chances of a successful crystallization of an RNA–protein complex, it is important to attempt crystallization of multiple, somewhat different constructs. Each construct should retain the key features of the studied complex (such as regions involved in RNA–protein interactions and other functionally

or structurally important parts), but nonessential (typically, peripheral) parts should be varied. In principle, both RNA and protein parts of the complex of interest can be modified; practically, in the vast majority of cases it is the RNA component that is subjected to such manipulations. Alternatively, homologous complexes from different organisms can be used. The use of these approaches dramatically increases the chances of crystallization while allowing the investigator to stay focused on a particular complex of interest. For the crystallization of the P3 RNA domain of yeast RNase MRP in a complex with proteins Pop6 and Pop7 (6), we used the available results of biochemical studies describing RNA–protein interactions in this complex (8) to identify parts of the RNA that could be modified without affecting its interactions with proteins, and multiple variants of RNA were screened before the successful crystallization was achieved (4).

28. Small drops dry out quickly: work the wells in small groups (finish and seal a group of 6–12 before proceeding with the next group).
29. Any disturbance may interfere with crystallization: use incubators that do not vibrate or shake. A typical source of vibration is an unbalanced fan; usual sources of occasional shaking are the compressor starting and stopping. We use incubators that run their compressors constantly.
30. Use polarization filters when checking for crystals: small crystals are typically easier to detect under polarized light.
31. High concentrations of salt may not be suitable for the purification of protein complexes because they may result in their dissociation; avoid exceeding 300–400 mM NaCl when purifying protein complexes.
32. Hydrophobic interaction chromatography may be unsuitable for some proteins resulting in their partial denaturation/loss of activity or precipitation.
33. Extraction of nucleic acids by polyethyleneimine may inactivate/denature some proteins.

Acknowledgements

We would like to thank Lydia Krasilnikova for her help with the preparation of this manuscript. This work was supported by NIH Grant GM085149 to A.S.K.

References

1. Steitz TA (2008) A structural understanding of the dynamic ribosome machine. Nat Rev Mol Cell Biol 9:242–253
2. Wahl MC, Will CL, Luhrmann R (2009) The spliceosome: design principles of a dynamic RNP machine. Cell 136:701–718
3. Altman S (2007) A view of RNase P. Mol Biosyst 3:604–607
4. Perederina A, Esakova O, Quan C, Khanova E, Krasilnikov AS (2010) Crystallization and preliminary X-ray diffraction analysis of the P3 RNA domain of yeast ribonuclease MRP in a complex with RNase P/MRP protein components Pop6 and Pop7. Acta Crystallogr F66:76–80
5. Esakova O, Krasilnikov AS (2010) Of proteins and RNA: the RNase P/MRP family. RNA 16:1725–1747
6. Perederina A, Esakova O, Quan C, Khanova E, Krasilnikov AS (2010) Eukaryotic ribonucleases P/MRP: the crystal structure of the P3 domain. EMBO J 29:761–769
7. Perederina A, Krasilnikov AS (2010) The P3 domain of eukaryotic RNases P/MRP: making a protein-rich RNA-based enzyme. RNA Biol 7:534–539
8. Perederina A, Esakova O, Koc H, Schmitt ME, Krasilnikov AS (2007) Specific binding of a Pop6/Pop7 heterodimer to the P3 stem of the yeast RNase MRP and RNase P RNAs. RNA 13:1648–1655
9. Studier FW (2005) Protein production by auto-induction in high-density shaking cultures. Protein Expr Purif 41:207–234
10. Milligan JF, Uhlenbeck OC (1989) Synthesis of small RNAs using T7 RNA polymerase. Methods Enzymol 180:51–62
11. Ferre-D'Amare AR, Doudna JA (1996) Use of cis- and trans- ribozymes to remove 5′ and 3′ heterogeneities from milligrams of in vitro transcribed RNA. Nucleic Acids Res 24: 977–978
12. Kao C, Zheng M, Rudisser S (1999) A simple and efficient method to reduce nontemplated nucleotide addition at the 3′ terminus of RNAs transcribed by T7 RNA polymerase. RNA 5:1268–1272
13. Milligan JF, Groebe DR, Witherell GW, Uhlenbeck OC (1987) Oligoribonucleotide synthesis using T7 RNA polymerase and synthetic DNA templates. Nucleic Acids Res 15(21):8783–8798

Chapter 10

Analysis of RNA Folding and Ligand Binding by Conventional and High-Throughput Calorimetry

Joshua E. Sokoloski and Philip C. Bevilacqua

Abstract

Noncoding RNAs serve myriad functions in the cell, but their biophysical properties are not well understood. Calorimetry offers direct and label-free means for characterizing the ligand-binding and thermostability properties of these RNA. We apply two main types of calorimetry—isothermal titration calorimetry (ITC) and differential scanning calorimetry (DSC)—to the characterization of these functional RNA molecules. ITC can describe ligand binding in terms of stoichiometry, affinity, and heat (enthalpy), while DSC can provide RNA stability in terms of heat capacity, melting temperature, and folding enthalpy. Here, we offer detailed experimental protocols for studying such RNA systems with commercially available conventional and high-throughput ITC and DSC instruments.

Key words: ITC, DSC, High-throughput, RNA

1. Introduction

RNA is capable of performing diverse biological roles in the cell due to its ability to fold into complex secondary and tertiary structures. Binding interactions figure prominently in RNA's biological function. RNA serves as a ligand for numerous regulatory proteins, both specific and nonspecific, and can itself be a binding site for small molecules and metal ions, such as in the case of riboswitches, ribozymes, and RNA-protein complexes like the spliceosome (1–4). Thus, understanding the binding and folding behavior of RNA is pivotal in elucidating the activity of noncoding RNA in the cell.

Isothermal titration calorimetry (ITC) and differential scanning calorimetry (DSC) offer direct, label-free methods of measuring the thermodynamics of ligand binding and folding of RNA, DNA, proteins, and lipid systems. Information that can be directly obtained via ITC experiments includes binding stoichiometries (n), dissociation constants (K_d), binding enthalpy (ΔH), and

Kenneth C. Keiler (ed.), *Bacterial Regulatory RNA: Methods and Protocols*, Methods in Molecular Biology, vol. 905,
DOI 10.1007/978-1-61779-949-5_10,

entropy (ΔS). Information from DSC experiments includes melting temperatures (T_M), folding enthalpy (ΔH), and heat capacity changes upon folding (ΔC_p). Over the past decade, there have been several excellent reviews on ITC and DSC, and we refer the reader to these for additional theory and methodology (5–9). In this chapter, we present detailed protocols for running RNA ITC and DSC experiments on MicroCal, Inc (Northampton, MA) instruments, guidelines for troubleshooting, and analysis of RNA-binding interactions with both conventional and high-throughput calorimeters.

1.1. Introduction to ITC

In ITC, a solution of ligand is injected into a solution of macromolecule, and the heat absorbed or released from this process is measured. The macromolecule solution is in a calorimetric sample cell, paired with a reference calorimetric cell containing either water or buffer, with both cells housed within an adiabatic jacket (Fig. 1a). Inserted into the sample cell is an injection syringe containing the

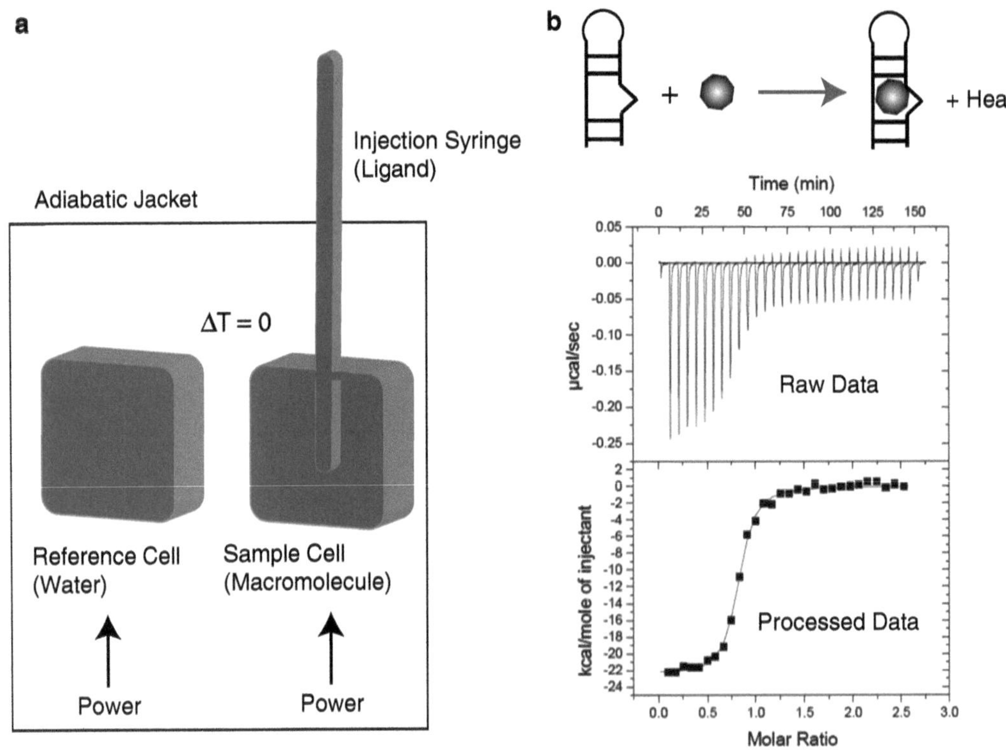

Fig. 1. Isothermal titration calorimetry (ITC). (**a**) Basic schematic of an isothermal titration calorimeter. There are two calorimetric cells: the reference cell containing water or buffer and the sample cell containing the macromolecule. An injection syringe contains the ligand that is to be titrated into the sample cell. The instrument supplies power to each cell to maintain them at a constant temperature. (**b**) A typical ITC experiment. Binding of a ligand (ball) to an RNA aptamer causes heat release or uptake by the system. This heat transfer is reflected in the raw differential power compensation data compared to the reference cell. In this example, the exothermicity results in a decrease of power in order to maintain a constant temperature in the sample cell. The processed integrated data are plotted as enthalpy per mole ligand versus the ligand to macromolecule mole ratio.

ligand solution. The difference between the terminology "ligand" and "macromolecule" is often semantic; for the purposes of this chapter, we refer to the chemical species in the syringe as the ligand and the interacting species in the sample cell as the macromolecule. The ITC instrument provides power to both cells to maintain them at a constant temperature. When the ligand is injected into the macromolecule, the interaction either releases (exothermic) or absorbs (endothermic) heat (there are rare cases where the interaction is thermoneutral and ITC cannot be used to study the system). Thus, the binding interaction acts as a heat source or sink causing the instrument to adjust the power supplied to the sample cell accordingly to keep the temperature constant. This differential power compensation per ligand injection event is the raw signal recorded in the experiment. Integration of this data vs. time gives the total heat transferred as a result of the injection. Plotting this heat per mol of injected ligand vs. the mol ratio of the ligand to the macromolecule then yields a binding curve from which the binding thermodynamic parameters can be calculated (Fig. 1b).

One chief advantage of ITC over most other biophysical methods for determining binding lies in it being a direct measurement of the heat involved in the binding interaction. There is therefore no need for a potentially perturbing label to be incorporated into either the ligand or the macromolecule. Also, the vast majority of binding interactions either release or absorb heat, making ITC a near-universal tool. RNA folding transitions, in particular, often release large amounts of heat because typical stacking interactions between nearest neighbors can have enthalpies up to ~−10 kcal/mol (10). The primary challenge in ITC experiments is sample quantity and quality. The amount of material needed for an ITC experiment is less than the amount needed for NMR spectroscopy or X-ray crystallography, but significantly more than a gel shift assay or fluorescence experiment. Typically, we find that we need 5–50 nmol of macromolecule and 50–750 nmol of ligand per run, although the details will depend on the K_d, stoichiometry, and enthalpy of the interaction, as well as the type of ITC instrument used. Of paramount importance to a successful ITC experiment is the purity of samples involved. Because it is so sensitive, ITC will detect virtually any binding interaction within the sample cell; thus, the macromolecule and ligand solutions should be free of contaminating species, which may otherwise compromise the accuracy of the ligand and macromolecule concentrations or that may undergo nonspecific binding interactions that obfuscate the signal from the interaction of interest. In addition, mismatch of buffer components between ligand and macromolecule solutions—even as low as 1%—can completely mask the binding signal. This chapter will provide guidelines for obtaining large amounts of RNA and for preparing the samples appropriately for ITC, including matching buffer components.

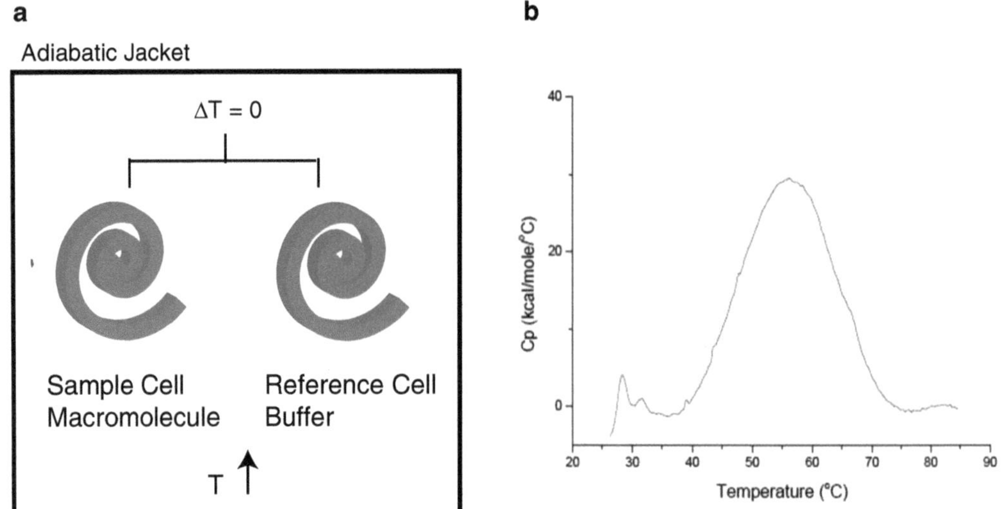

Fig. 2. Differential scanning calorimetry. (**a**) Basic schematic of a differential scanning calorimeter. There are two calorimetric cells: the reference cell containing the buffer and the sample cell containing the macromolecule. Power is supplied to each cell to raise the temperature with the instrument ensuring that the two cells are at exactly the same temperature. (**b**) A typical DSC experiment of a RNA aptamer unfolding.

1.2. Introduction to DSC

In DSC, two calorimetric cells—the sample cell containing the macromolecule and the reference cell containing just buffer—are heated (or cooled) through a temperature range, and the power necessary to effect this change is monitored (Fig. 2a). Since the macromolecule's secondary and tertiary structure add additional modes of energy storage, more power will have to be supplied to the sample cell in order to raise its temperature as compared to the reference cell. The differential power necessary to maintain a ΔT of 0 between the two cells is the physical basis for the unfolding thermodynamics measurement. The differential power represents the excess heat capacity of the macromolecular structure, and the unfolding process is represented by a peak in the sample cell signal as compared to the reference cell (Fig. 2b). Subtraction of the reference signal and analysis of the peak morphology and area lead to extraction of folding thermodynamic parameters.

1.3. Commercial ITC and DSC Instruments for Biological Calorimetry

The principal manufacturer of instrumentation for biological calorimetry is MicroCal, Inc., a subsidiary of GE Healthcare (Northampton, MA). There are two isothermal titration calorimeters from this company that are currently employed in industrial and academic research: the VP-ITC and the iTC_{200}. They differ in the volume and concentration required for experiments, the time required for a complete titration, and in the fact that the iTC_{200} can be incorporated into a robotic operation platform that can automatically load the instrument, perform the titration experiments, and clean the cell and syringe. If you have the choice between the two instruments, we recommend the AutoiTC_{200} for the screening of multiple ligands and conditions, and the VP-ITC for more detailed thermodynamic studies. The company also features two differential scanning calorimeters: the VP-DSC and the VP-Capillary DSC. We recommend the VP-capillary DSC because of features including full automation, a much smaller sample volume, and a higher maximum temperature. This chapter contains protocols and recommendations for operating the AutoiTC_{200}, the VP-ITC, and the VP-Capillary DSC.

2. Materials

2.1. Instrumentation

1. VP-ITC isothermal titration calorimeter from MicroCal, Inc. (GE Healthcare)

   OR

   AutoiTC_{200} automated isothermal titration calorimeter from MicroCal, Inc. (GE Healthcare).
2. ThermoVac degassing station from MicroCal, Inc. (GE Healthcare).
3. VP-Capillary DSC from MicroCal, Inc. (GE Healthcare).

2.2. Instrument Accessories

2.2.1. General

1. 5-mL plastic vials with caps available from MicroCal Inc.
2. Borosilicate Glass Culture Tubes (6 × 50 mm).
3. Microstir bars (7 × 2 mm).

2.2.2. Isothermal Titration Calorimetry

1. Hamilton Syringes: 2.5-mL syringe (18/8.5″/PT3) for VP-ITC, 0.5-mL syringe (22/3″/Special) for the AutoiTC_{200}. These are included with the instruments and spares can be purchased from the Hamilton Company.
2. Cleaning accessory and 5-mL filling syringe (included with MicroCal VP-ITC instruments).
3. 96-Deep-well plates (RNase/DNase free) from Nunc for AutoiTC_{200}.
4. 96-Well plate covers (EZ Pierce Zone Free Films) from Excel Scientific for AutoiTC_{200}.

2.2.3. Differential Scanning Calorimetry

1. 96-Deep-well plates (RNase/DNase free) from MicroLiter Analytical Services.
2. 96-Well MicroMat pre-slit plate covers from Sun Sri.

2.3. In Vitro Transcription of RNA from a Hemi-Duplex

1. 10× Transcription Buffer: 400 mM Tris (pH 8.0), 250 mM $MgCl_2$, 20 mM dithiothreitol, 10 mM spermidine.
2. 100 mM NTPs: separate solutions for each NTP (pH adjusted to 7.5–8.0 with 2 M Tris).
3. Hemi-duplex top-strand DNA oligo: 5′-TAATACGACTCA CTATAG-3′.
4. Hemi-duplex DNA template oligo: 5′-<Reverse Complement to your sequence>-TATAGTGAGTCGTATTA-3′, where the fixed region is reverse complementary to the sequence in the previous step.
5. Annealing Buffer: 10 mM Tris (pH 7.5), 10 mM NaCl, and 1 mM Na_2EDTA.
6. T7 RNA Polymerase: lab-made or purchased from Ambion, New England Biolabs, or Epicentre.
7. 1× TE buffer: 10 mM Tris (pH 7.5), 1 mM Na_2EDTA.
8. 5 M NaCl.

2.4. RNA Purification

1. 10× TBE Buffer: 1.0 M Tris, 0.83 M boric acid, 10 mM EDTA.
2. Polyacrylamide gel electrophoresis (PAGE) reagents: 20% acrylamide (29:1) in 7 M urea and 1× TBE; 7 M urea in 1× TBE (These two solutions are mixed to give the desired percent acrylamide in 7 M urea/1× TBE.); 10% Ammonium persulfate; TEMED.
3. 2× Formamide/EDTA loading buffer: 95% formamide, 0.1× TBE, 20 mM EDTA, 0.1% xylene cyanol, 0.025% bromophenol blue.
4. 1× TEN_{250} Buffer: 10 mM Tris pH 7.5, 1 mM EDTA, 250 mM NaCl.
5. 190 or 200 proof Ethanol.
6. Dialysis membranes with appropriate molecular weight cut-off (MWCO) for the macromolecule sample.

2.5. Instrument Maintenance

1. 1–5 mg/mL proteinase K in water or TE.
2. RNase ZAP from Ambion, Inc.
3. Contrad-70 from Decon Labs.
4. 10–50 mM NaCl solutions for instrument quality control.

3. Methods

3.1. In Vitro Transcription with a Hemi-Duplex

The easiest way to prepare RNA in significant quantities with the requisite purity is in vitro transcription. There are three ways to prepare the DNA template: (1) digested plasmid; (2) PCR product; and (3) a hemi-duplex. Wear gloves and use good practice for working with RNA; for a more complete description, we refer the reader to ref. (11). We describe only the last option here; protocols for preparing your DNA template via a digested plasmid or PCR product are described in ref. (11). Hemi-duplex in vitro transcription offers the advantage of speed and economy and is suitable for a range of RNA sizes that includes many functional RNA (~20–100 nt). DNA templates of the type described in Subheading 2.3 are synthesized routinely by suppliers such as IDT (Coralville, IA). For wild-type T7 phage RNA polymerase, the RNA construct should start with a G, and abortive events are decreased by having the sequence start with several Gs and/or As (12). If the transcript requires a non-G nucleotide at the 5′ position, the P266L T7 polymerase mutant can be used (13); this mutant also helps if abortive transcription events are severe enough to make the yield unacceptable. Commercially available transcription kits are available from Ambion (Austin, TX) and Epicentre (Madison, WI). However, for extensive ITC work, we strongly recommend producing and purifying in-house T7 polymerase in order to avoid the high cost of standard transcription kits.

The size of the transcription reaction required depends upon the number of ITC experiments to be performed. A 1 mL transcription is often enough to perform one to two ITC experiments. A 10–20 mL transcription should be performed for studies that will require numerous experiments. Regardless of the size of the transcription to be performed, small scale (10–20 μL) test transcriptions, body-labeled with an α(^{32}P)-NTP, should be performed in order to optimize the concentrations of NTP, DNA template, and especially the T7 polymerase. When troubleshooting the enzyme, important parameters to keep in mind are that the T7 reaction is highly salt-sensitive so that the lowest possible concentrations of sodium ions should be used, and that T7 polymerase requires storage in the presence of active DTT to ensure proper operation. Thus, fresh DTT should be added periodically to the batch of purified polymerase, or aliquots of polymerase should be frozen at −80°C until needed. The general procedure for a large-scale transcription is as follows:

1. The DNA template must be annealed to the T7 top-strand promoter DNA. Mix equal moles of the two oligonucleotides to a final concentration of 5–10 μM each, in annealing buffer.
2. To promote annealing, heat the solution for 2 min at 90°C and then cool immediately on ice for 10 min.
3. For a 1 mL in vitro transcription, combine 100 μL of 10× transcription buffer, 40 μL of each NTP, 600 μL of double deionized (dd)H_2O, 100 μL of DNA template, and ~40 μL of T7 polymerase (50,000 units/mL) in a polypropylene tube or vial. For larger reactions, scale these volumes accordingly. Final concentrations of this solution will be 4 mM each NTP,

700 nM DNA template, and 4% (v/v) T7 polymerase. No matter the size of the transcription, do not use glass containers, as RNA adheres to glass surfaces.

4. Immerse the reaction in a water bath at 37°C for 2–4 h. The reaction should turn cloudy as the transcription progresses due to precipitation of the magnesium pyrophosphate by-product.

5. Upon completion of a large-scale reaction of 5 mL or larger, it is often necessary to concentrate the transcription products via centrifugal concentrators (such as the Vivaspin line (GE Healthcare)) or ethanol precipitation in order to facilitate loading of the sample onto a gel for PAGE. If the latter is chosen, add 0.5 M EDTA to a final concentration of 20 mM in order to sequester the magnesium ions, and add 5 M NaCl to a final concentration of 250 mM. Then combine three volume equivalents of ice-cold 200 proof ethanol with the transcription solution and either place it on dry ice for 45–60 min or place it in a −20°C freezer for overnight. Centrifuge the precipitation mixture for 30 min at 17,500 × *g* in a table top centrifuge or for 30 min at 20,400 × *g* in a Beckman JA-20 rotor or equivalent for large volumes, and store the pellet at −20 or −80°C until PAGE purification of the RNA. For transcriptions on the scale of 5–20 mL, centrifuge the mixture in 50-mL polycarbonate Oak Ridge tubes at 20,400 × *g*.

3.2. PAGE Purification of the RNA

Purification of the RNA after transcription is accomplished by performing PAGE to isolate the pure sample RNA, precipitating the RNA, and then extensively dialyzing into the desired buffer. The standard procedure for large-scale PAGE purification is as follows:

1. Pour a PAGE gel of the appropriate percentage. For the large-scale transcriptions, the maximum size of gel that can be prepared in your lab should be used. A gel with combs and spacers 3 mm thick and sufficient height (usually upwards of 25 cm) for the RNA of interest to resolve is needed for a >10 mL transcription.

2. Dissolve the transcription product pellet in 1–2 mL ddH_2O and mix with an equal volume of 2× formamide/EDTA loading buffer.

3. Heat the sample at 90°C for 2 min to fully denature the RNA before loading onto the gel.

4. After PAGE, visualize the RNA with UV shadowing (or by an indirect method such as staining with acridine orange if you want to avoid UV exposure of your RNA, although the dye will have to later be removed) and cut out the band with a sterile blade.

5. Crush the band and mix 4–5× the volume of the crushed pieces with 1× TEN_{250}.
6. Rotate the mixture overnight at 4°C. A second incubation with fresh TEN_{250} can be done to improve yield.
7. The crush and soak solution from step 6 is then vacuum-filtered to remove the gel pieces and combined with three equivalents volumes of ice-cold ethanol to precipitate the RNA, as described in Subheading 3.1, step 5.

3.3. Dialysis of the RNA and Ligand

Dialysis of the RNA is extremely important. It removes impurities and ensures a buffer balance between the RNA and ligand. The choice of buffer is also extremely important for ITC. Because ITC measures every interaction within the sample cell, the buffer's heat of protonation should be minimal. Different buffers at the same pH with different heats of protonation can be used to determine if the binding event is protonation-coupled (14). Also, if multiple experimental temperatures are to be used, the buffer's pK_a should not have a high temperature dependence. For these reasons, cacodylic acid is often used as the buffer in calorimetric experiments. Phosphate and HEPES are also good choices, but Tris should be avoided entirely. For binding interactions with protein, the buffer should not contain DTT in order to avoid disruptions to the baseline signal. Beta-mercaptoethanol is recommended for use instead.

For dialysis, membranes with a MWCO ~ 1/3 the mass of your sample should be used. Dialysis can be accomplished through bag dialysis, button dialysis, or cassette dialysis. Spin column dialyzers can be used, but several runs may be necessary. A large batch of dialysis buffer should be prepared, and this exact same batch should be used for all aspects of the experiments. For interactions of RNA with proteins or another nucleic acid, both species should be dialyzed simultaneously against the same batch of buffer. For small molecule ligands, the buffer after dialysis should be used to dissolve the pure solid ligand. To avoid degradation of the RNA, it is recommended that the magnesium salt needed for proper folding of the RNA is left out of the dialysis buffer. Instead, prepare a concentrated stock solution of magnesium salt using the 1× dialysis buffer, and add it to the RNA and protein before the experiment. The dialysis buffer should be saved for dilution of the RNA stock and for rinsing of the instrument.

3.4. Concentration Determination

The thermodynamics quantities obtained via ITC are only as reliable as the concentrations for the samples. For RNA, the extinction coefficient at 260 nm for your sequence can be calculated via a variety of programs. These values descend from nearest neighbor considerations, which are only valid at 90°C where the RNA is fully denatured (15, 16). Therefore, an aliquot of the RNA sample prepared for ITC should have a UV spectrum measured at 90°C, using a spectrophotometer with temperature control available. The absorbance at 260 nm will give the RNA concentration, and a A_{260}/A_{280} ratio of 2.0 or higher will give an indication of the sample being free from protein or acrylamide contamination from the purification steps.

Calculating the concentration of the protein or ligand that interacts with the RNA can be more challenging if the extinction coefficient is not known. If there are tryptophan residues in the protein, then the extinction coefficient can be estimated using the program ProtParam (17) or by using the Pace protocol (18). Small molecule extinction coefficients may need to be calculated using mass and absorbance measurements. If the metal concentration needs to be precisely quantified, atomic absorption spectroscopy is the preferred method.

3.5. Designing the ITC Experiments

3.5.1. Type of Study

The AutoiTC$_{200}$ and VP-ITC can be used in synergy to screen interactions and characterize binding, or they can be employed individually. Due to a lower sample volume and faster equilibration and run times, the AutoiTC$_{200}$ is excellent for high-throughput screening studies, testing different ligands, temperatures, buffer composition, etc., either to optimize the binding interaction in one set of conditions or to measure how the thermodynamics change with these variables. This instrument can also be used for detailed binding thermodynamics, although this is the area where the VP-ITC works the best since it has a larger sample cell. Likewise, such screening experiments can also be done with the VP-ITC instrument; however, they will take significantly more time (a week vs. a day in most cases). Figure 3 provides sample data comparing the thermodynamic information received for the same titration on both instruments, which is further analyzed in the legend. No matter the instrument being used, the design of the ITC experiment follows the guidelines presented in the next two subsections. Advanced ITC experiments such as heat capacity curves, competition, and dissociation experiments are discussed below (see Note 1).

3.5.2. Basic Set-Up

The first basic step in designing the ITC experiment is to decide which species, the macromolecule or the ligand, goes into the sample cell and which goes into the syringe. This decision is based in large part upon consideration of solubility and aggregation. Because the goal of the experiment is to vary the ligand's concentration from lower than to higher than the macromolecule's concentration, the solution in the injection syringe will necessarily be much higher in concentration (10–15 fold) than the solution in the sample cell. Therefore, the molecule in the syringe must be very soluble in the buffer and be in the monomeric form, despite its high concentration. With RNA and many proteins, dimerization has to be considered. UV-monitored thermal melting experiments over a range of RNA concentrations can reveal dimerization through an

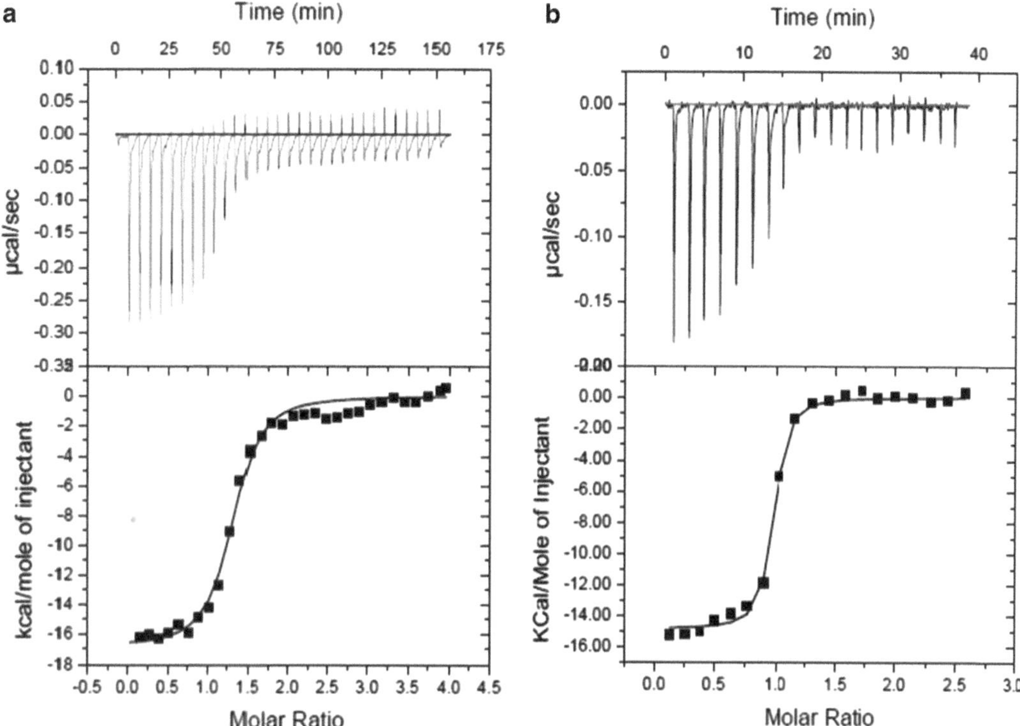

Fig. 3. Comparison of a ligand binding to an RNA aptamer under identical conditions on the VP-ITC (**a**) and the $AutoiTC_{200}$ (**b**). The thermodynamic information from the VP-ITC data is $n=1.2$, $K_d=81$ nM, and $\Delta H=-17$ kcal/mol, while for the $AutoiTC_{200}$ the data is $n=0.93$, $K_d=15$ nM, and $\Delta H=-15$ kcal/mol. The differences between n and ΔH values are within experimental error from repeat measurements of the titration and the difference between K_d values is attributed to the disparity in the number of data points in the transition region. It is for this reason that we recommend using the $AutoiTC_{200}$ to survey binding interactions and the VP-ITC to rigorously characterize the binding thermodynamics.

upwards shift in the T_M, and PAGE under native conditions can reveal dimerization as a slow mobility band. Such experiments are useful in determining the upper concentration limit for the ITC experiment so that the RNA stays monomeric. Given the above considerations, it is typically prudent to have the RNA within the sample cell of the instrument.

Given that the RNA and protein are monomers, the concentrations to be used in the ITC experiment can be determined by the using the following equations:

$$[\text{Macromolecule}] = \frac{c^* K_{D(\text{est})}}{n} \quad (1)$$

$$[\text{Ligand}] = n * 15 * [\text{Macromolecule}] \quad (2)$$

where [Macromolecule] is the concentration of the species in the calorimetric sample cell, K_d is the estimated dissociation constant for the interaction, n is the number of binding sites, and "c" is a value which reflects the quality of the ITC binding curve.

The value for c should be between 1 and 50, with 5 being a good value to use for a first experiment. If the c-value is too low, the binding curve will be too broad to reliably extract thermodynamic information. If the c-value is too high, the binding curve will appear as a step function, allowing only binding stoichiometry and enthalpy to be determined from the data. Equation 1 can be rearranged to calculate the c-value after an initial ITC run, and this information can be used to improve the results of subsequent experiments. For the estimate of the K_d, biochemical data such as gel shift assays or comparison to similar types of systems in the literature can be used, although the K_d is often best estimated from initial ITC runs. For the concentration of the ligand in the syringe, the value should be ~15 times the macromolecule concentration in the cell (Equation 1), which ensures that the mole ratio reaches a value of 3 at the end of the titration. This will ensure a complete binding curve for a one-site process even if the binding event is weak, although a ligand concentration of 10 times the macromolecule concentration will work for tight binders. For larger binding stoichiometries, multiply the ligand concentration according to the number of projected binding sites (Equation 1). Note that in no case does the value of the ligand have to be precisely 15 (or 10) times the macromolecule concentration; rather, all the concentrations, whatever they may be, have to be known precisely. MicroCal offers planning and prediction software called "ITC Expert," which is included with the instruments, that aids in the process of planning the basics of an ITC experiment.

3.5.3. Instrument Parameters

The next step is to choose the instrument parameters for the ITC experiment. These parameters fall into two categories: titration and calorimetric. For the titration parameters, the number, volume, and spacing of the injections need to be defined. For the VP-ITC instrument, the syringe holds up to 300 μL and can dispense a volume as small as 2 μL in a single injection. The manufacturer recommends that the first injection be 2 μL regardless of the other injection parameters and that this data point be deleted from the analysis, as the first injection can often be problematic (e.g., a small air bubble trapped at the end of the syringe, or a volume less than 2 μL). Our experience is similar, and we reject a small first injection. The injection scheme after that can be whatever suits the experiment, with typical injection volumes being between 5 and 10 μL. The default setting for the speed of each injection is 0.5 μL/s and there are few circumstances that require this to change. The filter period also should be left at 2 s. The spacing of time between injections, however, is a crucial factor in the experiment, as the signal must return to baseline prior to subsequent injection in order for the succeeding injection to have accurate integration data. Weak signal events require as little as 120 s to return to baseline, while very strong and/or slow thermal events may need as long as 600 s before the system equilibrates. A spacing of 300 s is a good value to try for the first experiment. A convenient feature of the operating software is that it allows the user to change the injection parameters "on the fly"; it is thus recommended to observe the first few injections to see if this spacing will be adequate.

The AutoiTC$_{200}$ has a much smaller volume for the injection syringe, at 40 μL. With this instrument, typical injection volumes are 2 μL, with the first injection point usually suitable for inclusion in data analysis. This injection volume allows for 19 data points in the binding curve, although the injection volume can be lowered to 1.5 or 1 μL if more points are desired, allowing that sufficient heat is generated. The injection speed default is still 0.5 μL/s. The spacing between injections is correspondingly lowered in the iTC$_{200}$ instrument, with 120 s being suitable for most binding events. Titrations are thus significantly faster with the iTC$_{200}$.

For the calorimetric parameters, the most important is the reference power, measured in μcal/s. This quantity sets the baseline for the experiment. The injection peak should never go below 0 μcal/s or above 30 μcal/s, as the instrument can no longer actively compensate for the heat transfer from the binding interaction. For the VP-ITC, the initial experiment should have a reference power of 15 μcal/s, which should cover a wide range of concentrations and injection volumes and can be adjusted accordingly in subsequent experiments. For exothermic interactions, if the signal is too small, reference power can be lowered; if it is too large, it can be raised. The reverse is true for endothermic binding. In this manner, the reference power functions as the instrumental gain. For the iTC$_{200}$, the standard reference power setting is 5 μcal/s.

The stirring speed is the last parameter to consider. The injection syringe, which has a propeller tip, revolves at high speed to aid mixing of the solution. For the VP-ITC, the normal stirring speed is 310 rpm, while for the iTC$_{200}$, the default setting is 1,000 rpm. These values rarely have to be adjusted.

3.5.4. Control Experiments

There are sources of thermal events that are in the background of every ITC experiment that will obscure the true thermodynamics of the binding interaction unless they are taken into account. Each injection itself will add a small quantity of mechanical heat; however, the heat of dilution of the ligand is usually more problematic. The ideal dilute approximation of thermoneutral mixing does not hold for biological systems, and the dilution of the ligand species into the sample cell will either release or absorb heat. To correct for this heat, a control experiment should be performed in which the ligand in buffer is injected into buffer. The data are processed assuming a fictional macromolecule concentration in the sample cell, in order to have the data points fall on the same *x*-coordinate as the actual binding experiment. Dilution curves usually have one of three outcomes: linear, linear with a slope of zero, or exponential.

The data points from the dilution control experiment can then be subtracted from the binding experiment to account for dilution and mechanical heat, leaving only the heat associated with the binding interaction. If the dilution has a linear profile with a slope of zero (i.e., a consistent injection heat for each injection across the titration), then the saturation data points from the binding experiment can be used for background subtraction, and no additional control dilution experiment is typically needed. Indeed, the control of injecting buffer into the macromolecule solution in the sample cell is rarely necessary, but can still be used in some cases to verify buffer matching of the macromolecule following dialysis. As discussed below, if the dilution control is a line with a slope of zero, the titration data can be corrected using the average value from the saturation portion of the titration, which is in fact a preferred method as it more closely reflects what is occurring during the titration itself.

3.6. Preparing the Samples for ITC

The RNA should be renatured before the ITC experiment. There are a multitude of renaturing/annealing procedures available. In order to facilitate comparison across other biochemical and biophysical methods, whatever protocol is followed for biochemical assays with the RNA of interest should be followed for the ITC experiment. One will need to prepare 2 mL of macromolecule solution for the VP-ITC and 400 μL for the AutoiTC$_{200}$. For the ligand solution, prepare at least 500 μL for the VP-ITC and 120 μL for the AutoiTC$_{200}$. The RNA sample should be degassed with stirring in the MicroCal Thermovac apparatus for at least 8 min immediately before the experiment. The ligand solution and an aliquot of buffer should be degassed as well. The AutoiTC$_{200}$ does not require samples to be degassed, but we find the results are better when they are treated in the same way as VP-ITC samples.

Following degassing, for the VP-ITC experiments the RNA should be incubated at or a few degrees below the experimental temperature for at least 5 min prior to loading the instrument. For the AutoiTC$_{200}$, the samples are stored in 96-well trays in a chamber within the instrument that should be kept within 5–10° below the experimental temperature. This is because both calorimeters work more efficiently when they are directly heating samples as opposed to indirectly cooling them.

3.7. AutoITC$_{200}$

Once the experiment has been planned, this information needs to be input to the operation interface for the AutoiTC$_{200}$ instrument. There are two programs that have to be running with the instrument. Following startup of the instrument, the "ITC200" and the "AutoItc200" programs should be opened. The "AutoItc200" program is the interface used to perform the experiment, as it in turn relays the necessary instructions to the "ITC200" program that directly operates the calorimeter. There are five tabs at the top of the program screen: "Sample Groups," "Experiments," "Data

Plots," "Instrument Set-up," and "System." "Sample Groups" is where each experiment is set-up. "Experiments" is the screen to verify and run the experiments. "Data Plots" gives the raw data in real time, while "Instrument Set-up" is where titration programs are made. "System" provides options for maintenance and cleaning of the instrument. For more information on this and all aspects of his protocol, consult the MicroCal instrument manual.

1. First, fill the reference with freshly degassed ddH_2O.
2. Using the tool provided, remove the protective Teflon ring around the sample cell port, exposing the reference cell port.
3. Using a 0.5-mL Hamilton syringe, remove the old water and replace it with fresh water with the appropriate burst fill technique and light tapping to remove any bubbles. This process should be done at least once a week.
4. Under the "Instrument Set-up" screen, enter in the parameters for the titration(s) using the guidelines from the previous section. These parameters can then be saved for current and future use.
5. Under the "Sample Groups" tab, there will be a series of subtabs that need to be filled. To start, in the "Dataset Naming" screen, enter the number of groups of experiments, as well as the number of samples within each group. A "group" is defined as the injection parameters (temperature, injection volume, spacing, etc.) that were created in the previous step. A sample is defined as one titration, and there can be multiple samples within one group. Also on this screen, you can enter the dataset naming scheme and any sample comments. A good measure to take to ensure that the instrument is working properly is to first run a water-into-water or buffer-into-buffer titration, which should give very small exothermic heats (see Note 3).
6. Next, on the "Method" screen, choose the automation, ITC run, and analysis methods for the group. There are numerous options for the automation method, with the most common being "Plates Standard B." There are other options including a cleaning step between each experiment, prerinsing the cell with buffer before the run, or saving the post-run cell contents afterwards. See the instrument manual for further details on these options. For the "ITC run" method, choose the injection program that you created in step 1. For "Analysis" method, choose either "saturation" for a ligand-to-macromolecule titration or "control" for a ligand-dilution run. These options are for the automatic data fitting that the instrument offers, although we recommend that fitting be done by the user.
7. On the "Concentration" screen, enter the concentrations of ligand and macromolecule. If they are unknown at this time, input estimated values—exact values can be used during the data analysis.

8. The next screen is "Plate Set-up," from which you can choose where on the plate the instrument will take samples for each group. The AutoiTC$_{200}$ instrument can accommodate four 96-well plates, with a minimum of two wells per titration–one for the syringe and one for the sample cell; thus, the instrument can perform up to 192 ITC experiments in succession. The instrument also holds five 35-mL tubes as reservoirs for sample cell macromolecule if the experiment requires it; with the macromolecule in these large reservoir tubes, only one well in the 96-well plate is now required for a single titration experiment, allowing for a new maximum instrument capability of 384 unattended experiments in succession. This is the preferred mode of operation if one is titrating a large number of ligands against a single macromolecule target, such as in drug screening. The wells on the plates will be color-coded as to whether it contains sample cell (salmon) or syringe material (light blue). Additional options are to use some of the wells for pre-rinsing (as described elsewhere) and some of the wells for recovering the titrated sample (post-run save spots). If you have pre-rinsing solutions (pink) or post-run save spots (yellow), they will be indicated as well on the diagram as to where to place them in the plate.
9. Using the set-up from the previous step, fill the 96-well plate accordingly. It is imperative that no material be left as droplets on the side of a well, as this will lead to filling errors. When complete, apply an adhesive plate cover. Then place the plate inside of the sample tray at the appropriate numbered position. Under the "System" primary tab, the temperature of the tray chamber can be set.
10. Next, within the "Experiments" primary tab, choose the command "Import Sample Groups." This will load the information input in the previous six steps into a spreadsheet on the screen. You can then review and edit the information as required. After checking the list of experiments, choose the "Validate" command. A screen will appear that will state how much ddH_2O, methanol, and cleaner solution will be required for the run and what volume of waste will be generated. These 0.5–1 L bottles are attached to the AutoiTC$_{200}$ via tubing in the back of the instrument and are generally placed next to the instrument. A tank of nitrogen must be attached to the instrument for cleaning purposes. Generally, the nitrogen should be of average purity and the pressure should be set to 20–25 psi, using the regulator supplied by MicroCal. This tank should be checked prior to a set of experiments to ensure that it will not run empty. A tank will typically last for about 200 individual runs.
11. The final step is to click on the start arrow at the top left corner of the program screen or to push the start arrow button on the

instrument housing, which initiates the experiment. The instrument will then clean the cell, injection syringe, and transfer pipettes, and then load the material for the first run. Once loaded, the instrument will set itself to the experimental temperature and go into baseline equilibration mode. Cell heating and baseline equilibrium should take no longer than 15 min. The equilibrium process and the subsequent raw titration data can be monitored in real time under the "Data Plots" tab. If you wish to stop the progression of experiments part way through, press the stop button; the instrument will finish the current experiment, clean, and then reset to the home position of the system's robotics. If there is a need to stop the run immediately, an abort option is available. When this button is pressed, the robotic system immediately freezes, but not the calorimeter. To stop syringe stirring and data collection, go into the "ITC200" program and press the stop button there. Then, return the system to the home position by going to the "System" tab on the "AutoItc200" program. However, aborting a run can be fraught with complications and should only be used in a last resort, such as in cases that could present some physical problem within the instrument, that could damage the syringe or transfer pipettes, or could lead to an unusually long baseline equilibration period.

3.8. VP-ITC

For the VP-ITC experiments, the samples are loaded manually. The program "VP Viewer 2000 ITC" is used to run the experiment. This program should be launched after the instrument has been turned on. Two screens will launch: the instrument control screen and an Origin real time data display screen. There are four tabs on the screen: "ITC Controls," "Thermostat/Calib.," "Setup/Maintenance," and "Constants." For a typical experiment, only "ITC Controls" and "Thermostat/Calib" are needed. For more information on this and all aspects of his protocol, consult the MicroCal instrument manual.

1. In the "Thermostat/Calib." tab of the "VP Viewer 2000 ITC" program, the thermostat temperature for the calorimeter should be set to a temperature 3–5°C lower than the experimental temperature, if it is not already.
2. Fill the reference cell (port located to the left of the sample cell in this instrument) with freshly degassed ddH_2O. The technique used to fill the cell is extremely important for the success of the experiment. The 5-mL Hamilton syringe is filled with 2 mL of water and placed down into the reference cell until the long needle touches the bottom of the cell. The syringe is then lifted up a few mm and the contents are injected with a slow continuous injection for the first mL and short bursts for the second mL. The syringe needle is then tapped very gently

against the bottom of the cell and gently swirled around the cell, tapping the walls to displace any bubbles from inside the cell. A bubble trapped in either the reference or sample cells can make the data unusable. The syringe is then withdrawn from the cell, and the tip is placed on the metal rim on the outside of the port where excess liquid is drawn up into the syringe. This method of overfilling and then withdrawing allows for reproducible filling of the cell with no air volume at the top. The white reference plug is then placed on top of the cell port. The reference cell should be emptied and refilled at least once a week.

3. Within the "ITC Control" tab, enter the experimental parameters decided upon in Subheading 4 and assign a filename. These parameters can be saved as an ITC run file for later use.
4. First, rinse the sample cell with buffer, or macromolecule solution if there is enough to spare, before the cell is filled with the macromolecule solution. The same filling technique described in step 1 should be used. Some of the excess macromolecule solution from the process of overfilling the sample cell can be saved and used to obtain a concentration that will account for any dilution from the buffer pre-rinse.
5. Place the ligand solution in either a 5-mL plastic tube or a 0.5-mL glass vial in the syringe mount holder such that the tip of the injection syringe is well immersed in solution. With the plunger in the open position (which can be enacted using the self-named command within the operating software), attach a needled syringe with Tygon tubing to the fill port of the syringe and draw solution up to the top of the syringe. Move the plunger to the closed position using the button on the VP Viewer program. The "Purge and Refill" command is then executed twice to ensure that no bubbles are in the syringe. The plunger will expel its contents into the vial and draw the solution back up, returning finally to the closed position.
6. With care taken not to bend it, place the syringe straight down into the sample cell, clicking it into place. If necessary, gently wipe off any ligand solution that is clinging to the outside of the syringe needle.
7. The run can now be started by clicking on the Start button. The instrument will then heat the cells to the run temperature and go into a pre-run thermostat mode for a few minutes. Then the instrument will proceed into baseline equilibrium, followed by a 1 min pre-titration delay. This entire process should take 15–20 min. If it goes any longer, it is advisable to stop the run and check the reference and sample cell filling.

Common ITC problems and their solutions are further discussed below (see Note 2). In addition, regular ITC maintenance is discussed below (see Note 3).

3.9. ITC Data Analysis

Analyzing the ITC data is usually accomplished using the MicroCal-specific Origin 7 software included with the instrument. This software will automatically load the raw data, integrate the injection peak areas, and generate a ΔH/mol ligand vs. mole fraction plot. For full details, consult the MicroCal data analysis manual that is included with the instrument and software. The AutoiTC$_{200}$ features the option of automatic analysis of the data fit to a one-site model, but we recommend performing the entire analysis procedure manually.

The first step in data analysis is to read in the raw ITC data file (with a .itc extension) using the "Read Data" command button on the "mRawITC" screen of the program. Origin will then integrate the injection peaks and launch a new "Delta H" screen. The following procedure is then used to extract thermodynamic information from either the VP-ITC- or AutoiTC$_{200}$-generated data set:

1. Using the "Concentration" button on the left side of the "Delta H" screen, enter in the exact concentrations of your samples if this has not been done during the experimental set-up. This has to be the first step, as entering the numbers later in the analysis procedure will undo everything that had previously been done to the data.
2. If the data are noisy, you may go back to the raw data screen and adjust the integration limits and baseline position for each individual peak. This adjustment of the computer integration should be done only in limited cases when there is a significantly noisy baseline, such as from injections of a bubble from the syringe. If the peak integration limit is adjusted, *all* peaks should have their integration limits adjusted in the same manner to keep the data analysis consistent.
3. The next step is to correct for dilution heat of the ligand. There are two ways of accomplishing this: subtracting a whole dilution data set from the titration data or subtracting an average heat of dilution value from every point. The proper course of action depends on the dilution control experiment. If the dilution control is a line with a slope of zero, the titration data can be corrected using a single average value. This average value can be obtained either from the dilution control data points or from the last five saturation points of the titration; the latter is preferred as there are often slight differences in dilution heat with and without the context of a macromolecule in solution. This data set correction is performed by going to the "Math" tab on Origin, selecting "Simple Math," and subtracting the appropriate number. If the dilution control data are a curve or a line with a nonzero slope, the entire dilution data set will have to be subtracted from the titration data set, which can be easily done with the "Subtract Reference Data" button from the "Delta H" screen after the reference data are read into the

same plot as a different layer. A variation of this correction is to subtract a best-fit straight line using the "Subtract Straight Line" option under the "Math" tab. This option has the advantage of introducing less scatter to the corrected data set.

4. Fitting of the titration data to a binding model can now be performed. There are several built-in options in the MicroCal Origin 7 program and the command buttons for them are on the left side of the "Delta H" screen. The "Single Set of Sites" model fits the data to a model where there are one or more equivalent binding sites. The "Two Sets of Sites" model fits the data to a situation in which there are two classes of binding sites on the macromolecule (for instance, a tight-binding specific site and a weak-binding nonspecific site). The "Sequential Binding" model fits the titration to a scenario wherein one site has to be saturated on the macromolecule before the second site is available. The "Competition Equation" is used for competing ligand experiments, and the "Dissociation Model" is used for dimer dilution dissociation experiments, as described in Note 1. Other models may be designed, written, and implemented using the Origin base fitting session as well. When the appropriate option is chosen, the program will do an initial fit, and then the number of Levenberg-Marquadt iterations to perform is chosen. In general, the algorithm is applied until the χ^2 is no longer reduced. When complete, the thermodynamic data will appear in a text box on the graph with the fit curve.
5. The common publication figure for ITC data is generated by selecting the "Final Figure" option under the ITC tab. The final figure consists of the raw titration data plotted in the upper panel and the processed fitted data in the lower panel. This figure can be altered and adjusted and saved as a pdf or image file.

Interpretation of the thermodynamic data extracted is ideally partnered with structural data to get the optimal understanding of the binding interaction. This is especially important for interpretation of the n value. An n-value of 0.5 could indicate that two macromolecules bind to one ligand or that the macromolecule is populating multiple folded states, with only 50% of macromolecule being in an active confirmation. Typically, other biophysical studies such as ultracentrifugation or gel electrophoresis are used to further interrogate this issue. The ΔH and ΔS values will offer information about the net gain or loss on intermolecular bonds during the interaction and/or the change in solvent accessible surface area.

3.10. Designing the DSC Experiment

3.10.1. Type of Study

The DSC experiment can reveal much about the folding behavior of the RNA or protein of interest, in either the absence or presence of a ligand. The standard experiment is to perform a temperature scan of the macromolecule under the desired buffer conditions to assess its tertiary and secondary structure stability. Salt and pH conditions can be varied to study their effects on the macromolecule folding. The VP-Capillary DSC is ideal for these types of screening experiments owing to its high-throughput qualities. With this automated instrument, only upwards temperature scans can be done, but the sample can be cooled to the starting temperature and rescanned. The conventional VP-DSC can perform both up and down temperature scans, but each run has to be conducted manually. For RNA with no divalent ions, the folded–unfolded transition should be reversible. However, presence of magnesium or other divalent ions may lead to significant degradation at the high temperatures and be unusable for repeated runs. The effect of ligand binding can also be measured by DSC by performing an experiment on a solution of the macromolecule-ligand bound complex. DSC experiments with varying ligand concentration can be used to provide an estimate of linked macromolecule conformational changes and the binding affinity, but ITC is superior for determining binding thermodynamics.

3.10.2. Basic Set-Up and Sample Preparation

The VP-Capillary DSC experiment requires 400 μL of macromolecule solution for the sample cell and 400 μL of dialysis buffer for the reference cell. These quantities are approximately one-half of what is required with the conventional VP-DSC instrument (750–800 μL). It is imperative that the macromolecule and reference solutions be perfectly buffer-matched. As with ITC, extensive dialysis is required to ensure that this condition is met. Any discrepancy in buffer or salt concentrations between solutions will lead to large data artifacts. The molar concentration of the RNA or protein needed will depend on the size of the macromolecule: larger molecules will have unfolding transitions involving more intermolecular bond breaking than smaller molecules and hence one can use a lower molar concentration. It is therefore more convenient to discuss sample requirements in units of mg/mL. Concentrations of macromolecule for Capillary DSC should be in the range of 0.1–10 mg/mL, with 0.5 mg/mL a good starting value. The upper limit of 10 mg/mL should not be exceeded at risk of extensive, damaging precipitation.

Much of the same buffer considerations of ITC apply to DSC. Buffers with low heat of ionization and a small temperature dependence of the pK_a are optimal. Cacodylate and phosphate are good for DSC, while Tris should never be used. In addition, DTT should be avoided, and β-mercaptoethanol used instead if a reducing agent if needed. A buffer-only experiment can be conducted to verify that there will be no anomalous peaks due to behavior of the buffer. The samples should be degassed akin to ITC samples to ensure no bubbles form during the experiments.

3.10.3. Instrument Parameters

There are three major parameters for the DSC experiment: temperature range, scan rate, and feedback mode. The temperature should start at 20–25°C and extend to at least 20° beyond the unfolding transition. The VP-Capillary DSC offers the advantage over the VP-DSC of extending the melting experiment beyond 100°C, as the cells are pressurized with a nitrogen source, thereby preventing boiling. This capability is invaluable when the macromolecule's T_M is in the 80–90°C range, which is the case for some stable RNAs. The instrument can go up to 120°C, but we find that 110°C is sufficient as an upper limit. In the case of RNA, degradation can be problematic at these temperatures, however, especially in Mg^{2+} solutions, and the quality of the RNA should be checked by PAGE after the experiment.

The scan-rate parameter will affect the sensitivity and resolution of the DSC data. Higher scan rates (the VP-Capillary DSC can go up to 200°C/h, the conventional equivalent has a maximum rate of 90°C/h) offer higher sensitivity and lower noise, but can lead to lost resolution or loss of equilibrium as compared to rates at a slow scan rate (<60°C/h). For RNA, we use scan rates in the range of 75–100°C/h, but slower scan rates should be tested to assure equilibration throughout the run. The filter period parameter is associated with the scan rate: the filter period should decrease as the scan rate increases. The following equation is used to calculate the ideal filter period required:

$$\text{Filter period} = \frac{0.2 * 3{,}600\ (\text{s/h})}{\text{Scan rate}\ (^{\circ}\text{C/h})} \tag{3}$$

Once this calculation is completed, the closest filter period available on the Origin operating software should be chosen.

The feedback mode chosen for the DSC experiment depends on the macromolecule under investigation. RNA or DNA DSC experiments should use low gain except at fast scan rates (>100°C/h) where mid gain should be used. For protein experiments, no gain should be used, however for lipids, high gain should be chosen.

3.10.4. Establishing a Thermal History

The DSC instrument will require several "warm-up" scans before the macromolecule of interest can be run. The first scan will almost always be unreliable until the calorimeter develops what is termed a "thermal history." Typically 4–6 runs of water/water, where both the reference and sample cells are filled with H_2O, DSC experiments are done first, followed by 2–3 buffer/buffer runs. These control runs should start overlapping as the instrument establishes its thermal history. Only then should the RNA or protein experiment be performed. The buffer/buffer runs that preceded the macromolecule scan can be used as a reference control for data analysis. Thus, a typical DSC experiment will involve these 6–9 preliminary scans in addition to the targeted macromolecule run.

3.11. Performing the DSC Experiment

The VP-Capillary DSC experiment is run using the Microcal-specific Origin software included with the instrument. The VP-Capillary DSC with its autosampling, higher scan rates, and higher temperature capabilities is ideal for RNA and therefore we limit our discussions in this section to this instrument, although many considerations apply to the conventional DSC as well. There are five tabs on the program, but only two tabs will be used to run the experiment: "Autosampler" and "DSC Controls":

1. Within the autosampler portion of the instrument, there are two buffer reservoir bottles and three cleaning solution bottles. Fill one of the buffer bottles with ddH_2O, and the other with the dialysis buffer (~250 mL minimum). Fill one of the cleaning bottles with water, and the other two may hold cleaning solution (5–10% Contrad-70 is recommended).
2. On the "Autosampler" tab of the Origin program, there are three side tabs. Open the "Tray set-up" tab.
3. First, input the number of DSC runs, and set the location on the 96-well tray. The spreadsheet to the right of the screen will summarize the scans to be performed. Notes can be added to provide details of each scan. In this spreadsheet, cleaning steps can be added to the list of instrument commands. Cleaning steps should occur after the last macromolecule run, or after each macromolecule scan if there is threat of precipitation. Three water/water runs should be done after a cleaning step to ensure that all cleaning solution is removed from the cells.
4. Prepare and degas the macromolecule solution and the dialysis buffer as described for ITC samples in Subheading 3.6, and then add to the appropriate wells of the 96-well tray, as laid out in the previous step.
5. Place the pre-slit tray cover on top, and place the tray in the cooling tower of the instrument in the chosen spot.
6. In the "DSC Controls" tab of the Origin program, input the scan parameters decided upon in the previous section. The prescan thermostat should be set to 10–15 min, and the post scan thermostat should be set to 0 min. The series of scans can be started by clicking on the "Start" icon at the top of the program screen.

Troubleshooting and maintenance of the DSC are discussed below (see Note 4).

3.12. DSC Data Analysis

Analysis of the DSC data is performed with the Origin software included with the instrument. The process of analyzing DSC data involves subtracting a buffer/buffer scan as a control, setting a baseline before and after each peak, and fitting the peak to the appropriate model. More detail can be found in the MicroCal DSC data analysis manual.

1. On the front screen of the DSC Origin analysis program, read in the data by clicking on the "Read Data" button on the left hand side. All the scans can be loaded at once if the Auto DSC data is chosen, or a single scan can be loaded. We recommend the use of the Auto DSC data to survey the results of the experiments and choose the buffer/buffer scan to be used as a control.
2. Once the scans to be analyzed are chosen, they are read into the Origin program with "Scan Rate Normalization" and "Delete Time Data" options checked. Two scans should be on the screen: the macromolecule scan and the chosen buffer/buffer control.
3. Subtract the control data from the macromolecule scan by selecting the "Subtract Reference" button on the left side of the screen.
4. Next, the control corrected data are normalized for macromolecule concentration by selecting the "Normalize Data" option on the left.
5. Construct the baseline for the data by selecting "Start Baseline Session" from the "Peak" menu. Automatic baselines will be drawn, which can be adjusted by choosing the "Adjust" option. The baselines should be set immediately before and after the peak, and their length should be 10–20°C.
6. By choosing the "Baseline" options, draw a baseline under the peak. There are several mathematical options for constructing this baseline, but the baseline drawn from choosing "Progress" works for most cases.
7. Once the baseline is drawn, baseline-correct the thermogram by selecting "OK." The data will go from being on a slant to being horizontal, as this baseline is subtracted from the raw data.
8. The "DSC" menu provides three models to fit the data. The two-state model assumes that no intermediates are populated during the unfolding process, while the non-two-state model does not make this assumption. The two-state-with-ΔC_p model assumes a two-state process with a nonzero ΔC_p. Start the fitting process by clicking on the peak, and use Levanberg-Marquandt algorithms to fit the data.

The data analysis will provide the T_M and the ΔH of unfolding. The non-two-state model gives two enthalpies, the calorimetric enthalpy (ΔH_{cal}) and the van't Hoff enthalpy (ΔH_{vH}). Comparison of these two quantities can offer much information about the system: the calorimetric enthalpy is generated from the area under the curve, while the van't Hoff enthalpy is derived from the shape of the peak. If the two quantities are equal, then there are probably no significantly populated folding intermediates and the system is likely two-state; however, if the ΔH_{vH} is less than the ΔH_{cal}, then there are likely intermediate states accessed during the transition, and if the ΔH_{vH} is greater than the ΔH_{cal}, then the macromolecule has likely oligomerized in solution. In summary, regardless of what is known about macromolecule behavior going into the experiment, we recommend that the non-two-state model always be tried to assess the two-state nature of the transition.

4. Notes

1. Advanced ITC experiments

   There are several more complex experiments beyond the basic case of two molecules interacting that can be performed using ITC. These include heat capacity (ΔC_p) curves, competition titrations, and dissociation experiments. Any of these experiments can be done using either the VP-ITC or the AutoiTC_{200}.

   Heat capacity provides useful information on the temperature dependence of the enthalpy. Such information is useful for understanding how ligand binding varies with temperature and for gaining insight into such phenomena as conformational changes of the RNA between the free and bound states (19–21). To measure the heat capacity of the interaction, the binding experiment is done at a series of temperatures. The heat capacity curve is then obtained by plotting the individual enthalpies vs. temperature:

$$\Delta H(T) = \Delta H(T_{ref}) + (T - T_{ref})\Delta C_p + (T - T_{ref})^2\left(\frac{d\Delta C_P}{dT}\right) \quad (4)$$

   The last term in Eq. 4, which reflects a temperature dependence of the ΔC_p itself, may or may not be needed to fit the data.

   Either instrument is capable of performing titrations in a range of 4–80°C, although it is preferable to keep the range of experiments within 5–65°C for most systems. For RNA, the upper limit of temperature for the heat capacity curve should be at least 10°C below the T_M in order to ensure the RNA is still properly folded. Elucidation of the heat capacity curve can offer important information regarding the role of the hydrophobic effect and surface area burial in the binding event (20, 22).

   Some binding interactions are so tight (K_d <<1 nM) that ITC cannot be used to measure the binding equilibrium directly, as the binding curve becomes a step function. A competition experiment can be then be done to weaken the apparent binding and indirectly calculate the binding affinity. The macromolecule is bound first to a weaker binding competitor, and the strong affinity ligand is then titrated into the bound complex solution. The bound complex can either be mixed together with pipette or produced via a separate ITC experiment. Using the enthalpy and binding affinity for the competitor ligand, the quantities for the strong binding ligand can be obtained using the data analysis described by Sigurskjold (23) or by using the Origin software. Such experiments are also useful when studying therapeutic drugs that are either direct agonist or antagonists, and when studying their corresponding competition for the active site with the cognate ligand.

ITC can also be used to measure dimerization (or higher oligomerization) as well. The thermodynamics of RNA duplex formation have been studied extensively with ITC by the Feig Lab (22, 24, 25). Typically, binding of two non-self-complementary oligonucleotides is studied, but homodimerization of a self-complementary oligonucleotide can also be studied by ITC utilizing a dissociation experiment. In this scenario, a self-complementary dimer at a concentration near its K_d is injected into buffer, and the sudden drop in concentration compels the dimer to dissociate into monomers. The ITC origin software can provide analysis for such data.

2. Common ITC problems and their solutions

The most common problems that could arise during the ITC run are usually due to one of three factors: cell cleanliness, air in the system, or variable environmental (lab) temperature. A dirty cell will give a noisy baseline or an odd morphology of the injection peaks. Cleaning should be done immediately after each run in order to prevent build-up of residue in the cell. The cleaning procedure is automatic with the AutoiTC$_{200}$ instrument, but the VP-ITC must be diligently cleaned manually. The standard cleaning protocol uses 100 mL of 5–10% Contrad-70, Top-Job, or 409 cleaner run through the sample cell using the cleaning attachment to the Thermo-Vac followed by 500–1,000 mL of ddH_2O. More rigorous cleaning can involve incubation overnight of cleaning solution in the sample cell at 55°C followed by ddH_2O washing. To remove any protein contamination, 1 mg/mL Proteinase K in water can be incubated in the instrument at 37°C overnight and flushed out in the morning.

Air bubbles within the sample cell or injection syringe will invariably lead to large disruptive thermal events to the ITC experiment. This problem is usually caused by poor sample-loading technique, insufficient degassing, or a mechanical problem with the syringe. Test runs of water injected into water, which should give very small exothermic heats (see Note 3), should be used to verify that the operator is loading the instrument properly. Filling problems in the AutoiTC$_{200}$ often stem from full or partial blockages in the lines, protein build-up in the injection syringe, or damage to the fill port adaptor.

Temperature fluctuations in the room during the ITC experiment will lead to drifting baselines. This relatively simple problem is often the hardest to solve as it may require relocation of the instrument to another site. Some considerations to optimize environmental conditions are to avoid proximity to windows, overhear air vents, and doors. The AutoiTC$_{200}$'s enclosure offers a level of protection to variable room temperatures, but the instrument is not immune to a bad laboratory environment with direct sunlight or air vent exposure. Another problem with long runs can be if air conditioning or heating levels change during the night. One consideration can be to locate the instrument in an NMR or crystallography facility, which often has very stable temperature conditions.

Other problems with the ITC experiment may be due to design or preparation of the sample. A saturation injection heat value that is much larger than the dilution control injection heats is an indication of buffer mismatch and can be solved by expanding dialysis times and/or volumes. On the other hand, lack of binding seen in a titration may reflect an inappropriately folded or degraded RNA, or perhaps a thermoneutral binding event. Degradation can be assessed by removing some sample post-run, 5′ end labeling with ^{32}P using polynucleotide kinase and γ-^{32}P ATP, and performing denaturing PAGE. This degradation could be due to high divalent concentrations, long experiment times, high experiment temperatures, or any combination of these factors. In the worst-case scenario, there may be RNase contamination in the instrument or syringe accessories, which should be handled with RNase Zap (Ambion, Inc.) treatment or related decontamination procedure. In the lack of evidence for degradation, misfolding of the RNA can be addressed through construct redesign or change in renaturation procedures that may rescue binding activity.

3. Regular ITC maintenance

The most important aspect of maintenance for both instruments is thorough cleaning immediately following a run. Proteinase K cleaning following protein experiments will help prevent protein precipitate build-up that could lead to clogs. RNase Zap treatment can be used to treat the instrument and other implements to prevent RNase-related problems. For the AutoiTC$_{200}$, the solution stock reservoirs and the nitrogen tank should be checked before each set of runs and refilled as needed. The fill port adaptor may need to be replaced on a

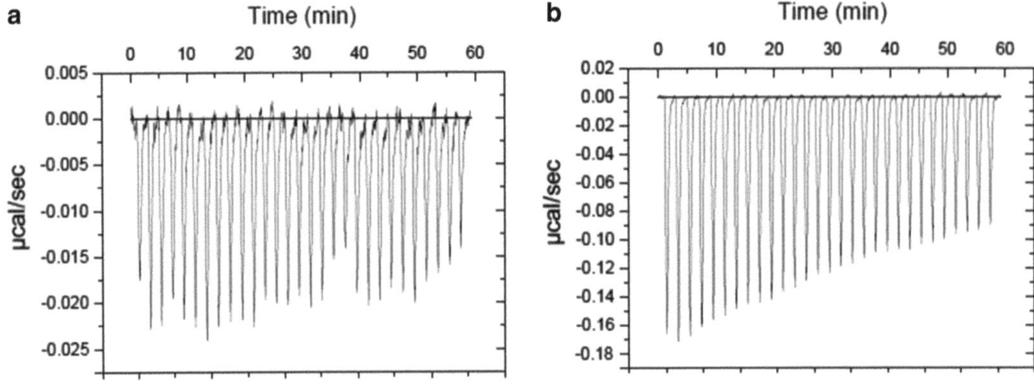

Fig. 4. Examples of quality control titrations to verify ITC performance. (**a**) A typical water into water injection on the VP-ITC. (**b**) 10 mM NaCl titrated into water on the VP-ITC.

yearly basis depending on the number of runs the instrument has performed, and the VP-ITC will have to have its plunger tip changed every 6 months. For both instruments, the injection syringe may need to be changed if it becomes bent or worn, although this is a rare occurrence, and the cause of such damage should be immediately investigated.

To check performance of the instrument, there are two ITC runs that are recommended (Fig. 4). The first is titration of water into water. This should show only the mechanical heat of injection and should be very small (<0.05 μcal/s for both instruments). This experiment will verify your filling technique, indicate if the reference cell needs to be changed, and test if the room conditions are acceptable for an ITC experiment. The second experiment that serves as a test of the instrument's operation is to take a 10–50 mM salt solution (NaCl, $MgCl_2$, etc.) and titrate it into water (26). This titration will give peaks of a size similar to those encountered in a biochemical experiment and will verify cleanliness of the cell, as well as proper operation of the instrument. Either experiment is recommended to be performed immediately prior to a series of ITC experiments to ensure that the instrument is in optimal condition before valuable biological samples are consumed. We have made it common practice to use a water-into-water or salt-into-water titration prior to every AutoiTC$_{200}$ run for these reasons.

4. DSC troubleshooting and maintenance

By far the most common problem encountered in DSC is the accumulation of macromolecule residue in the cell. Cleaning of the cells, the injection plumbing, and the loading syringe should therefore be performed before and after every series of scans using the "Maintenance" subtab located under the "AutoSampler" tab on the Origin operating software. Incubation with Contrad-70 at high temperatures (50–60°C) should suffice for most RNA and protein runs. Denatured proteins are often insoluble, so residue build up in the capillary cell can cause extreme problems. Weak solutions of nitric acid may often be required to remediate serious precipitate. Consult the operator manual and MicroCal for advice in this case.

Acknowledgments

We like to thank Dr. Andrew Feig for helpful discussions about ITC cleaning, maintenance, and troubleshooting, and Drs. Verna Frasca and George Makhatadze for general helpful discussion about sample preparation and data analysis. We also would like to thank Matthew Coppock for comments on the manuscript, and NSF MRI 0922974 for support of biological calorimetry at Penn State University.

References

1. Smith AM, Fuchs RT, Grundy FJ, Henkin TM (2010) Riboswitch RNAs: regulation of gene expression by direct monitoring of a physiological signal. RNA Biol 7(1):104–110
2. Ferre-D'Amare AR (2010) The glmS ribozyme: use of a small molecule coenzyme by a gene-regulatory RNA. Q Rev Biophys 43(4): 423–447
3. Scott WG, Martick M, Chi YI (2009) Structure and function of regulatory RNA elements: ribozymes that regulate gene expression. Biochim Biophys Acta 1789(9–10):634–641
4. Zhang J, Lau MW, Ferre-D'Amare AR (2010) Ribozymes and riboswitches: modulation of RNA function by small molecules. Biochemistry 49(43):9123–9131
5. Gilbert SD, Batey RT (2009) Monitoring RNA-ligand interactions using isothermal titration calorimetry. Methods Mol Biol 540:97–114
6. Bruylants G, Wouters J, Michaux C (2005) Differential scanning calorimetry in life science: thermodynamics, stability, molecular recognition and application in drug design. Curr Med Chem 12(17):2011–2020
7. Ladbury JE (2010) Calorimetry as a tool for understanding biomolecular interactions and an aid to drug design. Biochem Soc Trans 38(4):888–893
8. Salim NN, Feig AL (2009) Isothermal titration calorimetry of RNA. Methods 47(3):198–205
9. Spink CH (2008) Differential scanning calorimetry. Methods Cell Biol 84:115–141
10. Serra MJ, Turner DH (1995) Predicting thermodynamic properties of RNA. Methods Enzymol 259:242–261
11. Bevilacqua PC, Brown TS, Chadalavada D, Parente AD, Yajima R (2003) Kinetic analysis of ribozyme cleavage. In: Johnson KA (ed) Kinetic analysis of macromolecules: a practical approach. Oxford University Press, Oxford, pp 49–74
12. Milligan JF, Uhlenbeck OC (1989) Synthesis of small RNAs using T7 RNA polymerase. Methods Enzymol 180:51–62
13. Guillerez J, Lopez PJ, Proux F, Launay H, Dreyfus M (2005) A mutation in T7 RNA polymerase that facilitates promoter clearance. Proc Natl Acad Sci U S A 102(17):5958–5963
14. Kaul M, Pilch DS (2002) Thermodynamics of aminoglycoside-rRNA recognition: the binding of neomycin-class aminoglycosides to the A site of 16S rRNA. Biochemistry 41(24): 7695–7706
15. Borer PN (1975) Optical properties of nucleic acids, absorption and circular dichroism spectra. In: Fasman GD (ed) Handbook of biochemistry and molecular biology nucleic acids. CRC Press, Cleveland, OH, p 589
16. Richards EG (1975) Use of tables in calculation of absorption, optical rotatory dispersion and circular dichroism of polyribonucleotides. In: Fasman GD (ed) Handbook of biochemistry and molecular biology: nucleic acids. CRC Press, Cleveland, OH, p 597
17. Gasteiger E, Hoogland C, Gattiker A, Duvaud S, Wilkins MR, Appel RD, Bairoch A (2005) protein identification and analysis tools on the ExPASy server. In: Walker JM (ed) The proteomics protocols handbook. Humana Press, Totowa, NJ, pp 571–607

18. Pace CN, Vajdos F, Fee L, Grimsley G, Gray T (1995) How to measure and predict the molar absorption coefficient of a protein. Protein Sci 4(11):2411–2423
19. Spolar RS, Record MT Jr (1994) Coupling of local folding to site-specific binding of proteins to DNA. Science 263(5148):777–784
20. Prabhu NV, Sharp KA (2005) Heat capacity in proteins. Annu Rev Phys Chem 56:521–548
21. Gilbert SD, Stoddard CD, Wise SJ, Batey RT (2006) Thermodynamic and kinetic characterization of ligand binding to the purine riboswitch aptamer domain. J Mol Biol 359(3):754–768
22. Mikulecky PJ, Feig AL (2006) Heat capacity changes associated with DNA duplex formation: salt- and sequence-dependent effects. Biochemistry 45(2):604–616
23. Sigurskjold BW (2000) Exact analysis of competition ligand binding by displacement isothermal titration calorimetry. Anal Biochem 277(2):260–266
24. Mikulecky PJ, Takach JC, Feig AL (2004) Entropy-driven folding of an RNA helical junction: an isothermal titration calorimetric analysis of the hammerhead ribozyme. Biochemistry 43(19):5870–5881
25. Takach JC, Mikulecky PJ, Feig AL (2004) Salt-dependent heat capacity changes for RNA duplex formation. J Am Chem Soc 126(21): 6530–6531
26. Tellinghuisen J (2007) Calibration in isothermal titration calorimetry: heat and cell volume from heat of dilution of NaCl(aq). Anal Biochem 360(1):47–55

Part IV

RNA–Protein Interactions

Chapter 11

Use of Aptamer Tagging to Identify In Vivo Protein Binding Partners of Small Regulatory RNAs

Colin P. Corcoran, Renate Rieder, Dimitri Podkaminski, Benjamin Hofmann, and Jörg Vogel

Abstract

Small regulatory RNAs (sRNAs) are short, generally noncoding RNAs that act posttranscriptionally to control target gene expression. Over the past 10 years there has been a rapid expansion in the discovery and characterization of sRNAs in a diverse range of bacteria. Paradigm shifts in our understanding of the breadth of posttranscriptional control by sRNAs were achieved in a number of pioneering studies that involved immunoprecipitation of a known RNA chaperone, the near-ubiquitous Hfq, followed by sequencing to identify novel putative regulators and targets. To perform the converse experiment, we previously developed a method which uses an aptamer-tagged sRNA to allow purification of in vivo assembled RNA–protein complexes and subsequent identification of bound proteins. We successfully implemented this protocol using the Hfq-associated sRNA, InvR, tagged with a tandem repeat of the commonly used MS2-aptamer. Incorporation of the aptamer had no effect on sRNA stability or activity. InvR-MS2 could be effectively purified along with associated proteins, such as Hfq, using maltose binding protein fused to the MS2 coat protein (MBP-MS2) immobilized on an amylose column. Mass-spectroscopy was also used to identify previously uncharacterized protein partners. These results have been described previously (Said et al., Nucleic Acids Res 37:e133, 2009) and thus the figures presented here are intended solely as an illustrative guide to complement this detailed step-by-step protocol.

Key words: Aptamers, Small RNAs, RNA-binding proteins, Affinity purification, RNA–protein interaction, Hfq, MS2, InvR

1. Introduction

The best studied and largest group of bacterial regulatory RNAs are the small regulatory RNAs (sRNAs) that modulate mRNA translation and stability by base pairing with limited sequence homology to target mRNAs (1–3). There are very few commonalities among

Kenneth C. Keiler (ed.), *Bacterial Regulatory RNA: Methods and Protocols*, Methods in Molecular Biology, vol. 905, DOI 10.1007/978-1-61779-949-5_11, © Springer Science+Business Media, LLC 2012

sRNAs, which vary dramatically in size (from 50 to 500 nucleotides), sequence, and secondary structure. One unifying factor is their association with the Sm-like RNA chaperone Hfq, which is generally required for both stability of the sRNA and annealing to target mRNAs (4). Some studies pioneering sRNA identification used Hfq co-immunoprecipitation and subsequent analysis (either by microarray array or pyrosequencing) of bound RNA to simultaneously identify numerous novel sRNAs and mRNA targets (5, 6). While this method has provided extensive insights into the post-transcriptional regulons of many bacteria, the discovery required prior knowledge of Hfq as an RNA chaperone and a pleiotropic regulator of gene expression (7). While Hfq is widely conserved in bacteria, its functional significance can differ dramatically between species (8). Equally, Hfq homologs have yet to be identified in many bacteria suggesting that alternative RNA binding proteins may fulfill the functional role of this generally pleiotropic regulator. Another major class of sRNAs do not base pair with target mRNAs but instead act to directly modulate protein function (9, 10). It would therefore be hugely beneficial to be able to identify novel sRNA-associated protein partners without a priori knowledge but instead by direct isolation and identification of protein bounds to an sRNA of interest.

There are a number of approaches available to identify RNA-associated proteins. For example, in vitro transcribed (IVT) RNAs of interest can be coupled to a column after which cell lysate is added allowing enrichment of proteins which bind tightly to the IVT RNA (11). This approach has a number of potential limitations since it relies on the correct folding of the immobilized IVT RNA and interaction of the RNA and protein in a far from physiological environment. Other approaches have utilized oligonucleotides complementary to the RNA of interest immobilized on columns to "catch" in vivo assembled RNA–protein complexes from cell lysates (12–14). While this approach has the benefit of purifying in vivo assembled RNA–protein complexes, it requires prior knowledge of the sRNA structure within the complex since it is essential to use RNA oligonucleotides complementary to accessible, single stranded regions on the target RNA. A third strategy, which overcomes the limitations of the previous two methods, involves the incorporation of short aptamer sequences into the sRNA, which allows the purification of in vivo formed RNA–protein complexes by passage of the cell lysate over immobilized protein that binds specifically to the aptamer.

RNA aptamers are small structured RNAs that are bound with high affinity by specific protein partners (e.g., eIF4a, lambda-N, MS2) (15). RNA aptamers have been used previously to identify novel RNA–protein interactions, for example by using maltose binding protein fused to the MS2 protein (MBP-MS2) immobilized on an amylose column. In a recent study from our group (16), we identified the lambda phage coat protein MS2 (17) as the most suitable aptamer for purifying in vivo formed RNA–protein complexes and utilized the MBP-MS2 fusion system to purify a number of sRNA and associated proteins. Specifically, it was the one of three tested aptamers that did not impair sRNA function or stability.

In this protocol, we describe in detail the approach developed in Said et al. (16) for the purification of in vivo assembled sRNA–protein complexes. The successful implementation of this protocol is illustrated using the Hfq-associated sRNA, InvR (18). We detail the system for tagging of InvR with a tandem MS2-aptamer, methods to ensure stability and functionality of the sRNA after incorporation of the ~40 nucleotide aptamer, and purification of the MS2-tagged sRNA along with associated proteins, as exemplified by the successful co-purification of Hfq.

2. Materials

Unless otherwise stated, autoclaved, deionized water (dH_2O) was used to make all solutions. Standard equipment used in molecular biology is essential for this protocol, e.g., incubators, gel electrophoresis apparatus and a bench top centrifuge.

2.1. Components Required for Culturing of Bacteria

1. Bacterial strains: wild-type with epitope-tagged *hfq* gene (e.g., *hfq*::3xFLAG; JVS-00944) (18) and the congenic strain carrying a chromosomal deletion for the respective sRNA (e.g., *hfq*::3xFLAG Δ*invR*; JVS-03618) (16). Store in LB containing 10% dimethyl sulfoxide (DMSO) at −80°C. Standard culture conditions were used for all experiments. Overnight cultures grown from single colonies were diluted 1:100 fresh LB medium supplemented with appropriate antibiotics. Cultures (typically 20 mL cultures in a 100-mL conical flask) were incubated at 37°C and 220 rpm to an OD_{600} of 2, at which point cells were harvested according to the individual protocols.
2. LB medium: 1% (w/v) tryptone, 0.5% (w/v) yeast extract, and 1% (w/v) sodium chloride. Autoclave (120°C, 20 min).
3. Antibiotics (1,000× stocks): ampicillin (100 mg/mL), kanamycin (50 mg/mL). Filter through a 0.45 μm membrane and store at −20°C.
4. Static and rotary incubators.

2.2. Components Required for Aptamer Tagging of sRNAs and Plasmid Transformation

1. Plasmid encoding the sRNA of interest.
2. Aptamer encoding DNA oligonucleotides (100 pmol/μL). Details for primer design are given in Subheading 3.1.
3. Phusion High-Fidelity DNA polymerase, 2,000 U/mL (Finnzymes), provided with 5× Phusion HF Buffer.

4. dNTPs: 10 mM stock solution (10 mM dATP, 10 mM dGTP, 10 mM dCTP, 10 mM dTTP).
5. 6× DNA loading dye: 10 mM Tris–HCl (pH 7.6), 0.03% bromophenol blue, 0.03% xylene cyanol FF, 60% glycerol, 60 mM EDTA.
6. 10× TAE Buffer: 242 g Tris base, 57.1 mL acetic acid, 100 mL 0.5 M EDTA, made to 1 L with dH_2O.
7. 1% (w/v) agarose, in 1× TAE Buffer.
8. Gel extraction kit (Qiagen).
9. Restriction Enzymes: DpnI, 20,000 U/mL. Other restriction enzymes may be required depending on the tagging method (see Fig. 2).
10. T4 DNA Ligase, 5,000 U/mL provided with buffer 10× T4 DNA Ligation Buffer: 400 mM Tris–HCl pH 7.8, 100 mM $MgCl_2$, 100 mM DTT, 5 mM adenosine triphosphate (ATP).
11. Calcium chloride competent *Escherichia coli* Top10 (cloning strain) and appropriate research strain (e.g., electrocompetent *Salmonella typhimurium*).
12. Temperature controlled water bath.
13. NucleoSpinPlasmid QuickPure.
14. Electroporation cuvettes.
15. MicroPulser.

2.3. Components for RNA Extraction and Northern Blotting

1. Stop solution; 95% EtOH, 5% Phenol.
2. Liquid nitrogen.
3. TRIzol Reagent (Invitrogen). Store at 4°C.
4. Phase Lock Gel (PLG) heavy tubes.
5. Chloroform.
6. Isopropanol.
7. Nanodrop spectrophotometer (NanoDrop Technologies).
8. Loading buffer II: 95% Formamide, 18 mM EDTA, and 0.025% each of sodium dodecyl sulfate (SDS). Xylene Cyanol and Bromophenol Blue. Store at −20°C.
9. 6% Polyacrylamide (PAA)/8.3 M urea gel: 15 mL 40% acrylamide/bisacrylamide solution (19:1), 49.8 g urea, 10 mL 10× TBE, make to 100 mL in dH_2O.
10. 10% Ammonium persulfate (APS).
11. Tetramethylethylenediamine (TEMED).
12. Vertical gel system provided with glass plates: 20×20 cm; spacers and comb: thickness 2 mm.
13. Whatman paper (thin).

14. Hybond-XL membranes.
15. Transfer apparatus (Tank electroblotter).
16. Antisense oligonucleotides (100 pmol/μL) to the respective sRNA e.g., InvR (JVO-0222) 5′-GATAAATGCAACGTAAGAGACAAATG-3′; MS2 aptamer (JVO-3562) 5′-GTGTCTGAAAAACGTACCCTGAT-3′; 5S rRNA (JVO-0322) 5′-CTACGGCGTTTCACTTCTGAGTTC-3′. Store oligonucleotides at −20°C.
17. T4 Polynucleotide kinase, 10,000 U/mL provided with 10× Reaction Buffer A: 500 mM Tris–HCl pH 7.6, 100 mM $MgCl_2$, 50 mM DTT, 1 mM spermidine.
18. γ-[^{32}P]ATP, 10 mCi/mL, supplied in 10 mM Tris–HCl, 2 mM DTT.
19. Sephadex G-25 spin column.
20. UV lamp (emission wavelength 302 nm).
21. HB-1000 Hybridizer, Hybridization oven.
22. Glass hybridization tubes.
23. Rapid Hybridization Buffer.
24. 20× SSC: 175.3 g NaCl, 88.2 g sodium citrate, dissolve in 800 mL dH_2O, set pH to 7 with HCl, fill to 1 L with dH_2O.
25. Sodium dodecyl sulfate (SDS).
26. Cassette, Phosphorimager film.
27. Phosphorimager.
28. Image analysis software. e.g., AIDA (Raytest) or Image J (U.S. National Institute of Health; http://rsb.info.nih.gov/ij/).

2.4. Components for Western Blotting

1. Vertical gel system provided with glass plates: 20 × 10 cm; spacers and comb: thickness 0.8 mm 40% acrylamide/bisacrylamide solution (37.5:1, Roth).
2. 10% APS.
3. Tetramethylethylenediamine (TEMED).
4. Separation Buffer: 60.57 g Tris base pH 8.8, 20 mL 10% SDS, final volume 500 mL.
5. Stacking Buffer: 60.57 g Tris base pH 6.8, 20 mL 10% SDS, final volume 500 mL.
6. 10× SDS Running Buffer: 30.3 g Tris base, 144 g glycine, 10 mL 1% SDS, final volume 1 L.
7. Protein loading buffer.
8. Prestained protein size marker.
9. Polyscreen, PVDF Transfer Membrane.
10. Methanol.

11. 10× Transfer Buffer: 60 g Tris base, 288 g glycine, made up to 2 L with dH_2O.
12. 1× Transfer Buffer: 200 mL 10× Transfer Buffer, 400 mL methanol (20% (v/v) final concentration), 1.4 L dH_2O.
13. Semi-dry electro blotter.
14. Bovine serum albumin (BSA).
15. TBST: 20 mM Tris base pH 7.4, 150 mM NaCl, 0.1% Tween 20.
16. Monoclonal anti-FLAG antibody: 1:1,000 in 3% (w/v) BSA/TBST.
17. Anti-rabbit antibody conjugated to horseradish peroxidase: 1:5,000 in 3% (w/v) BSA/TBST.
18. Anti-GroEL antibody: 1:20,000 in 3% (w/v) BSA/TBST.
19. Anti-rabbit antibody conjugated to horseradish peroxidase: 1:5,000 in 3% (w/v) BSA/TBST.
20. Western Lightning Chemoluminescent Reagent.
21. Detection apparatus e.g., Typhoon LAS 4000 (GE Healthcare).

2.5. Components for Affinity Purification of Tagged sRNAs and Interacting Proteins

1. Buffer A: 20 mM Tris–HCl pH 8.0, 150 mM KCl, 1 mM $MgCl_2$, 1 mM DTT.
2. French Press.
3. Amylose resin and bio-Spin disposable chromatography columns.
4. Yeast tRNA: 10 mg/mL.
5. Maltose.
6. Phenol-chloroform-isoamyl alcohol, 25:24:1 (v/v).
7. Ethanol: 3 M sodium acetate (pH 6.5) solution, 30:1 (v/v).
8. Acetone.

3. Methods

3.1. Aptamer Tagging of a Plasmid Borne sRNA

In principle sRNAs can be tagged at any position except after the transcriptional terminator. Nevertheless, the choice of position might be crucial to maintain stable expression and functionality of the sRNA. Hfq-dependent sRNAs are typically modular in nature, consisting of an mRNA targeting domain, an Hfq binding domain, and a 3′ transcriptional terminator (4). Incorporation of an RNA aptamer into an sRNA could inadvertently affect one of these three components, which are usually essential for sRNA function. For example, incorporation of the RNA within the mRNA targeting region will most likely result in loss of regulation by the tagged sRNA without affecting stability. On the other hand, disruption of either the Hfq-binding domain or the terminator stem-loop may cause a dramatic reduction in sRNA stability with resulting deleterious effects on the regulatory capabilities of the molecule. Efforts should therefore be made to avoid disrupting these domains when incorporating an aptamer. If little is known about the sRNA of interest, mRNA targeting regions can often be identified by their high levels of sequence conservation. Also, Hfq is known to bind A/U rich regions flanking a stem-loop structure (4) and thus, regions of this nature should not be disrupted by insertion of an RNA aptamer. If required, the terminator can be used as a scaffold for the MS2 aptamer by simply replacing the apical loop with a single MS2 aptamer, which should result in minimal disruption to the sRNA characteristics (see Note 1).

Throughout the following section we will describe in detail the aptamer tagging of a plasmid-borne sRNA (InvR; pNS-9) (16) (Fig. 1) under control of the $P_{LlacO-1}$ promoter, which is constitutive in cells without a LacI repressor protein. The insertion of the MS2-aptamer tag at the 5′-end of InvR is exemplified in Fig. 2a. In general, any plasmid-borne sRNA can be tagged following this method. Alternatively, general plasmids are also provided (pNS-18, -19, -20, -21) harboring different aptamer sequences (1× boxB, 4× boxB, eIF4A, and an alternative form of MS2, termed MS2*) flanked by unique NheI and XbaI cleavage sites and followed by the

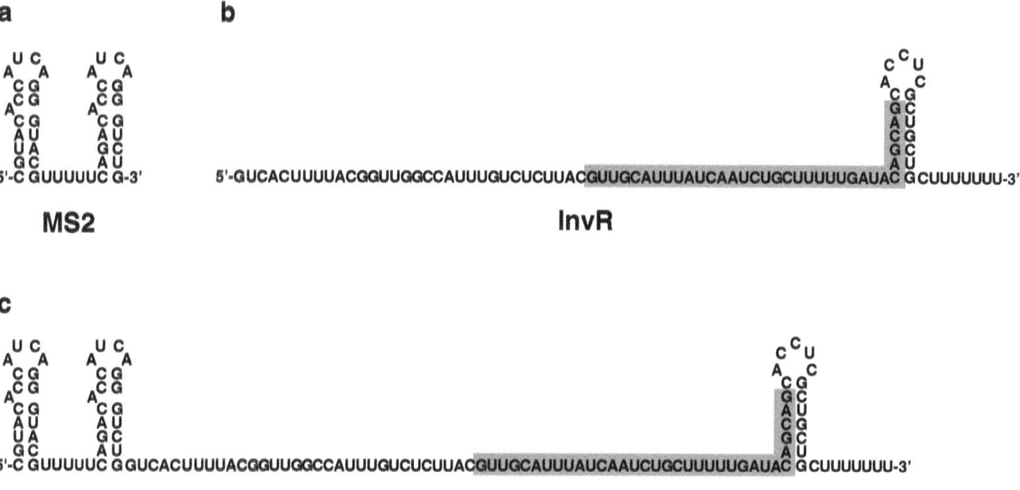

Fig. 1. Sequences and secondary structures of (**a**) a tandem repeat of the MS2 aptamer, (**b**) InvR, and (**c**) InvR-MS2. Nucleotides that are involved in target recognition by InvR are indicated by a *gray* background. InvR and InvR-MS2 are 83 nt and 133 nt in length, respectively.

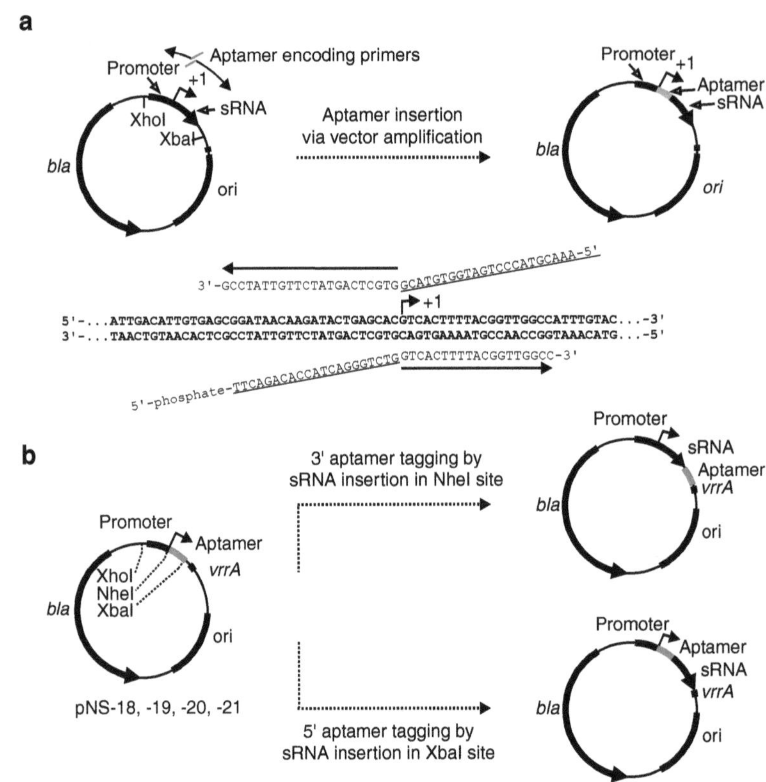

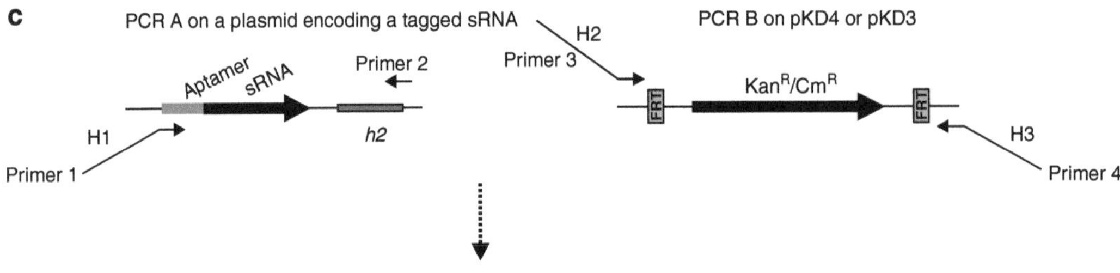

Fusion PCR of PCR product A and PCR product B using primer 1 and primer 4.

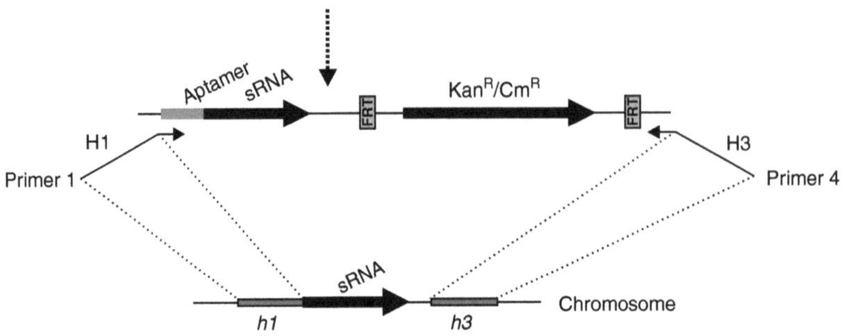

Fig. 2. Strategies for aptamer tagging of sRNAs. (**a**) Strategy I: primers containing the aptamer sequence (*gray/underlined*) 5′ to the priming sequence (homologous to the sRNA) allows for insertion of the aptamer at any position. (**b**) Strategy II: sRNA fragments can be amplified by PCR and introduced into the general aptamer cloning vectors, pNS-18, pNS-19, pNS-20, and

vrrA transcription terminator (Fig. 2b). These vectors allow aptamer tagging of an sRNA of interest at either the 5′- or 3′-end. The cloned sRNA will then be transcribed constitutively from the $P_{LlacO-1}$ promoter. Alternatively, an XhoI site further upstream in the plasmid facilitates the insertion of the sRNA gene with its native promoter (16). In some cases it might be necessary to express the sRNA from its native chromosomal location. Therefore we have developed a strategy for the chromosomal aptamer tagging of sRNAs (Fig. 2c). The general plasmids and strategy for chromosomal aptamer tagging of sRNAs are described in detail elsewhere (16).

1. Design divergent primers, allowing whole plasmid amplification, that flank the desired region of insertion for the MS2 aptamer (Fig. 2). Primers should contain 5′-overhanging regions (see Note 2) which, when combined, encode a tandem repeat of the MS2 aptamer sequence (Fig. 1a). The primer sequences used for MS2 tagging of InvR at the 5′-end are shown in Fig. 2a (also see Note 3).
2. Perform PCR amplification using 50 ng purified plasmid DNA in a reaction mixture containing 10 μL 5× HF buffer, 0.5 μL 10 mM dNTP solution, 0.5 μL of each primer (100 pmol/μL), and 0.5 μL Phusion High-Fidelity DNA polymerase (2 U/μL) in a final volume of 50 μL. Set up the thermo-cycling reaction as follows: initial denaturation at 98°C for 30 s followed by 30 cycles of denaturation at 98°C for 10 s, annealing at 60°C for 30 s, extension at 72°C for 1.5 min, and final extension at 72°C for 5 min (see Note 4).
3. Ensure amplification of the correct sized product (pNS-9, 2,412 nt; pNS-10, 2,455 nt, pNS-17 2,130 nt) by electrophoresis of 5 μL of the PCR product on a 1% TAE/agarose gel.
4. Add 1 μL of DpnI to the remaining PCR reaction and incubate for 1 h at 37°C (see Note 5).
5. Load the digestion reaction on 1% TAE/agarose gel, visualize by brief exposure to UV light and excise the band of correct size. Purify the DNA with Gel Extraction Kit.
6. For ligation of the amplified plasmid, dilute 40 ng of the purified PCR product to 17 μL with water, add 2 μL 10× T4 DNA ligase buffer and 1 μL T4 DNA ligase, and incubate for 1 h at room temperature.

Fig. 2. (continued) pNS-21, using unique NheI and XbaI sites. This strategy permits fusion of the aptamer to either the 5′- (insertion at XbaI site) or 3′-end (NheI) of the sRNA. Cloned fragment will be expressed from a $P_{LlacO-1}$ promoter located immediately upstream of the NheI site. Alternatively, cloning in the XhoI–NheI region permits 3′ tagging of the sRNA and expression from its native promoter. (**c**) Aptamer-tagged sRNAs can be chromosomally integrated at their native locus using a variation of the λ-Red method for one-step inactivation of chromosomal genes (30). For this, the aptamer-tagged sRNA is fused to the kanamycin resistance cassette of pKD4 (30) via overlap-extension PCR (31). This generates a single DNA fragment with flanking regions of homology for the chromosomal locus, thus allowing efficient integration. Diagrams not to scale.

7. Transform competent *E. coli* Top10 cells with 1 μL of the ligation reaction using a suitable method. We typically transform ligations using the calcium chloride method. Briefly, harvest exponentially growing cells by centrifugation (5 min, 4,000 × *g*, 4°C) and resuspend in 100 mM $CaCl_2$/10% glycerol (chilled). Wash cells three times (50 mL each) in this solution before final resuspension in 1/20 the original volume. Use calcium chloride competent cells immediately or store at −80°C. For transformation, incubate 1 μL of the ligation reaction with 30 μL of competent cells on ice for 30 min. Heat-shock in a temperature controlled water bath (30 s, 42°C), cool on ice (2 min), and add of 1 mL LB. Incubate reaction mixtures for 1 h at 37°C before concentration by centrifugation (5 min, 4,000 × *g*) and plating with appropriate antibiotic selection. Successful integration of the aptamer tag should be initially verified by colony PCR using flanking primers (for pNS-9 derivatives, use pZE-A (5′-GTG CCA CCT GAC GTC TAA GA-3′) and pZE-T1 (5′-CGG CGG ATT TGT CCT ACT-3′)), and electrophoresis on a 2% agarose/TAE gel (expected amplicon sizes for pNS-9, pNS-10 are 462 bp and 505 bp respectively). Plasmids containing insertions of the appropriate size were purified using the NucleoSpin Plasmid QuickPure kit and sequenced to ensure correct incorporation of the aptamer.

3.2. Analysis of Expression and Stability of Aptamer-Tagged sRNAs

As described earlier, it is essential to confirm the stability and functionality of tagged sRNAs before attempting the affinity purification protocol. We use Northern blotting for this as it allows simultaneous analysis of the expression and integrity of the sRNA (see Note 6). Here we use general culture conditions (see Subheading 2.1) that are suitable for purification of a constitutively expressed sRNA but which may need to be altered depending on the promoter used. Typically sRNAs are induced during stress responses and thus are often upregulated upon entry to stationary phase (1, 2). Therefore, harvesting cells in early stationary phase (e.g., OD_{600} of 2) will often result in a good yield of an aptamer-tagged sRNA under the control of its native promoter. Empirical testing is required to determine optimal expression conditions for each sRNA. The stability of the aptamer-tagged sRNA should always be compared directly with that of the untagged sRNA. Also, the ~40 nt increase in size of the tagged sRNA (corresponding to the tandem MS2 aptamer) should be clearly visible compared to the untagged sRNA. Comparison of strains expressing InvR and InvR-MS2 compared with the strain expressing a vector control ensures the specificity of the oligonucleotide probe and also should allow easy visualization of the regulatory capabilities of the tagged sRNA (see Subheading 3.3).

1. Transform plasmids expressing a control nonsense RNA composed predominantly of the *rrnB* terminator (pNS-17), InvR (pNS-9), and InvR-MS2 (pNS-10) into electrocompetent *Salmonella* (JVS-03618; *hfq*::3xFLAG, Δ*invR*) using a MicroPulser (Biorad). Briefly, prepare electrocompetent cells as described for the calcium chloride method (Subheading 3.1, step 7) with the important variation that cells are washed and resuspended in H_2O rather than 100 mM $CaCl_2$/10% glycerol and used immediately. For transformation, mix 1 μL (~50 ng) of purified plasmid with 50 μL of electrocompetent cells and transfer to a prechilled electroporation cuvette (2 mm gap width). Apply a voltage of 2.5 kV and immediately add 1 mL LB broth. Allow cells to recover for 1 h at 37°C before concentration by centrifugation (5 min, 4,000×*g*) and plating with appropriate antibiotic selection (all plasmids used in this study conferred ampicillin resistance). Prepare permanent stocks from an overnight culture grown from a single colony by mixing 900 μL of culture with 100 μL or DMSO in a cryotube. Maintain permanent stocks at −80°C.
2. Inoculate LB medium (supplemented with appropriate antibiotics) 1:100 with fresh overnight cultures.
3. Grow cultures to an OD_{600} of 2 and transfer 1.5 mL of the culture to a 2-mL tube containing 300 μL of stop-solution.
4. Snap-freeze in liquid nitrogen. If required, samples can be stored indefinitely at −80°C at this step (see Note 7).
5. Thaw samples on ice, harvest cells by centrifugation (5 min, 16,000×*g*, 4°C) and resuspend the pellet in 1 mL TRIzol reagent on ice.
6. Transfer the TRIzol/RNA solution to pre-spun (30 s, 16,000×*g*) Phase-Lock-Tubes.
7. Add 400 μL chloroform and shake immediately for 10 s (do *not* vortex).
8. Let stand for 2–5 min at room temperature.
9. Separate the organic and inorganic phases by centrifugation (15 min, 16,000×*g*, 4°C).
10. Take the aqueous (upper) phase (~500 μL), transfer it to a new tube containing 450 μL isopropanol, immediately mix by inversion (do *not* vortex), and incubate at room temperature for 30 min.
11. Pellet the precipitated RNA by centrifugation (30 min, 16,000×*g*, 4°C).
12. Remove supernatant and wash the pellet by addition of 700 μL 70% EtOH (do *not* pipette up and down or attempt to dissolve pellet by other means).

13. Centrifuge samples (10 min, 16,000 × *g*, 4°C) and remove the supernatant.
14. Air-dry the pellet by incubation at room temperature with an open lid for 10 min.
15. Resuspend the RNA in 25 μL of RNase-free water by incubation at 65°C (10 min, preferably with mixing, 600 rpm). Do not resuspend by pipetting. Store at −20°C.
16. Determine the concentration of total RNA using a Nanodrop or other suitable apparatus (expected yield, ~2 μg/μL).
17. Prepare a 6% PAA/8.3 M urea gel using 70 mL of the acrylamide premix (see Subheading 2.3) to which add 700 μL of 10% APS and 70 μL TEMED. Immediately pour the gel mixture in between two 20 × 20-cm glass plates, separated by 2-mm spacers and insert a comb containing the appropriate number of wells (see Note 8).
18. The amount of total RNA required for detection depends on the abundance of the target RNA and the efficiency of the detection method. Typically 5 μg of total RNA is sufficient to detect the majority of sRNAs using a 5′-radiolabelled oligonucleotide probe (see step 23 and Note 9). Mix 5 μg total RNA 1:1 with Loading buffer II.
19. Denature for 5 min at 95°C and cool the samples on ice for 5 min.
20. Load samples (see Note 10) and electrophorese for 1–2 h at 300 V (room temperature).
21. Transfer RNA (see Note 11) onto Hybond XL membrane (1 h/50 V in 1× TBE, 4°C) and covalently link it to the membrane by UV treatment (0.120 mJ).
22. Place the membrane in a glass hybridization tube, add 15 mL RotiQuickHyb buffer, and incubate for 30 min with rotation at 42°C.
23. A γ-[^{32}P]ATP labeled oligonucleotide (~20 nt) complementary the sRNA of interest is commonly used for identification of RNA species. The labeling reaction is as follows: 11 μL of dH_2O, 1 μL of oligonucleotide (10 pmol/μL) (JVO-0222 was used for detection of InvR), 2 μL of 10× Reaction Buffer A, 1 μL of T4 phosphonucleotidekinase (PNK), 3 μL of γ-[^{32}P] ATP (10 μCi/μL). After incubation for 1 h at 37°C, inactivate T4 PNK by heating at 95°C for 5 min.
24. Remove unincorporated γ-[^{32}P]ATP using Sephadex G-25 spin columns, according to the manufactures guidelines. Briefly, resuspend the resin by vortexing. Open the lid and break the base of the column to allow flow-out of the storage solution during centrifugation (1 min, 735 × *g*). Add the 20 μL labeling reaction to the center of the resin and place the column in a

new 1.5-mL tube. Elute the labeled, purified, oligonucleotide by centrifugation (2 min, 735 × *g*). Store at −20°C.

25. Boil the labeled probe for 5 min at 95°C.
26. Cool on ice for 2 min.
27. Add 2 μL of the probe to the QuickHyb solution (not directly on the membrane) and proceed with incubation for 1 h with rotation at 42°C.
28. Wash the membrane sequentially using 50 mL 5×, 1×, and 0.5× SSC solutions containing 0.1% SDS for 15 min each with rotation at 42°C.
29. After washing, allow the blot to dry and place it in a clear plastic "pocket." We use the Typhoon Phosphorimaging system for detection although X-ray film and developing solutions can also be used. With our system, gels are typically exposed overnight to obtain a strong signal.
30. For quantitative comparison of samples, we detect the highly abundant and constitutively expressed 5S rRNA, which acts as a loading control. First, remove the labeled oligonucleotides used for detection of the sRNA by adding boiling water/0.1% SDS to the membrane and shaking for 5 min. Repeat the hybridization and detection procedure described above (starting with preincubation of the membrane as described in step 22) using a labeled oligonucleotide (JVO-322) targeting 5S rRNA (expected size, ~120 nt).

Analysis of the Northern blot (Fig. 4a) revealed that InvR-MS2 accumulated as a single dominant species of the expected size and at levels comparable to untagged InvR. If the tagged sRNA does not accumulate as a single homogenous species of the expected size, alternative tagging approaches should be attempted, as described briefly earlier and in detail elsewhere (16). While the InvR-MS2 sRNA was stably expressed as a homogenous species, it was still necessary to ensure that the function was not abrogated (described below).

3.3. Functional Analysis of Tagged sRNAs

There are multiple methods to analyze the functionality of an sRNA. Naturally, this requires prior knowledge of a target and its regulation. Assuming that the stability of target mRNA is altered by the active sRNA, the simplest test is to re-probe the membrane used to confirm sRNA stability for the known target of the sRNA. The regulatory effect should be clear when comparing the sRNA deletion strain with isogenic strains complemented with the tagged or untagged sRNA. However, since not all sRNA-mRNA interactions result in an alteration of mRNA stability, we use a standard protocol involving Western blotting (see Note 6) of target proteins that allows quantification of target protein levels and reliable evaluation of sRNA activity. For detection of InvR target regulation, an anti-porin serum was used, which allows visualization of the specific downregulation of OmpD levels by InvR, while levels of other major porins (such as OmpF and OmpC) are unaltered. If an antibody is not available for the target of interest, a GFP reporter system has been developed that allows rapid fusion of target proteins and detection by fluorescence or using a commercially available GFP-monoclonal antibody (19, 20).

1. Either thaw the protein samples on ice (see Subheading 3.2, Note 7) or grow fresh cultures using culture conditions as described in Subheading 3.2.
2. Harvest 500 μL of the culture by centrifugation (5 min, 16,000 × *g*, 15°C).
3. Resuspend the pellet in 100 μL 1× loading buffer and immediately heat for 5 min at 95°C to lyse cells and denature proteins.
4. As standard we used 12–16% PAA gels for protein analysis. Here a 15% PAA gel was chosen to allow clear separation of OmpF and OmpC (~39 and ~40 kDa, respectively). To prepare a 15% SDS-PAA gel, mix 3.75 mL of separation buffer, 3.75 mL of acrylamide/bisacrylamide, 100 μL of 10% SDS, and 2.5 mL of dH_2O. Add 75 μL of 10% APS and 7.5 μL of TEMED to catalyze the polymerization reaction. Swirl to homogeneity and pour the gel in between the glass plates with 0.8 mm spacers, held by a gel cassette, to within ~1 cm of the top and immediately overlay with isopropanol.
5. After polymerization of the separation gel remove the isopropanol completely using absorbent tissue (or Whatman paper). To prepare the stacking gel, mix 1.25 mL stacking buffer, 0.75 mL acrylamide/bisacrylamide, 7.82 mL dH_2O, and 10 μL 10% SDS followed by 150 μL 10% APS and 15 μL TEMED. Swirl the mixture to homogeneity, pour over the polymerized stacking gel and insert the comb immediately (see Note 8).
6. After polymerization, fill the electrophoresis apparatus with 1× SDS running buffer, flush the wells of the gel with buffer, and load the samples. Load a pre-stained protein marker as a size reference. We typically electrophorese protein gels at 30 mA for 1–2 h or until good separation of the pre-stained protein markers has occurred.
7. For transfer of protein samples, activate the PVDF membrane by incubation for 90 s in methanol in a fume hood, followed by 5 min in dH_2O and 5 min in 1× Transfer buffer.
8. Transfer for 2 h at 2 mA/cm^2 of the membrane at 4°C (see Note 11).

9. After transfer, "block" the membrane with 3% (w/v) BSA/TBST (or 3% low-fat milk powder/TBST) for 1 h at room temperature or overnight at 4°C.
10. To confirm activity of InvR and InvR-MS2, we detected the major OMPs using porin antiserum (1:20,000 in 3% (w/v) BSA/TBST, provided by Rajeev Misra, Arizona State University, USA), which detects porins such as OmpA/C/D/F for 1 h at room temperature under agitation.
11. Wash the membrane in TBST five times for 6 min each.
12. Incubate with anti-rabbit conjugated horseradish peroxidase (GE Healthcare, 1:5,000 in 3% (w/v) BSA/TBST) for 1 h at room temperature.
13. Wash again in TBST six times for 10 min each.
14. Detect proteins by incubation with Western Lightning Reagent (5 min, room temperature) and visualization using a Typhoon LAS 4000 CCD camera, or other suitable system. We typically detect the constitutively expressed chaperone GroEL (Sigma Aldrich, 1:20,000 in 3% (w/v) BSA/TBST) to ensure equal loading of all lanes.

Analysis of the Western blot (Fig. 4a) shows that InvR-MS2 down-regulated OmpD protein synthesis comparably to wild-type InvR. This shows that InvR-MS2 is an active regulator and is functionally unaffected by addition of the RNA aptamer. We therefore proceeded with the affinity purification protocol. If the function of the sRNA is abrogated, alternative tagging approaches should be attempted. Insertion of the aptamer may disrupt the targeting region of the sRNA, either by direct disruption or potentially by base-pairing with the targeting region. sRNAs typically require single stranded targeting regions to bind effectively with target mRNAs and thus pairing of sequences within the targeting region with the MS2 aptamer will most likely reduce function without affecting the stability of the sRNA. To address loss-of-function issues, the MS2 aptamer should be incorporated into multiple positions within the sRNA, as discussed in Subheading 3.1 and detailed previously (16). When necessary it is possible to incorporate the MS2 aptamer into at the transcriptional terminator of the sRNA, which should not affect sRNA targeting (see Note 1).

3.4. Affinity Purification of Tagged sRNA–Protein Complexes

In order to test whether the newly encoded aptamer sequences enable the purification of in vivo expressed sRNAs with associated protein partners, we adapted a previously established protocol for the isolation of in vitro assembled spliceosomal ribonucleoprotein complexes (RNPs) (21), which employs the MS2-MBP fusion protein (Fig. 3). In this protocol *Salmonella* expressing MS2-tagged sRNAs from plasmids are harvested by centrifugation, followed by rapid lysis. The cleared lysate is then directly applied to an amylose

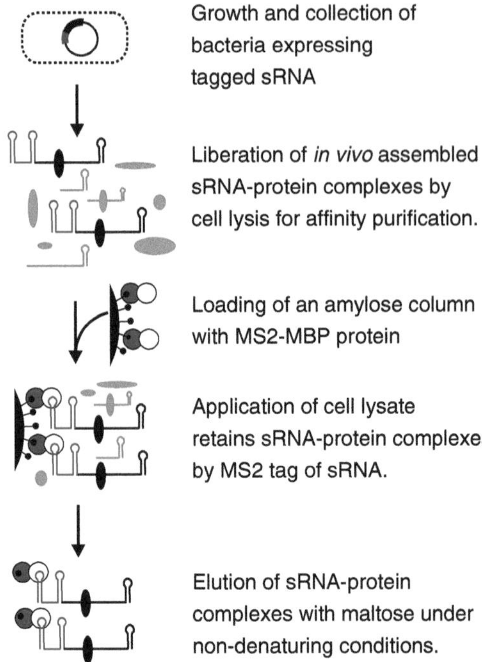

Fig. 3. Experimental strategy to purify MS2 aptamer-tagged sRNAs and in vivo formed RNA–protein complexes under non-denaturing conditions from bacterial lysates. *Light gray lines* represent MS2 aptamers fused to the respective sRNA (*black*). *Gray circles* on the amylose resin denote maltose binding protein fused to the MS2-coat protein (*white circles*) non-covalently bound to the resin (*black circles*), which immobilizes MS2-tagged sRNAs.

column that is non-covalently coupled with MS2-MBP protein (via amylose binding of the MBP moiety) (see Note 12). Upon binding of the MS2-tagged RNA species to MS2-MBP, the column is extensively washed to remove nonspecifically bound molecules. Finally, the tagged sRNA and associated protein(s) are eluted with a maltose-containing buffer that disrupts the MBP–amylose interaction. Purification of the untagged sRNA was attempted in a parallel experiment to determine the degree of nonspecific binding to the column.

1. Follow standard culture conditions (see Subheading 2.1); chill cells equivalent to 50 OD_{600} on ice for 20 min before harvesting for 20 min at 2,900 × *g* at 4°C.
2. Wash cells in 1 mL of Buffer A (see Note 13) and centrifuge for 5 min at 11,200 × *g* at 4°C. At this point pellets can be snap-frozen in liquid nitrogen and stored at –80°C.
3. Resuspend the pellets in 2 mL of Buffer A.
4. Break cells by three applications of 800 psi to the cells in a French press (chill the lysate on ice between each step).

5. Clear the lysate by centrifugation for 30 min at 16,000 × *g* at 4°C and collect the soluble fraction.
6. Prepare the affinity column by applying 50 μL amylose resin to Bio-Spin disposable chromatography columns. Importantly, perform all steps in affinity chromatography at 4°C.
7. Wash the column three times with 2 mL of Buffer A.
8. Apply 100 pmol of MS2-MBP (see Notes 14–16) diluted in 1 mL of Buffer A to the amylose resin (see Note 17).
9. Wash the column with 2 mL of Buffer A.
10. Load the cleared lysate onto the column (see Notes 18 and 19) and wash three times with 2 mL of Buffer A. Collect aliquots equivalent to 2 OD_{600} of the lysate, flow-through and wash fractions for RNA extraction and analysis by Northern blotting, as described in Subheading 3.2. Additionally, collect aliquots equivalent to 0.5 OD_{600} for protein analysis by SDS-PAGE, as described in Subheading 3.3.
11. Elute RNA–protein complexes from the column with 900 μL of Buffer A containing 12 mM maltose.
12. Add 1 volume phenol-chloroform-isoamyl alcohol to the eluate and mix by strong shaking for 20 s.
13. Separate water and organic phase by centrifugation for 30 min at 16,000 × *g* at 15°C.
14. For RNA precipitation, transfer upper aqueous phase to a new tube (see Note 20) add 20 μg yeast tRNA, and mix with 3 vol EtOH:NaAc (30:1) solution. Retain the organic phase for protein precipitation starting with step 19.
15. Precipitate the RNA by incubation for 2 h to overnight at −20°C and concentrate by centrifugation for 30 min at 16,000 × *g* at 4°C.
16. Wash the RNA pellets by adding 500 μL chilled 70% ethanol (see Note 21) and centrifuge again for 10 min at 16,000 × *g* at 4°C.
17. Carefully remove the supernatant (see Note 22) and air-dry the pellet by incubation (room temperature) with an open lid for 10 min.
18. Resuspend the RNA pellet in 25 μL of RNase-free water and store at −20°C. Use 5–10 μL for Northern blotting (see Note 23).
19. For protein precipitation, add 3 volumes of acetone to the lower organic phase, mix and incubate from 2 h to overnight at −20°C.
20. Pellet the proteins by centrifugation for 30 min at 16,000 × *g* at 4°C.

21. Wash the pellet using 500 μL of ice-cold acetone (do not pipette up and down) and centrifuge for 10 min at 16,000 × *g* at 4°C.
22. Remove the acetone very carefully (see Note 24) and repeat the wash step.
23. Air-dry the sample for 10 min at room temperature.
24. Dissolve the pellet in 50 μL of 1× protein loading buffer, denature by heating for 5 min at 95°C, and store at −20°C. Use 10–20 μL for SDS-PAGE followed by Western blot analysis for detection of specific proteins of interest or for silver-staining to visualize all precipitated proteins (see Note 25). In this example, we detected Hfq$^{3\times FLAG}$ by Western blotting using a monoclonal anti-FLAG antibody and a horseradish peroxidase conjugated anti-rabbit antibody.

Northern blot analysis of the eluted fractions (Fig. 4b, lane 5) indicates that the MS2 tag permits the specific enrichment and purification of InvR-MS2 compared to the untagged sRNA. Quantification of the Northern blot signals in the lysate (input) and eluted fractions suggested that 9% of InvR-MS2 RNA was recovered (see Note 26). In contrast, only 0.14% of InvR RNA (which is expressed from the plasmid at the same level as InvR-MS2; Fig. 4a) was recovered, arguing that InvR-MS2 was enriched 65-fold as compared to untagged InvR. InvR is bound tightly by Hfq both in vivo and in vitro and this strong association has allowed enrichment of InvR with Hfq during co-immunoprecipitation of Hfq from *Salmonella* lysates (5, 22). Conversely, if the MS2-MBP affinity approach indeed recovered sRNAs in their native complexes, Hfq should not only be detectable in the eluted fraction of InvR-MS2 but should also be considerably enriched when compared to the eluate from untagged InvR. Figure 4c shows that the Hfq protein was strongly enriched when comparing InvR-MS2 to the InvR. To ensure that Hfq enrichment was specific to its association with InvR and not the MS2 tag, we tested a construct expressing the MS2 tag fused to the *rrnB* terminator (pRR-05). Even though this ~80 nucleotide RNA was stable in vivo and was recovered by MS2-MBP affinity purification in amounts comparable to InvR-MS2 (Fig. 3b), it showed no enrichment of Hfq indicating that Hfq was purified specifically through interaction with the InvR component of InvR-MS2, rather than by interaction with the MS2 or with the column-bound MS2-MBP protein.

This protocol for the specific purification of in vivo formed RNA–protein complexes can be readily applied to identify proteins associated with small noncoding RNAs. As a note of caution, the approach by itself does distinguish between productive or nonproductive complexes, so we advise the isolated complex be tested for function, availability of a suitable assay permitting.

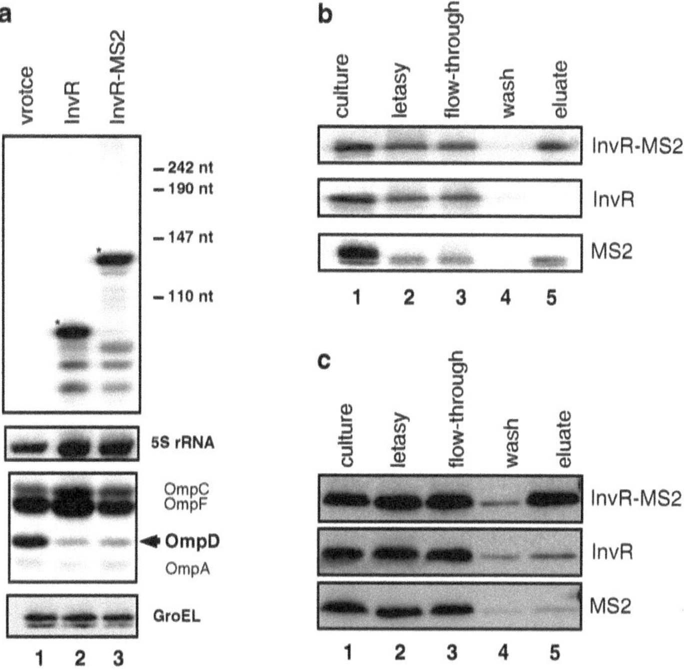

Fig. 4. Analysis of stability, function, and purification of InvR-MS2. (**a**) *Salmonella* Δ*invR* (JVS-03008) carrying an "empty" control vector (lane 1, pNS-17), or plasmids expressing InvR (lane 2, pNS-9) or InvR-MS2 (lanes 3, pNS-10). Detection was performed with an oligonucleotide probe, JVO-0222. *Asterisks* indicate bands of expected full-length sRNAs. Positions of a co-migrating size marker are shown to the *right* of the blot. The *second panel* shows probing for 5S rRNA (JVO-0322) to confirm equal loading. The *third panel* shows the relevant section of a Western blot probed with a general anti-porin serum to show specific downregulation of OmpD levels by both InvR and InvR-MS2 sRNAs. The *bottom panel* shows the same Western blot probed for GroEL protein to confirm equal loading. (**b**) Affinity purification of in vivo expressed MS2-tagged InvR. InvR-MS2 and control constructs (expressing InvR or MS2 alone) were expressed in *Salmonella hfq*$^{3\times FLAG}$Δ*invR* (JVS-03618). RNA samples were prepared from intact cells before lysate preparation (lane 1), from the cleared lysate shortly before loading onto the column (lane 2), from the flow-through fraction collected after loading on a column preincubated with MS2-MBP protein (lane 3), from the combined column wash fractions (lane 4) and of the eluate (lane 5). Note that the amount of RNA loaded in lanes 1–4 each corresponds to 1 OD_{600} of bacterial culture, but in lane 5 the eluate corresponds to 10 OD_{600} of culture. RNA was analyzed by Northern blotting using InvR or MS2 specific probes. (**c**) Co-purification of Hfq with InvR-MS2. Protein fractions of the samples described in (**b**) were analyzed by Western blotting using a monoclonal antibody directed against the FLAG-epitope tag of the Hfq$^{3\times FLAG}$ protein. Samples corresponding to 0.05 OD_{600} of bacterial culture were loaded for lanes 1–4. Sample corresponding to 10 OD_{600} was loaded for the eluate (lane 5).

Please note these results have been described elsewhere (16) thus the figures shown here are purely for illustrative purposes.

4. Notes

1. For very small RNAs or those with undefined regulatory nucleotides, the aptamer can be inserted on top of the intrinsic transcription terminator. We have successfully purified a small RNA in which the terminator loop has been replaced with nucleotides 5–20 of a single MS2-aptamer. Naturally, it is important to ensure that this does not have deleterious effects on stability of the tagged sRNA.
2. One of the primers must be phosphorylated at the 5′-end to allow ligation by T4 ligase. Typically this is generated by endonucleolytic cleavage using a restriction enzyme although when DNA is to be ligated without digestion, one primer should be phosphorylated either during synthesis or later using PNK, as described by the manufacturer.
3. Amplification of the plasmid using primers that flank the sRNA and encoding the MS2 aptamer encoded in 5′-overhangs allows construction of a control plasmid that exclusively expresses the MS2-aptamer.
4. Phusion polymerase is highly processive and is ideal for inverse PCR-based applications. Modify the annealing temperature as appropriate to your primers. Finnzymes recommend using the Tm calculator (https://www.finnzymes.fi/tm_determination.html) to find appropriate annealing temperature. Modify extension time of the reaction depending on the size of the amplicon (15–30 s/kb).
5. A control should be performed to ensure efficient digestion by DpnI. Typically, we digest 50 ng of plasmid DNA in parallel with the experimental reaction. A mock digestion (incubation with buffer but without enzyme) can also be performed as a comparison for digestion efficiency. This mock digestion also acts as a positive control for the transformation reaction, since the undigested plasmid DNA should transform with high efficiency.
6. Researchers unfamiliar with techniques such as Western blotting and Northern blotting can find more detailed protocols elsewhere (23, 24).
7. Samples for protein analysis by Western blotting can also be taken at this point (see Subheading 3.3).
8. In the preparation of any gel for electrophoresis it is important to ensure bubbles are not introduced as air bubbles can inhibit the polymerization reaction and can cause aberrant migration of samples.
9. If detection using 5′ radiolabelled olidonucleotides is ineffective, body-labelled RNA probes (riboprobes), which contain

homology to the full sRNA coding sequence (excluding the terminator structure) can be used. To generate a riboprobe, a 5′ extension encoding the T7 promoter sequence is included in the antisense primer homologous to the 3′ end of the sRNA preceding the transcriptional terminator. This oligonucleotide is used in combination with a sense oligonucleotide designed against the 5′ end of the sRNA. Importantly, column purify the resulting PCR product before using a Maxiscript kit (Ambion) in presence of α-[^{32}P] UTP, but do not gel purify the DNA template as we find this interferes with the in vitro transcription reaction. The resulting RNA molecules contain multiple labeled nucleotides (5′-radiolabelled oligonucleotides only contain one each) resulting in significantly increased signal per bound molecule. When using a riboprobe, hybridization and wash steps are performed at 70°C. Also the initial wash step is performed using 2×, not 5×, SSC buffer. Otherwise the protocols are identical. It should be noted that due to the high affinity of the riboprobe for the target RNA, riboprobes can generally not be removed from the membrane (e.g., by addition of boiling water/0.1% SDS).

10. It is essential to wash the slots immediately before loading to remove unpolymerized acrylamide and urea, which may affect migration of the RNA.
11. It is standard procedure in electrophoretic transfer (both for Northern and Western blotting) to soak three thin Whatman papers in the appropriate transfer buffer and lay them on either side of the gel and membrane. At this stage take care to prevent introduction of air bubbles in between the different layers, which can cause aberrant electrophoretic transfer.
12. The non-covalent interactions involved in the affinity chromatography are essential for the success of the protocol. If the protocol fails, it is important to ensure that MS2-MBP protein binds to the beads under the buffer conditions used. Secondly, it must be confirmed that the tagged RNA associates with the column bound MS2-MBP protein under the buffer conditions used. To facilitate this, an IVT tagged RNA can be used. The use of radioactively-labeled RNA allows easy analysis of the efficiency of the step during the experiment.
13. Since the ionic strength of the buffer conditions can have a strong effect on association and stability of RNA–protein complexes (25–27), the effects of KCl were tested by increasing the concentration of KCl in all buffers from 150 mM to 1 M. 1 M KCl permitted better co-purification of Hfq$^{3\times FLAG}$ with InvR-MS2 RNA but did not necessarily increase binding of other proteins (16).
14. We tested whether the incubation time with MS2-MBP fusion protein influences the recovery of tagged sRNAs from the cell

lysate. Preincubation of InvR-MS2 with MS2-MBP for up to 70 min had no effect on recovery and thus, at least in this case was unnecessary (16).

15. For affinity purification of MS2-tagged sRNAs, a maltose-binding protein fused N-terminal to MS2 coat protein (MS2-MBP) was used to immobilize sRNA–protein to the column (16). The protein was purified as described previously (28).
16. The effect of varying MS2-MBP protein concentration on the recovery of Hfq co-purified with InvR-MS2 was tested by loading lysates directly on columns immobilized with increasing concentrations (10, 50, 100, 200 pmol) of MS2-MBP. Under the conditions used here, 10 pmol of MS2-MBP is fully sufficient to recover Hfq along with InvR-MS2 RNA. Increasing MS2-MBP concentration does result in a notable increase in purified Hfq, although high amounts of MS2-MBP (200 pmol) protein seem also to increase unspecific binding of Hfq. Therefore 100 pmol MS2-MBP is recommended for use in this protocol.
17. Application of MS2-MBP two to three times may increase loading of the column and subsequent purification of binding partners.
18. The lysate can be applied two to three times to increase binding of the complexes to the column.
19. It may be beneficial to change the order of affinity chromatography by incubating the MS2-MBP protein first with the lysate and followed by loading on to the column. Here it is important to test that the MS2-MBP protein still binds efficiently to the column after incubation in the lysate. Incubation temperature may also be critical in identifying certain RNA–protein interactions, and variations in this step should be considered.
20. We leave some drops of the aqueous phase together with the organic phase as this seems to support protein precipitation.
21. This step is to remove excess salt which co-precipitates with the RNA. Do not attempt to redissolve the pellet by vortexing or pipetting. Simply inverting the tube a number of times is sufficient as a wash step.
22. Typically we repeat this step to ensure all of the supernatant is removed before air drying.
23. If no RNA can be detected it is important to test if complexes are being efficiently eluted, for example by attempting to purify RNA from the resin before and after addition of the elution buffer.
24. Pellets are not fixed and can easily be disturbed.
25. Eluted fractions equivalent to 5×10^{10} *Salmonella* cells (or 25 mL of a stationary phase bacterial culture) were analyzed

by silver-staining after SDS-PAGE (16). The amount of associated proteins was sufficient to perform a first proteome analysis by liquid chromatography tandem mass spectrometry (LC-MS/MS) of the tagged InvR-MS2-protein complexes. A protocol for this and detailed list of proteins that specifically associate with InvR-MS2 are provided elsewhere (16).

26. Regardless of the enrichment, comparison of the flow-through fractions of the tagged vs. untagged RNAs showed that generally <20% of the MS2-tagged sRNAs bind to the column. It is possible that recovery of the tagged sRNA is influenced by the type of complex formed in vivo and that the 20% recovered represents a distinct subset of complexes in which the sRNA is a component. For example, the addition of RNaseE in the complex (29) may obscure the MS2 tag and prevent efficient recovery.

References

1. Papenfort K, Vogel J (2010) Regulatory RNA in bacterial pathogens. Cell Host Microbe 8:116–127
2. Waters LS, Storz G (2009) Regulatory RNAs in bacteria. Cell 136:615–628
3. Corcoran CP, Papenfort K, Vogel J (2011) In: Hess WR, Marchfelder A (eds) Regulatory RNAs in prokaryotes. Springer, New York, pp 15–50
4. Vogel J, Luisi BF (2011) Hfq and its constellation of RNA. Nat Rev Microbiol 9:578–589
5. Sittka A, Lucchini S, Papenfort K, Sharma CM, Rolle K, Binnewies TT, Hinton JC, Vogel J (2008) Deep sequencing analysis of small noncoding RNA and mRNA targets of the global post-transcriptional regulator. PLoS Genet 4:e1000163
6. Zhang A, Wassarman KM, Rosenow C, Tjaden BC, Storz G, Gottesman S (2003) Global analysis of small RNA and mRNA targets of Hfq. Mol Microbiol 50:1111–1124
7. Sharma CM, Vogel J (2009) Experimental approaches for the discovery and characterization of regulatory small RNA. Curr Opin Microbiol 12:536–546
8. Chao Y, Vogel J (2010) The role of Hfq in bacterial pathogens. Curr Opin Microbiol 13:24–33
9. Babitzke P, Romeo T (2007) CsrB sRNA family: sequestration of RNA-binding regulatory proteins. Curr Opin Microbiol 10:156–163
10. Wassarman KM (2007) 6S RNA: a small RNA regulator of transcription. Curr Opin Microbiol 10:164–168
11. Caputi M, Mayeda A, Krainer AR, Zahler AM (1999) hnRNP A/B proteins are required for inhibition of HIV-1 pre-mRNA splicing. EMBO J 18:4060–4067
12. Schnapp G, Rodi HP, Rettig WJ, Schnapp A, Damm K (1998) One-step affinity purification protocol for human telomerase. Nucleic Acids Res 26:3311–3313
13. Kurth I, Cristofari G, Lingner J (2008) An affinity oligonucleotide displacement strategy to purify ribonucleoprotein complexes applied to human telomerase. Methods Mol Biol 488:9–22
14. Wassarman KM, Storz G (2000) 6S RNA regulates *E. coli* RNA polymerase activity. Cell 101:613–623
15. Bunka DH, Stockley PG (2006) Aptamers come of age—at last. Nat Rev Microbiol 4:588–596
16. Said N, Rieder R, Hurwitz R, Deckert J, Urlaub H, Vogel J (2009) In vivo expression and purification of aptamer-tagged small RNA regulators. Nucleic Acids Res 37:e133
17. Witherell GW, Gott JM, Uhlenbeck OC (1991) Specific interaction between RNA phage coat proteins and RNA. Prog Nucleic Acid Res Mol Biol 40:185–220
18. Pfeiffer V, Sittka A, Tomer R, Tedin K, Brinkmann V, Vogel J (2007) A small non-coding RNA of the invasion gene island (SPI-1) represses outer membrane protein synthesis from the *Salmonella* core genome. Mol Microbiol 66:1174–1191
19. Urban JH, Vogel J (2007) Translational control and target recognition by *Escherichia coli* small RNAs in vivo. Nucleic Acids Res 35:1018–1037

20. Urban JH, Vogel J (2009) A green fluorescent protein (GFP)-based plasmid system to study post-transcriptional control of gene expression in vivo. Methods Mol Biol 540:301–319
21. Zhou Z, Reed R (eds) (2003) Purification of functional RNA-protein complexes using MS2-MBP. Wiley, New York
22. Sittka A, Sharma CM, Rolle K, Vogel J (2009) Deep sequencing of *Salmonella* RNA associated with heterologous Hfq proteins in vivo reveals small RNAs as a major target class and identifies RNA processing phenotypes. RNA Biol 6:266–275
23. Lee C (2007) Western blotting. Methods Mol Biol 362:391–399
24. Lopez-Gomollon S (2011) Detecting sRNAs by Northern blotting. Methods Mol Biol 732:25–38
25. Talbot SJ, Altman S (1994) Kinetic and thermodynamic analysis of RNA-protein interactions in the RNase P holoenzyme from *Escherichia coli*. Biochemistry 33:1406–1411
26. Lohman TM, Overman LB, Ferrari ME, Kozlov AG (1996) A highly salt-dependent enthalpy change for *Escherichia coli* SSB protein-nucleic acid binding due to ion-protein interactions. Biochemistry 35:5272–5279
27. Baumann C, Otridge J, Gollnick P (1996) Kinetic and thermodynamic analysis of the interaction between TRAP (trp RNA-binding attenuation protein) of *Bacillus subtilis* and trp leader RNA. J Biol Chem 271:12269–12274
28. Jurica MS, Licklider LJ, Gygi SR, Grigorieff N, Moore MJ (2002) Purification and characterization of native spliceosomes suitable for three-dimensional structural analysis. RNA 8:426–439
29. Aiba H (2007) Mechanism of RNA silencing by Hfq-binding small RNAs. Curr Opin Microbiol 10:134–139
30. Datsenko KA, Wanner BL (2000) One-step inactivation of chromosomal genes in *Escherichia coli* K-12 using PCR products. Proc Natl Acad Sci U S A 97:6640–6645
31. Lee J, Shin MK, Ryu DK, Kim S, Ryu WS (2010) Insertion and deletion mutagenesis by overlap extension PCR. Methods Mol Biol 634:137–146

Chapter 12

Gel Mobility Shift Assays to Detect Protein–RNA Interactions

Alexander V. Yakhnin, Helen Yakhnin, and Paul Babitzke

Abstract

The gel mobility shift assay is a powerful technique for detecting and quantifying protein–RNA interactions. While other techniques such as filter binding and isothermal titration calorimetry (ITC) are available for quantifying protein–RNA interactions, gel shift analysis provides the added advantage that you can visualize the protein–RNA complexes. In the gel shift assay, protein–RNA complexes are typically separated from the unbound RNA using native polyacrylamide gels in Tris/borate/EDTA buffer, although an alternative Tris-glycine buffering system is superior in many situations. Here, we describe both gel shift methods, along with strategies to improve separation of protein–RNA complexes from free RNA, which can be a particular challenge for small RNA binding proteins.

Key words: Protein–RNA interaction, Gel mobility shift assay, Electrophoretic mobility shift assay, EMSA, RNA binding protein, CsrA, TRAP

1. Introduction

Gel mobility shift (gel shift) assays, also referred to as electrophoretic mobility shift assays (EMSA), allow rapid detection and quantification of protein–RNA interactions (1–5). In vitro generated RNA is typically end-labeled at the 5′ end with [γ-^{32}P]ATP, while the protein is unlabeled. Following an incubation step to allow protein–RNA complex formation, complexes are separated from unbound (free) RNA by native (nondenaturing) polyacrylamide gel electrophoresis (PAGE). Visualization of bound RNA in the complex and free RNA is accomplished using a phosphorimager. The intensity of the bound and free RNA species are then quantified using commercially available software (e.g., ImageQuant from Molecular Dynamics). The fraction of bound RNA is plotted as a function of protein concentration to derive the apparent equilibrium binding constant (K_d), which is defined as the concentration

Kenneth C. Keiler (ed.), *Bacterial Regulatory RNA: Methods and Protocols*, Methods in Molecular Biology, vol. 905, DOI 10.1007/978-1-61779-949-5_12,

of protein in which 50% of the RNA is bound. Thus, the K_d value is a measure of the affinity that the protein has for its RNA binding target (3).

The use of a Tris/borate/EDTA (TBE) buffering system (1, 2, 6) has been widely used for gel shift analysis. While this buffering system works well for a variety of applications, we found that an alternative Tris-glycine buffering system, commonly used for separating proteins under denaturing conditions (7), offers superior results in many situations (3–5, 8–10). Thus, we recommend that both buffering systems be tested to determine which system works best for the particular application.

Another important consideration is the length of the RNA to be used in the analysis, to ensure that PAGE adequately resolves the bound and free RNA species. In general, the separation of bound and free RNA species improves as the length of RNA is decreased. This consideration is critical for small RNA binding proteins, such as CsrA of *Escherichia coli* (MW = 13.7 kDa), and less so for larger proteins such as TRAP of *Bacillus subtilis* (MW = 91.6 kDa). Gel shift analyses with these two proteins will be used as examples throughout this chapter.

2. Materials

It is critical that RNase-free conditions be maintained throughout the procedure. All glassware, metal spatulas, and stir bars used for preparing solutions should be baked in a 250°C oven for 4 h to destroy RNases. As diethylpyrocarbonate (DEPC) destroys RNases, all solutions for protein–RNA binding reactions should be prepared with DEPC-treated and autoclaved water (see Note 1). Prepare gel casting and running buffer solutions with distilled water.

2.1. RNA Binding by CsrA

1. DEPC-treated water: Add 1 mL of DEPC to 0.5 L of distilled water in a 1-L bottle. Mix, incubate overnight at room temperature, and autoclave for 30 min the next day. Store at room temperature.
2. 2 M Tris–HCl, pH 7.5: Weigh 24.2 g Tris base and transfer to a 125-mL glass beaker containing a stir bar. Add DEPC-treated water to a volume of ~75 mL. Mix on a magnetic stirrer and adjust the pH to 7.5 with HCl. Add DEPC-treated water to a final volume of 100 mL. Store at room temperature.
3. 2 M $MgCl_2$: Weigh 20.3 g magnesium chloride hexahydrate and transfer to a 50-mL plastic conical tube. Add DEPC-treated water to a final volume of 50 mL. Store at room temperature.

4. 4 M KCl: Weigh 29.8 g potassium chloride and transfer to 125-mL glass beaker. Add DEPC-treated water to a final volume of 100 mL. Store at room temperature.
5. 0.5 M EDTA, pH 8.0: Weigh 18.6 g EDTA, disodium salt, and transfer to a 125-mL glass beaker containing a stir bar. Add DEPC-treated water to a volume of ~75 mL. Mix on a magnetic stirrer and adjust the pH to 8.0 with 10 M sodium hydroxide. Add DEPC-treated water to a final volume of 100 mL. Store at room temperature.
6. TE buffer: 10 mM Tris–HCl, pH 7.5, 1 mM EDTA. Mix 0.5 mL of 2 M Tris–HCl, pH 7.5, 0.2 mL of 0.5 M EDTA, and 99.3 mL of DEPC-treated water.
7. 1 M Dithiothreitol (DTT): Dissolve 154 mg DTT in a 1.5-mL microcentrifuge tube with 1 mL of DEPC-treated water. Store at −20°C.
8. 50% Glycerol (v/v): Weigh 31.5 g glycerol (25 mL) and transfer to a 50-mL plastic conical tube. Add 25 mL DEPC-treated water and mix. Store at room temperature.
9. 5% Xylene cyanol: Dissolve 50 mg xylene cyanol in a 1.5-mL microcentrifuge tube with 1 mL DEPC-treated water. Store at −20°C.
10. 10 mg/mL yeast RNA: Dissolve 10 mg yeast RNA in a 1.5-mL microcentrifuge tube with 1 mL of DEPC-treated water. Store at −20°C (see Note 2).
11. 10× CsrA binding buffer: Mix 50 μL of 2 M Tris–HCl, pH 7.5 (100 mM final concentration), 50 μL of 2 M $MgCl_2$ (100 mM final concentration), 250 μL of 4 M KCl (1 M final concentration), and 650 μL of DEPC-treated water. Store at room temperature (see Note 3).
12. CsrA dilution buffer: Mix 5 μL of 2 M Tris–HCl, pH 7.5 (10 mM final concentration), 2 μL of 1 M DTT (2 mM final concentration), 200 μL 50% glycerol (10% final concentration), and 793 μL DEPC-treated water. Store at −20°C.
13. RNasin (Promega)
14. 5′ End-labeled RNA: (see Note 4).

2.2. RNA Binding by TRAP

1. DEPC-treated water: See Subheading 2.1, step 1.
2. 1 M Tris-acetate, pH 8.0: Weigh 12.1 g Tris base and transfer to a 125-mL glass beaker containing a stir bar. Add DEPC-treated water to a volume of ~75 mL. Mix on a magnetic stirrer and adjust the pH with acetic acid. Add DEPC-treated water to a final volume of 100 mL. Store at room temperature.
3. 2 M magnesium acetate: Weigh 21.4 g magnesium acetate tertrahydrate and transfer to a 50-mL plastic conical tube.

Add DEPC-treated water to a final volume of 50 mL. Store at room temperature.

4. 0.5 M EDTA, pH 8.0: See Subheading 2.1, step 5.
5. TE buffer: See Subheading 2.1, step 6.
6. 1 M DTT: See Subheading 2.1, step 7.
7. 50% Glycerol (v/v): See Subheading 2.1, step 8.
8. 5% Xylene cyanol: See Subheading 2.1, step 9.
9. 10 mg/mL *E. coli* tRNA: Dissolve 10 mg tRNA (*E. coli* MRE600) (Roche) in a 1.5-mL microcentrifuge tube with 1 mL of DEPC-treated water. Store at –20°C (see Note 2).
10. 20 mM L-tryptophan: Dissolve 41 mg L-tryptophan in a 15-mL plastic conical tube with 10 mL of DEPC-treated water. Store at –20°C.
11. 6× TRAP binding buffer: Mix 300 μL of 1 M Tris-acetate, pH 8.0 (300 mM final concentration), 12 μL of 2 M magnesium acetate (24 mM final concentration), 30 μL of 1 M DTT (30 mM final concentration), 120 μL of 10 mg/mL tRNA (1.2 mg/mL final concentration), 12 μL of 5% xylene cyanol (0.6 mg/mL final concentration), and 526 μL of DEPC-treated water. Store at –20°C (see Notes 3 and 5).
12. TRAP dilution buffer: Mix 10 μL of 1 M Tris-acetate, pH 8.0 (10 mM final concentration), 2 μL of 1 M DTT (2 mM final concentration), 200 μL 50% glycerol (10% final concentration), and 788 μL DEPC-treated water. Store at –20°C.
13. 5′ End-labeled RNA: (see Note 4).

2.3. Tris-Glycine Polyacrylamide Gel Components

1. Gel buffer: 1.5 M Tris–HCl, pH 8.8. Weigh 36.3 g Tris base and transfer to a 250-mL glass beaker containing a stir bar. Add water to a volume of ~150 mL. Mix on a magnetic stirrer and adjust pH with HCl. Add water to a final volume of 200 mL. Store at 4°C.
2. 5× Tris-glycine gel running buffer: 135 mM Tris, 960 mM glycine, 5 mM EDTA, pH 8.3. Weigh 16.3 g Tris base, 72 g glycine, and 1.8 g EDTA, disodium salt. Transfer to a 1-L glass beaker containing a stir bar. Add ~600 mL water. Mix on a magnetic stirrer. Add water to a final volume of 1 L. The pH does not require adjustment. Store at room temperature.
3. 1× Tris-glycine gel running buffer: Mix 200 mL of 5× Tris-glycine gel running buffer with 800 mL of water. Store at room temperature.
4. 40% Acrylamide:bisacrylamide, 37.5:1 solution. Store at 4°C.
5. Ammonium persulfate, 10% solution in water. Store at –20°C.

6. *N*, *N*, *N*′, *N*′-tetramethylethylenediamine (TEMED). Store at 4°C.
7. Fixing solution, 30% methanol and 10% acetic acid: Mix 0.3 L of methanol, 0.1 L of glacial acetic acid, and 0.6 L of water. Store at room temperature.

2.4. TBE Polyacrylamide Gel Components

1. 5× TBE buffer: 465 mM Tris, 445 mM boric acid, 10 mM EDTA, pH 8.3. Weigh 56.3 g of Tris base, 27.6 g of boric acid, and 3.7 g of EDTA, disodium salt. Transfer to a 1-L glass beaker containing a stir bar. Add water to a volume of ~600 mL. Mix on a magnetic stirrer. Add water to a final volume of 1 L. pH does not require any adjustment. Store at room temperature.
2. 0.5× TBE buffer: 47 mM Tris, 45 mM boric acid, 1 mM EDTA, pH 8.3. Mix 100 mL of 5× TBE buffer with 900 mL of water. Store at room temperature.
3. 40% Acrylamide:bisacrylamide solution: See Subheading 2.3, step 4.
4. Ammonium persulfate (10%): See Subheading 2.3, step 5.
5. TEMED: See Subheading 2.3, step 6.
6. Fixing solution: See Subheading 2.3, step 7.

3. Methods

3.1. Gel Casting for Tris-Glycine System

1. 6% Polyacrylamide gel: Mix 1.8 mL of 40% acrylamide: bisacrylamide, 1.2 mL of 50% glycerol, 2.4 mL of 1.5 M Tris–HCl (pH 8.8), 6.6 mL of water, 12 μL of TEMED, and 120 μL of 10% ammonium persulfate. This mixture is sufficient for two gels. Cast gels using 8 × 10-cm gel plates with 0.5-mm spacers. Insert a 10-well comb in each gel. Allow the gels to polymerize for 30 min. Transfer gels into a vertical gel running apparatus and fill the buffer reservoirs with 1× Tris-glycine gel running buffer (see Notes 6 and 7).
2. 10% Polyacrylamide gel: Mix 3.0 mL of 40% acrylamide: bisacrylamide, 1.2 mL of 50% glycerol, 2.4 mL of 1.5 M Tris–HCl (pH 8.8), 5.4 mL of water, 12 μL of TEMED, and 120 μL of 10% ammonium persulfate. This mixture is sufficient for two gels. Cast gels using 8 × 10-cm gel plates with 0.5-mm spacers. Insert a 10-well comb in each gel. Allow the gels to polymerize for 30 min. Transfer gels into a vertical gel running apparatus and fill the buffer reservoirs with 1× Tris-glycine gel running buffer.

3.2. Gel Casting for TBE System

1. 6% Polyacrylamide gel: Mix 1.8 mL of 40% acrylamide: bisacrylamide, 0.6 mL of 50% glycerol, 1.2 mL of 5× TBE buffer (0.5× final concentration), 8.4 mL of water, 12 μL of TEMED, and 120 μL of 10% ammonium persulfate. This mixture is sufficient for two gels. Cast gels using 8×10-cm gel plates with 0.5-mm spacers. Insert a 10-well comb in each gel. Allow the gels to polymerize for 30 min. Transfer gels into a vertical gel running apparatus and fill the buffer reservoirs with 0.5× TBE buffer (see Notes 6 and 7).
2. 10% Polyacrylamide gel: Mix 3 mL of 40% acrylamide: bisacrylamide, 0.6 mL of 50% glycerol, 1.2 mL of 5× TBE buffer (0.5× final concentration), 7.2 mL of water, 12 μL of TEMED, and 120 μL of 10% ammonium persulfate. This mixture is sufficient for two gels. Cast gels using 8×10-cm gel plates with 0.5-mm spacers. Insert a 10-well comb in each gel. Allow the gels to polymerize for 30 min. Transfer gels into a vertical gel running apparatus and fill the buffer reservoirs with 0.5× TBE buffer.

3.3. CsrA-RNA Gel Shift Analysis (Fig. 1)

1. Make up to 8 twofold serial protein dilutions in CsrA dilution buffer. The concentration of RNA should be at least tenfold lower than the lowest protein concentration used in the binding reaction.

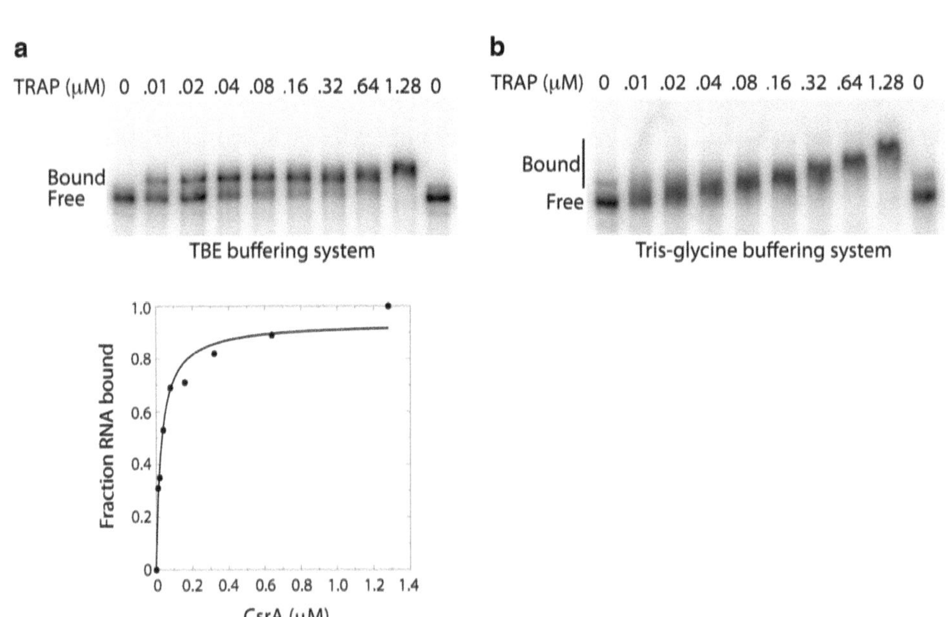

Fig. 1. Gel mobility shift analysis of CsrA-*csrA* RNA interaction. 5′ end-labeled *csrA* RNA (0.1 nM) was incubated with the concentration of CsrA shown at the top of each lane. Samples were loaded onto 10% polyacrylamide gels with a 37.5:1 acrylamide:bisacrylamide ratio. Positions of bound and free RNA are shown. (**a**) TBE buffering system. (**b**) Tris-glycine buffering system. The TBE buffering system gave superior results in this particular case. The simple binding curve for these data is shown below the gel. The K_d for this reaction was calculated to be 30 nM CsrA.

2. Label ten Eppendorf tubes (0.65 mL) 1 through 10. Add 2 μL of protein dilution buffer to tubes 1 and 10. Add 2 μL of each protein dilution into tubes 2 through 9. As the bound and free RNA species run close to one another with the small CsrA protein, it is helpful to have unbound controls (no CsrA) in the first and last lanes of the gel.
3. Dilute 5′ end-labeled RNA (gel purified) to 1 nM in TE and heat to 85°C for 3 min. Slow cool for 10 min at room temperature (see Note 8).
4. Prepare binding mixture (80 μL). Mix 10 μL of 10× CsrA binding buffer, 10 μL of 1 nM RNA (0.1 nM final concentration), 15 μL of 50% glycerol, 2 μL 1 M DTT, 1 μL of RNasin, 2 μL of 10 μg/μL yeast tRNA, 10 μL 5% xylene cyanol, and 30 μL of DEPC-treated water. Mix well. Add 8 μL to each tube containing protein dilution buffer or protein (see Notes 5 and 9).
5. Incubate tubes at 37°C for 30 min in a heat block and then immediately load 2–5 μL aliquots onto a polyacrylamide gel (see Note 10).
6. Gel running for TBE system: Pre-run the gel using 0.5× TBE for 10–15 min at 200 V (constant). Wash the wells with gel running buffer using a syringe and needle. Load the gel and run for 1–2 h at 200 V (constant).
7. Gel running for Tris-glycine system: Do not pre-run the gel. Wash the wells with 1× Tris-glycine gel running buffer using a syringe and needle. Load the gel and run at 12 mA (constant) and 150 V (maximum) for 2–3 h (see Note 11).
8. Transfer the gel to Whatman paper, cover with plastic wrap and dry in a gel dryer. Place the dried gel overnight in a Phosphorimager cassette. Scan and quantify the bound and free RNA species using ImageQuant software (Molecular Dynamics) (see Notes 12 and 13).
9. Fit the binding data to simple and cooperative binding equations (see Note 14).

3.4. TRAP-RNA Gel Shift Analysis (Fig. 2)

1. Make up to 9 twofold serial dilutions of protein in TRAP dilution buffer. The concentration of RNA should be at least tenfold lower than the lowest protein concentration used in the binding reaction.
2. Label ten Eppendorf tubes (0.65 mL) 1 through 10. Add 2 μL of protein dilution buffer to tube 1. Add 2 μL of each protein dilution to tubes 2 through 10.
3. Dilute 5′ end-labeled *B. subtilis trp* leader RNA (gel purified) to 1 nM in TE and heat to 85°C for 3 min. Cool slowly for 10 min at room temperature (see Note 15).

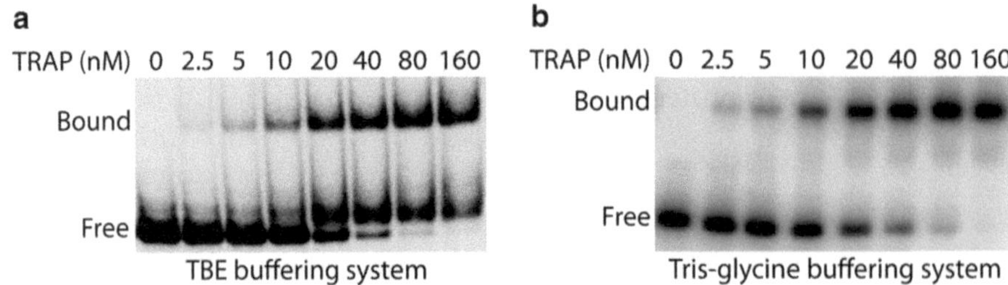

Fig. 2. Gel mobility shift analysis of TRAP-*trp* leader RNA interaction. 5′ end-labeled *trp* leader RNA (0.1 nM) was incubated with the concentration of TRAP shown at the top of each lane. Samples were loaded onto 6% polyacrylamide gels with a 37.5:1 acrylamide:bisacyrylamide ratio. Positions of bound and free RNA are shown. (**a**) TBE buffering system. (**b**) Tris-glycine buffering system. The Tris-glycine buffering system gave superior results in this particular case. The K_d for this reaction was calculated to be 17 nM TRAP.

4. Prepare binding mixture (80 μL). Mix 16.7 μL of 6× TRAP binding buffer, 10 μL of 1 nM RNA (0.1 nM final concentration), 5 μL of 20 mM L-tryptophan, and 48.3 μL of DEPC-treated water. Mix well. Add 8 μL to each tube containing protein dilution buffer or protein (see Note 16).
5. Incubation: Same as for CsrA-RNA interaction (see Note 10). See Subheading 3.3, step 5.
6. Gel running for TBE system: Same as for CsrA-RNA interaction. See Subheading 3.3, step 6.
7. Gel running for Tris-glycine system: Same as for CsrA-RNA interaction (see Note 11). See Subheading 3.3, step 7.
8. Gel handling and quantification (see Notes 12 and 13). See Subheading 3.3, step 8.
9. Fit the binding data to simple and cooperative binding equations (see Note 14).

4. Notes

1. Establishing and maintaining RNase-free conditions is a critical consideration. Gloves should be worn at all times to prevent the introduction of RNases from your fingers. Do not use pipeters that have been used for procedures containing RNase A (e.g., plasmid miniprep procedures). While DEPC inhibits RNases, it is critical to ensure that DEPC is destroyed by autoclaving. It is recommended that chemicals and solutions used for RNA work be used exclusively for that purpose.
2. A nonspecific RNA competitor must be included in the binding reaction in large excess over labeled RNA to prevent nonspecific protein–RNA interactions. Although tRNA is

routinely used as a competitor, a pilot experiment must be performed to determine whether tRNA is suitable for any given protein. For example, tRNA interferes with CsrA binding to its targets. Instead, total yeast RNA was found to be an effective nonspecific competitor for CsrA-RNA binding studies.

3. Binding conditions, such as pH and the concentration of monovalent and divalent salts, must be determined empirically.
4. PCR products containing a T7 RNA polymerase promoter are used as DNA templates. RNA is synthesized in vitro with a commercially available transcription kit by following the manufacturer's protocol. Gel-purified RNA is dephosphorylated with calf intestinal alkaline phosphatase and subsequently 5′-end labeled with (γ-^{32}P)-ATP and T4 polynucleotide kinase (4). Labeled RNAs are gel purified and suspended in TE buffer.
5. Inclusion of glycerol and xylene cyanol in the binding reaction simplifies gel loading. Glycerol also protects diluted proteins from denaturation. Both CsrA and TRAP remain active in their binding buffers at low concentrations. However, inclusion of acetylated BSA at 100 μg/mL (final concentration) to binding reactions may be required for less stable proteins.
6. The concentration of the polyacrylamide gel is an important consideration for gel shift analysis. The size of the protein and the RNA length must be taken into consideration to obtain adequate resolution of bound and free RNA species. In general, separation of these species is not a problem for a large protein such as TRAP (MW = 91.6 kDa). In this case, we typically use 100–200 nucleotide transcripts and obtain good separation in 6% gels. However, adequate separation with small RNA binding proteins such as CsrA (MW = 13.7 kDa) is more challenging. In general, the separation with CsrA improves as the transcript size is decreased (e.g., 100–16 nucleotide). However, care must be taken to ensure that unbound (free) short RNAs do not run too close to the gel front. Thus, for small RNA binding proteins we suggest trying a variety of RNA lengths and/or acrylamide concentrations (e.g., 8, 10 and 15%) in gels. Resolution may also be improved using different acrylamide:bisacrylamide ratios (e.g., 37.5:1 vs. 19:1).
7. We use the Hoefer SE 250 Mighty Small II 10 × 8 cm electrophoresis units. Other similar vertical slab gel systems can be substituted.
8. CsrA binds to hundreds of *E. coli* mRNAs (6, 8, 10, 11).
9. As CsrA and TRAP do not contain cysteine residues, reducing agents such as DTT are not required. However, DTT is required to maintain activity of RNase inhibitor (RNasin). Therefore, DTT should be included in the binding reaction if RNasin is used. RNasin may be omitted if proper RNase-free conditions are maintained.

10. The length of time required to establish an equilibrium in which association and dissociation of the protein–RNA complex are balanced must be determined empirically. A time course of binding (e.g., 10, 30, 60 min) should be performed using the same protein dilution. Equilibrium is obtained when the fraction bound no longer increases with time.
11. Each protein–RNA pair requires appropriate pH and salt concentrations for interaction. Loading of large reaction volumes (more than 10 μL) containing salt concentrations above 150 mM may disturb gel running and/or the shape of the bound and free RNA bands. In general, gels run in TBE buffer are more sensitive to salt perturbation than gels run in Tris-glycine buffer. Loading smaller volumes of binding reactions containing lower salt concentrations provides better separation and sharper bands.
12. High percentage polyacrylamide gels do not stick well to Whatman paper. Gels ≤10% will stick to Whatman paper. Be careful so as not to tear or distort the shape of the gel during this process.
13. Fractionation of the binding reaction in native gels may result in dissociation of protein–RNA complexes during gel running. The newly released RNA will run between the free and bound species, sometimes running as a smear. Because this newly released RNA was part of a protein–RNA complex at the time of gel loading, it should be considered as bound RNA for calculation of the apparent K_d value.
14. Calculation of the apparent equilibrium binding constant (K_d) (3, 12): provided that the concentration of RNA used in the binding reaction was at least tenfold lower than the lowest protein concentration used, you can assume that the concentration of free protein does not change during the binding reaction. In this case the binding data can be fit to the simple binding equation:

$$f = A \times P_0 / (K_d + P_0),$$

where f is the fraction of RNA bound, A is the maximal fraction of RNA bound, P_0 is the concentration of total protein, and K_d is the equilibrium binding constant. If you are unable to use an RNA concentration that is at least tenfold lower than the lowest protein concentration, a more complicated quadratic equation must be used to determine K_d (13). To determine whether the binding reaction is cooperative, it is advisable to also fit the data to the Hill equation:

$$f = (A \times P_0 / S_{0.5})^n / (1 + P_0 / S_{0.5})^n,$$

where f is the fraction of RNA bound, A is the maximal fraction of RNA bound, P_0 is the concentration of total protein, $S_{0.5}$ is the equilibrium binding constant, and *n* is the cooperativity coefficient.

15. TRAP binds to 11 (G/U)AG repeats in the *B. subtilis trp* leader (3).
16. L-tryptophan activates TRAP such that it can bind to the *B. subtilis trp* leader (3).

Acknowledgement

This work was supported by NIH grant GM059969.

References

1. Fried M, Crothers DM (1981) Equilibria and kinetics of *lac* repressor-operator interactions by polyacrylamide gel electrophoresis. Nucleic Acids Res 9:6505–6525
2. Garner MM, Revzin A (1981) A gel electrophoresis method for quantifying the binding of proteins to specific DNA regions: application to components of the *Escherichia coli* lactose operon regulatory system. Nucleic Acids Res 9:3047–3060
3. Yakhnin AV, Trimble JJ, Chiaro CR, Babitzke P (2000) Effects of mutations in the L-tryptophan binding pocket of the Trp RNA-binding attenuation protein of *Bacillus subtilis*. J Biol Chem 275:4519–4524
4. Dubey AK, Baker CS, Romeo T, Babitzke P (2005) RNA sequence and secondary structure participate in high-affinity CsrA-RNA interaction. RNA 11:1579–1587
5. Yakhnin H, Pandit P, Petty TJ, Baker CS, Romeo T, Babitzke P (2007) CsrA of *Bacillus subtilis* regulates translation initiation of the gene encoding the flagellin protein (*hag*) by blocking ribosome binding. Mol Microbiol 64:1605–1620
6. Yakhnin H, Yakhnin AV, Baker CS, Sineva E, Berezin I, Romeo T, Babitzke P (2011) Complex regulation of the global regulatory gene *csrA*: CsrA-mediated translational repression, transcription from five promoters by $E\sigma^{70}$ and $E\sigma^{S}$, and indirect transcriptional activation by CsrA. Mol Microbiol 81:689–704
7. Laemmli UK (1970) Cleavage of structural proteins during the assembly of the head of bacteriophage T4. Nature 227:680–685
8. Baker CS, Morozov I, Suzuki K, Romeo T, Babitzke P (2002) CsrA regulates glycogen biosynthesis by preventing translation of *glgC* in *Escherichia coli*. Mol Microbiol 44:1599–1610
9. Yakhnin H, Yakhnin AV, Babitzke P (2006) The *trp* RNA-binding attenuation protein (TRAP) of *Bacillus subtilis* regulates translation initiation of *ycbK*, a gene encoding a putative efflux protein, by blocking ribosome binding. Mol Microbiol 61:1252–1266
10. Baker CS, Eöry LA, Yakhnin H, Mercante J, Romeo T, Babitzke P (2007) CsrA inhibits translation initiation of *Escherichia coli hfq* by binding to a single site overlapping the Shine-Dalgarno sequence. J Bacteriol 189:5472–5481
11. Edwards AN, Patterson-Fortin LM, Vakulskas CA, Mercante JW, Potrykus K, Vinella D, Camacho MI, Fields JA, Thompson SA, Georgellis D, Cashel M, Babitzke P, Romeo T (2011) Circuitry linking the Csr and stringent response global regulatory systems. Mol Microbiol 80:1561–1580
12. Yang M, Chen X, Militello K, Hoffman R, Fernandez B, Baumann C, Gollnick P (1997) Alanine-scanning mutagenesis of *Bacillus subtilis trp* RNA-binding attenuation protein (TRAP) reveals residues involved in tryptophan binding and RNA binding. J Mol Biol 270:696–710
13. Schaak JE, Yakhnin H, Bevilacqua PC, Babitzke P (2003) A Mg^{2+}-dependent RNA tertiary structure forms in the *Bacillus subtilis trp* operon leader transcript and appears to interfere with *trpE* translation control by inhibiting TRAP binding. J Mol Biol 332:555–574

Chapter 13

RNase Footprinting of Protein Binding Sites on an mRNA Target of Small RNAs

Yi Peng, Toby J. Soper, and Sarah A. Woodson

Abstract

Endoribonuclease footprinting is an important technique for probing RNA–protein interactions with single nucleotide resolution. The susceptibility of RNA residues to enzymatic digestion gives information about the RNA secondary structure, the location of protein binding sites, and the effects of protein binding on the RNA structure. Here we present a detailed protocol for using RNase T2, which cleaves single stranded RNA with a preference for A nucleotides, to footprint the protein Hfq on the *rpoS* mRNA leader. This protocol covers how to form the RNP complex, determine the correct dose of enzyme, footprint the protein, and analyze the cleavage pattern using primer extension.

Key words: Footprinting, Primer extension, RNA, RNA protein interactions, RNase T2, Sequencing gel

1. Introduction

In bacteria, noncoding small RNAs (sRNAs) play a critical role in posttranscriptional gene regulation (1). These sRNAs bind target mRNAs directly to either activate or repress translation. The major class of sRNAs requires the ubiquitous and abundant RNA binding protein Hfq as a cofactor for regulation (2). In the best studied cases, such as OxyS inhibition of *fhlA* (3, 4), Spot42 inhibition of *galK* (5), and DsrA activation of *rpoS* (6, 7) it is clear that Hfq binding to both the sRNA and its mRNA target is important for correct regulation. Unlocking the mechanism behind the formation and function of these RNPs is a growing area of research.

"Footprinting" methods are important tools for probing the structures and composition of RNPs (8), including those containing sRNAs or their mRNA targets (9–11). These methods assay the ability of a specific protein or cellular extract to protect an RNA

Kenneth C. Keiler (ed.), *Bacterial Regulatory RNA: Methods and Protocols*, Methods in Molecular Biology, vol. 905,
DOI 10.1007/978-1-61779-949-5_13, © Springer Science+Business Media, LLC 2012

of interest from cleavage or modification by chemicals or endonucleases (12). An important advantage of footprinting methods is that they can have single nucleotide resolution, allowing for a very accurate determination of where proteins are binding and how that binding affects the RNA.

A variety of endonucleases with different cleavage specificities have been widely used to probe the structure and protein binding properties of RNAs (13). Different endonucleases have different nucleotide specificities, and probing all the residues in an RNA of interest requires using combinations of multiple enzymes. Commonly used enzymes for structure probing include RNase A, which cleaves 3′ of single stranded U and C nucleotides; RNase T1, which cleaves 3′ of single stranded G nucleotides; RNase T2, which cleaves 3′ of all single stranded nucleotides with a preference for A nucleotides; and RNase V1, which, in contrast to the enzymes above, cleaves double-stranded RNA (13). As these enzymes are much larger than chemical footprinting reagents such as dimethylsulfate or hydroxyl radical, their cleavage pattern is less affected by dynamic "breathing" of the RNA structure.

Here we describe nuclease footprinting of the Hfq protein on the *rpoS* mRNA leader, which is a regulatory target of the sRNA DsrA (14, 15). RNase T2 was chosen for this example because Hfq has a preference for single stranded A and U rich RNA, and the putative Hfq binding site in the *rpoS* leader is an $(AAN)_4$ repeat element (6). However, the protocol can be easily adapted for other ribonucleases.

For RNAs around 100 nucleotides or less, it is common to footprint 5′-end labeled RNAs and resolve the cleavage products directly with a polyacrylamide sequencing gel. An example of this approach is the use of RNase T1 and RNase T2 to footprint the sRNA DsrA (87 nucleotides) and a short (140 nucleotide) section of the *rpoS* mRNA leader (10). Because of the length of the *rpoS* RNA we used in the following protocol (324 nucleotides), we analyzed the RNase T2 cleavage pattern by extension of a ^{32}P-labeled primer to generate cDNAs that were then resolved on a sequencing gel (Fig. 1). The methods we present for 5′-end labeling, determining the RNase T2 dose response, and footprinting of the Hfq–*rpoS* RNP complex are suited for direct analysis of radiolabeled RNA or indirect analysis of the cleavage pattern by primer extension.

The primer extension protocol we present here uses ^{32}P-labeled primers and resolves the cDNAs with a traditional polyacrylamide sequencing gel. However, by using fluorescent dye-labeled primers, the cDNAs may be analyzed with a DNA sequencer (16, 17). The DNA sequencer provides longer reads (≥400 bases) and more automated quantitation of footprinting products.

Correct dosing is vital for RNase footprinting experiments. Useful data requires that each RNA that is cleaved by RNase be cleaved no more than once, but that there are sufficient cleavage events for robust detection. These conditions are met when 80–85%

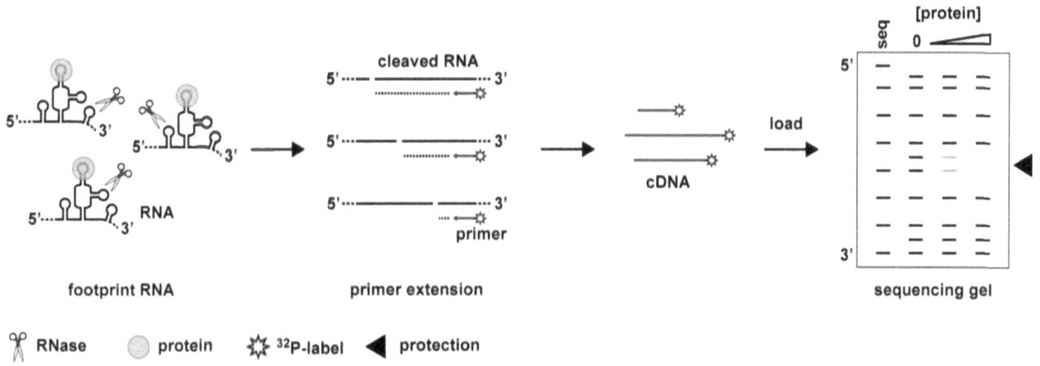

Fig. 1. Protocol for RNase footprinting analyzed by primer extension. The accessible single stranded regions of assembled RNPs are exposed to RNase under single-cleavage conditions. cDNAs are generated by primer extension of the RNase-treated RNA using ^{32}P-labeled primers. The cDNAs are resolved on a polyacrylamide sequencing gel and regions of protection from RNase cleavage are identified.

of the RNA remains undigested (18). Therefore, the first step in an endoribonuclease footprinting experiment is establishing the dose-response for the particular RNA and RNase being used.

Here we describe methods to label primers and use primer extension to generate cDNA for analysis on sequencing gels, how to generate a dose-response curve to determine the optimal concentration of RNase T2, and how to footprint a protein on an RNA with RNase T2.

2. Materials

All solutions should use RNase-free deionized water purified to 18 MΩ of resistivity.

2.1. RNP Footprinting Components

1. 5× Tris sodium potassium (TNK) RNA folding buffer: 50 mM Tris–HCl pH 7.5, 250 mM NaCl, 250 mM KCl.
2. Protein storage buffer: 50 mM Tris–HCl pH 7.5, 250 mM NH_4Cl, 1 mM EDTA, 10% glycerol (v/v).
3. Tris EDTA (TE) buffer: 10 mM Tris–HCl pH 7.5, 1 mM EDTA.
4. 50 μM Hfq_6 protein in storage buffer (see Note 1).
5. 1.25 μM in vitro transcribed *rpoS* RNA in TE (see Note 2).
6. RNase T2 (see Note 3).
7. 0.05 mg/mL yeast carrier tRNA.
8. Tris buffered phenol.
9. 24:1 Chloroform:isoamyl alcohol.
10. 3 M Sodium acetate pH 5.3.

2.2. Primer Labeling Components

1. 10× Polynucleotide kinase buffer: 700 mM Tris–HCl pH 7.5, 100 mM $MgCl_2$, 50 mM DTT (see Note 4).
2. 3.3 μM gel-purified custom DNA primer (see Note 5).
3. γ ^{32}P-ATP.
4. T4 Polynucleotide kinase.

2.3. Primer Extension Components

1. Superscript III first strand buffer (provided with enzyme).
2. 0.1 mM DDT (provided with enzyme).
3. 10 mM dNTP mix (10 mM of each dNTP in H_2O).
4. Individual 2.5 mM ddNTP solutions (in H_2O).
5. 200 nM ^{32}P-labeled primer.
6. 1.25 μM *rpoS* RNA for sequencing.
7. RNase T2 treated *rpoS* RNA samples.
8. Superscript III reverse transcriptase (Invitrogen).

2.4. Polyacrylamide Sequencing Gel Components

1. 10× Tris borate EDTA (TBE) buffer: 1 M Tris, 830 mM boric acid, 10 mM EDTA.
2. 40% 29:1 acrylamide:bisacrylamide solution: mix 116 g of acrylamide powder, 4 g bisacrylamide powder and bring the volume up to 300 mL with RNase-free water (see Note 6).
3. Urea pellets.
4. 45 cm × 35 cm glass plate, 43 cm × 35 cm offset glass plate, 0.2 mm spacers, comb.
5. Medium binder clips (see Note 7).
6. 10% APS.
7. TEMED.
8. 2× Formamide/TBE dye: 9.50 mL deionized formamide, 400 μL 10× TBE, 50 μL 2% bromophenol blue (BP) dye, 50 μL 2% xylene cyanol (XC) dye (see Note 8).

3. Methods

All reactions should use RNase-free deionized water purified to 18 MΩ of resistivity.

3.1. Primer Labeling

1. Mix 3 μL of 3.3 μM gel-purified primer, 1 μL of 10× PNK buffer, and 5 μL of γ ^{32}P-ATP in a 1.5 mL microcentrifuge tube. Vortex and centrifuge briefly (see Note 9).
2. Add 1 μL (10 U) of T4 polynucleotide kinase. Mix by gentle tapping and centrifuge briefly.

3. Incubate 30 min at 37°C.
4. Incubate reaction for 10 min at 65°C to inactivate the enzyme. Add 40 μL of H_2O. Vortex and centrifuge briefly.
5. Purify with a size exclusion column such as a TE-10 column (Clontech).
6. Store labeled primer at −20°C.

3.2. RNase T2 Dose Response

1. For each reaction, mix 2 μL of TNK buffer, 2 μL of protein storage buffer, 1 μL of TE, and 1 μL of H_2O in a 1.5 mL microcentrifuge tube.
2. Renature 1.25 μM *rpoS* transcript by heating at 80°C for 1 min followed by 5 min at room temperature.
3. Add 2 μL of renatured 1.25 μM *rpoS* RNA to salt mix from step 1. Incubate RNA 5–10 min at room temperature (see Note 10).
4. Create serial dilutions of RNase T2. It is best to start with a decadal series, for instance: 0.0, 0.005, 0.05, 0.5, and 5 U/μL RNase T2. Keep the RNase dilutions on ice (see Note 11).
5. Add 2 μL of each RNase T2 dilution to a reaction tube from step 3. Incubate at room temperature for 1 min.
6. Quench the reactions by adding 10 μL of buffered phenol, and immediately vortexing vigorously (see Note 12).
7. For each reaction, add 189 μL of H_2O, 1 μL of 0.05 mg/mL yeast carrier tRNA, and 190 μL of phenol to bring the volume to a level suitable for extraction. Vortex vigorously and centrifuge at maximum *g* for 5 min to separate layers (see Note 13).
8. For each reaction, transfer the aqueous layer to a clean 1.5 mL microcentrifuge tube and add 200 μL of 24:1 chloroform: isoamyl alcohol. Vortex vigorously and centrifuge as in step 7.
9. For each reaction, transfer aqueous layer to a clean 1.5 mL microcentrifuge tube and precipitate the RNA with 20 μL of 3 M sodium acetate and 600 μL of cold 100% ethanol, overnight at −20°C (see Note 14).
10. Centrifuge the precipitated RNA from step 9 in a microcentrifuge at maximum *g* for 30 min at 4°C. Decant the supernatant.
11. Wash the pellets with 500 μL of cold 70% ethanol and centrifuge at maximum *g* again for 5 min. Decant the supernatant.
12. Dry the pellets under vacuum (e.g., SpeedVac Concentrator or equivalent), then resuspend the pellets in 10 μL of H_2O.
13. Perform primer extension reactions on 2 μL from each nuclease digestion reaction, and run the cDNAs on a sequencing gel according to the protocols in Subheadings 3.3 and 3.4.

3.3. Primer Extension

Primer extension reactions should be assembled on ice.

1. Make a sequencing cocktail by mixing 10 μL of γ ^{32}P-labeled primer from Subheading 3.1, 5 μL of 10 μM dNTP mix, 2 μL of 1.25 μM *rpoS* RNA, and 29 μL of H_2O. This cocktail will go into four sequencing reactions and a pausing control, each containing 0.5 pmol RNA (see Note 15).
2. Add 3.8 μL of each ddNTP to a separate 0.5 mL thick walled PCR tube. Add 3.8 μL of H_2O to a fifth tube for the pausing control (see Note 16).
3. Add 9.2 μL of sequencing cocktail to each sequencing tube and the pausing control. Vortex and centrifuge briefly.
4. Make a reaction cocktail by mixing 2 μL of γ ^{32}P-labeled primer, 1.5 μL of 10 mM dNTP mix, and 7.5 μL of H_2O for each sample to be reverse transcribed.
5. Add 2 μL of experimentally treated RNA from either Subheading 3.2 or 3.6 to a 0.5 mL thick-walled PCR tube.
6. Add 11 μL of reaction cocktail to each tube from step 5. Mix by vortexing and centrifuge briefly.
7. Anneal the sequencing and experimental reactions from steps 3 and 6 by incubating at 65°C for 5 min and then returning to ice.
8. Make a reverse transcription (RT) cocktail by mixing 4 μL of 5× first strand buffer, 1 μL of 0.1 M DTT, 1.75 μL of H_2O, and 0.25 μL of Superscript III reverse transcriptase for each primer extension reaction to be performed (total number of sequencing and experimental reactions, including the pausing control).
9. Add 7 μL of RT cocktail to each reaction. Mix by tapping and centrifuge briefly.
10. Incubate reactions at 55°C for 30 min.
11. Precipitate the reactions overnight at −20°C with 2 μL of 3 M sodium acetate and 60 μL of cold 100% ethanol.
12. Centrifuge, wash, and dry the precipitated cDNA as in steps 10–12 in Subheading 3.2 (see Note 17).

3.4. Run Sequencing Gel

1. Make 100 mL of 8% acrylamide gel solution by mixing 48 g of urea pellets, 10 mL of 10× TBE, and 20 mL of 40% 29:1 acrylamide:bisacrylamide. Bring up to 100 mL with H_2O, dissolve the urea by incubating at 65°C, then degas the solution for 10 min in a side-arm flask using an aspirator (see Note 18).
2. Add 350 μL of 10% APS and 35 μL of TEMED to the gel solution, mix gently, and use a 60 mL syringe to inject it between clamped sequencing gel plates separated by 0.2 mm spacers. Insert comb (see Note 19).

3. Mount the sequencing gel on a gel electrophoresis apparatus and add 1 L of 1× TBE running buffer. Pre-run the gel at 55 W for 30 min.
4. Resuspend the cDNA from Subheading 3.3, step 12 in 5 μL of formamide/TBE dye. Vortex and centrifuge briefly.
5. Heat cDNA samples at 95°C for 30 s then load them onto the 8% sequencing gel.
6. Run the gel at 55 W until the bromophenol blue is at the bottom of the gel (about 2 h).
7. Separate the plates and pull the gel off the plates with piece of 3 mm Whatman filter paper cut to size (see Note 20).
8. Cover the gel with plastic wrap and dry it under vacuum.
9. Expose the gel to a phosphor screen, and then scan it with a phosphorimager (see Note 21).

3.5. Dose-Response Data Analysis

1. After scanning the gel, use image analysis software such as ImageQuant (Molecular Dynamics) or NIH Image J (19) to quantify the intensity (pixel volume) of the full-length primer extension product (FL). Separately quantify the volume of all other bands in the lane, but excluding full-length product and the unextended primer. These bands are cDNA fragments derived from RNase T2 cleavage products (P) (Fig. 2a).
2. For each lane, calculate the fraction full-length cDNA with the following formula: FL/(FL+P). This is taken to represent the amount of uncleaved RNA.
3. Divide the fraction of full-length cDNA for each RNase T2 concentration by the fraction of full-length cDNA in the zero T2 lane to obtain the normalized fraction of full-length cDNA.
4. Plot the logarithm of the normalized fraction full-length against RNase T2 concentration (a semi-log plot; Fig. 2b). Fitting the data with an exponential decay function will give a straight line. Use the line to determine the nuclease concentration that gives ~80% full-length (uncleaved) RNA. For example, in Fig. 1b, this would correspond to 3 U/μL RNase T2. Use this concentration in future experiments.

3.6. Footprinting Hfq Protein on the rpoS mRNA

1. Make an Hfq hexamer dilution series consisting of: 50, 25, 15, 5, and 0.5 μM Hfq_6 in protein storage buffer. Add 2 μL of each dilution or 2 μL of buffer to seven separate 1.5 ml microcentrifuge tubes (one no RNase T2 control, and six reactions containing 0, 0.1, 1, 3, 5, and 10 μM Hfq that will be digested with RNase T2). Keep protein stocks and dilutions on ice.
2. Assemble a cocktail of *rpoS* RNA as in Subheading 3.2, steps 1–3, omitting the protein storage buffer.

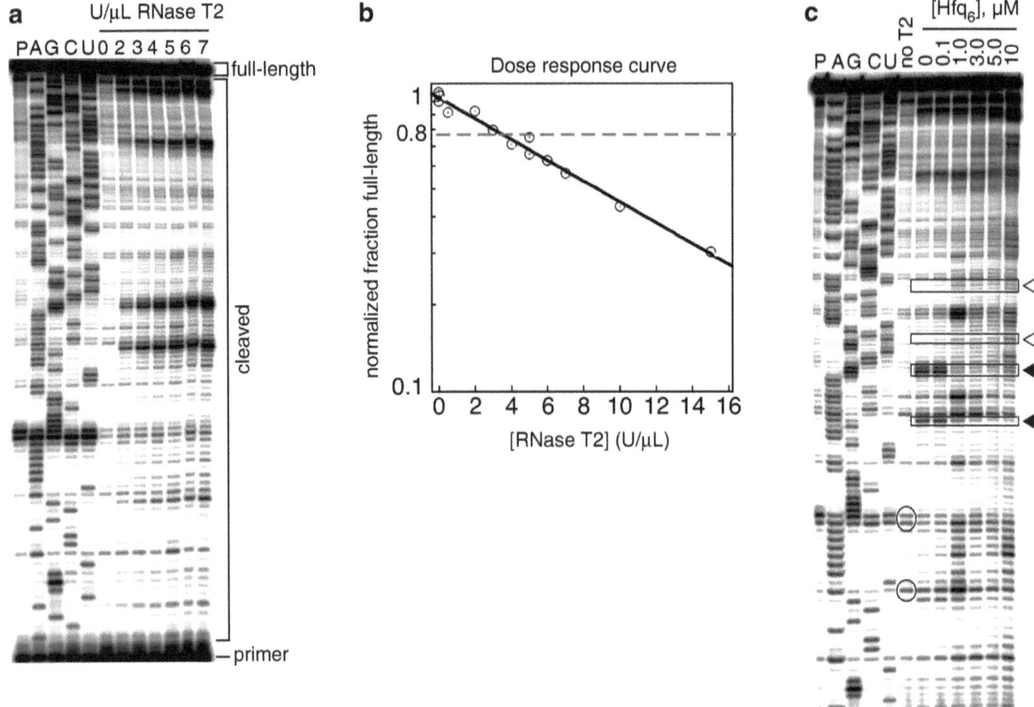

Fig. 2. RNase T2 footprinting of Hfq–*rpoS* complex. (**a**) Representative sequencing gel showing the RNase T2 dose response for the *rpoS* mRNA leader. An RNA-only pausing control (P) is followed by ddNTP generated sequencing lanes (A, G, C, and U). The next seven lanes are folded *rpoS* RNA reacted for 1 min with increasing concentration (U/μL) of RNase T2. The positions of cDNAs generated from the full-length or cleaved RNAs, as well as the primer, are indicated. (**b**) Semi-log plot of the dose response, calculated by dividing the counts in the full-length band by the total counts in the cleaved bands, omitting the primer. Full-length percentages are normalized to the zero RNase T2 lane. The data are fit with an exponential decay function. The RNase concentration that gives ~80% full-length (*dashed line*) is used in future experiments. (**c**) Footprinting Hfq on the *rpoS* leader. Folded *rpoS* RNA is incubated with increasing concentrations of Hfq hexamer for 5–10 min then reacted with 3 U/μL RNase T2 for 1 min. *Boxed bands* indicate representative nucleotides that are protected (*filled triangles*) or deprotected (*empty triangles*) with increasing Hfq_6.

3. Add 6 μL of RNA cocktail to each protein dilution and the no RNase T2 control and incubate for 5–10 min at room temperature.
4. Add 2 μL of 3 U/μL RNase T2 (as determined by the dose-response experiment) to each reaction tube except the no RNase T2 control (add 2 μL of H_2O to the control) and incubate for 1 min at room temperature.
5. Quench the RNase with phenol and extract the reactions with phenol and chloroform as in Subheading 3.2, steps 6–10 (see Note 22).
6. Reverse transcribe 2 μL of each sample and run the cDNA on a sequencing gel according to the protocols in Subheadings 3.3 and 3.4.

7. Scan the gel and analyze to find changes in cleavage levels as a function of Hfq. Bands whose intensity decreases as the concentration of Hfq_6 increases are considered to be protected, for example the bands marked with filled triangles in Fig. 2c. Protections can be due to either direct Hfq binding or structural changes in *rpoS* RNA as a result of Hfq binding. Hfq binding can also increase cleavage in some bands, for example those marked with empty triangles in Fig. 2c. These enhancements are due to structural changes in *rpoS*. Dark bands in the no RNase T2 control lane (the bands circled in Fig. 2c for example) indicate strong natural pause sites inherent to the RNA and those regions of the sequence should not be analyzed for Hfq-dependent changes. To quantify changes in band intensity, use ImageQuant or Image J as in Subheading 3.5, or use the program Semi-Automated Footprinting Analysis (SAFA) (20).

4. Notes

1. Hfq hexamers used in these experiments were purified by passing lysate containing untagged protein over a Co^{2+} column (21, 22).
2. RNAs were transcribed with T7 RNA polymerase from plasmid or PCR DNA templates containing T7 promoter sequences according to the protocol found in (6) and purified by denaturing gel electrophoresis according to (23).
3. RNase T2 has optimal activity at pH 4.5 (13). However, it is active and stable at physiological pH (13). The experiments detailed here were conducted at pH 7.5 and the RNase T2 was diluted into the pH 7.5 TNK buffer.
4. To keep the DTT fresh, remake this buffer often and avoid large numbers of freeze/thaw cycles.
5. We find it most cost effective to gel purify primers in the lab. We run 1–5 nmol of primer on a 10% acrylamide preparative gel, visualize the band by UV shadowing, cut it out of the gel, and elute the primer by rocking in TEN (10 mM Tris–HCl pH 7.5, 1 mM EDTA, 250 mM NaCl) 16–24 h at 4°C. The eluted primer is removed from the gel slice and precipitated.
6. Acrylamide is a neurotoxin, so wear a mask, gloves, and a lab coat when working with powdered acrylamide, and take care to avoid spilling or dispersing the powder.
7. These are sold by most office supply stores.

8. Deionizing the formamide in the loading dye improves the quality of band resolution in the sequencing gel. To deionize formamide, add 1–2 g of AG501-X8 Resin (Bio-Rad) to 10 mL formamide and shake 4–16 h at 4°C.
9. To label a smaller RNA for direct readout of RNase cleavage, use 40 pmol of in vitro transcribed RNA, dephosphorylated with calf intestinal phosphatase.
10. Two μL of 1.25 μM RNA is 2.5 pmol of RNA. The protocol presented here uses 0.5 pmol of RNA per primer extension reaction, so that one cleavage reaction will contain enough material for five primer extensions. Primer extension reactions should use 0.25–1.0 pmol of RNA. In the case of a smaller RNA that doesn't require primer extension to read out the cleavages, use 1 μL of ^{32}P-labeled RNA, with the concentration depending on binding properties of the protein of interest but containing at least 10^5 cpm/μL.
11. More accurate dose information will be obtained by repeating the dose response as well as using the initial results to refine the range of RNase T2 concentrations tested.
12. If a labeled RNA is being cleaved and analyzed directly without primer extension, quench the reaction with 2 μL of 10 mM aurin tricarboxcylic acid (ATA), add an equal volume of formamide/TBE dye, load the RNA directly on to the gel, and proceed from step 17 of Subheading 3.3. Do not use ATA if proceeding to primer extension because the ATA will inhibit the reverse transcriptase and is hard to remove from the reaction.
13. Adding carrier tRNA increases the amount of total RNA in the reaction, improving the precipitation efficiency.
14. For a faster precipitation step, use 30 min at −80°C.
15. When directly analyzing labeled RNA, sequencing is generated by nuclease digestion of denatured RNA.
16. The pausing control differs from the no RNase reaction in that the RNA being reverse transcribed is never incubated in reaction conditions and never phenol/chloroform extracted. It is a useful check on the quality of input RNA and any effects of processing, but does not need to be quantified for background subtraction.
17. Pellets can be stored 1–2 days at −20°C. Once they are resuspended in loading buffer, proceed directly to the sequencing gel.
18. The gel solution must be cooled to room temperature or below, or else polymerization will be too fast to cast a usable gel. In our experience, degassing for 10 min cools the solution sufficiently.

19. To prevent leaking gel solution, tape the bottom and sides of the plates before clamping them with binder clips. The comb should also be clamped to ensure proper well formation.
20. A thin spatula works well for separating the plates. Remove the spacers then insert the spatula blade from the bottom and slowly apply leverage. After covering the gel with plastic wrap it is usually necessary to trim away excess filter paper and plastic.
21. Depending on the number of counts loaded and the sensitivity of the phosphorimager, gels can be exposed from 3 to 24 h. It is useful to save the gel after scanning in case rescanning is required.
22. Process the no RNase T2 control the same as the reactions that are digested with RNase T2.

Acknowledgement

We would like to acknowledge Richard Lease, Subrata Panja, and Sarah Soper for providing reagents, technical advice, and useful discussions.

References

1. Gottesman S, Storz G (2011) Bacterial small RNA regulators: versatile roles and rapidly evolving variations. Cold Spring Harb Perspect Biol. 2011 Dec 1;3(12). pii: a003798. doi: 10.1101/cshperspect.a003798
2. Valentin-Hansen P, Eriksen M, Udesen C (2004) The bacterial Sm-like protein Hfq: a key player in RNA transactions. Mol Microbiol 51:1525–1533
3. Zhang A, Wassarman KM, Ortega J, Steven AC, Storz G (2002) The Sm-like Hfq protein increases OxyS RNA interaction with target mRNAs. Mol Cell 9:11–22
4. Salim NN, Feig AL (2010) An upstream Hfq binding site in the fhlA mRNA leader region facilitates the OxyS-fhlA interaction. PLoS One 5:e13028
5. Moller T, Franch T, Hojrup P, Keene DR, Bachinger HP, Brennan RG, Valentin-Hansen P (2002) Hfq: a bacterial Sm-like protein that mediates RNA-RNA interaction. Mol Cell 9:23–30
6. Soper TJ, Woodson SA (2008) The rpoS mRNA leader recruits Hfq to facilitate annealing with DsrA sRNA. RNA 14:1907–1917
7. Soper TJ, Doxzen K, Woodson SA (2011) Major role for mRNA binding and restructuring in sRNA recruitment by Hfq. RNA 17: 1544–1550
8. Ruskin B, Green MR (1985) Specific and stable intron-factor interactions are established early during in vitro pre-mRNA splicing. Cell 43:131–142
9. Brescia CC, Mikulecky PJ, Feig AL, Sledjeski DD (2003) Identification of the Hfq-binding site on DsrA RNA: Hfq binds without altering DsrA secondary structure. RNA 9:33–43
10. Lease RA, Woodson SA (2004) Cycling of the Sm-like protein Hfq on the DsrA small regulatory RNA. J Mol Biol 344:1211–1223
11. Geissmann TA, Touati D (2004) Hfq, a new chaperoning role: binding to messenger RNA determines access for small RNA regulator. EMBO J 23:396–405
12. Parker KA, Steitz JA (1989) Determination of RNA-protein and RNA-ribonucleoprotein interactions by nuclease probing. Methods Enzymol 180:454–468
13. Ehresmann C, Baudin F, Mougel M, Romby P, Ebel JP, Ehresmann B (1987) Probing the structure of RNAs in solution. Nucleic Acids Res 15:9109–9128

14. Lease RA, Cusick ME, Belfort M (1998) Riboregulation in *Escherichia coli*: DsrA RNA acts by RNA:RNA interactions at multiple loci. Proc Natl Acad Sci U S A 95: 12456–12461
15. Majdalani N, Cunning C, Sledjeski D, Elliott T, Gottesman S (1998) DsrA RNA regulates translation of RpoS message by an anti-antisense mechanism, independent of its action as an antisilencer of transcription. Proc Natl Acad Sci U S A 95:12462–12467
16. Mitra S, Shcherbakova IV, Altman RB, Brenowitz M, Laederach A (2008) High-throughput single-nucleotide structural mapping by capillary automated footprinting analysis. Nucleic Acids Res 36:e63
17. Vasa SM, Guex N, Wilkinson KA, Weeks KM, Giddings MC (2008) ShapeFinder: a software system for high-throughput quantitative analysis of nucleic acid reactivity information resolved by capillary electrophoresis. RNA 14: 1979–1990
18. Brenowitz M, Senear DF, Shea MA, Ackers GK (1986) Quantitative DNase footprint titration: a method for studying protein-DNA interactions. Methods Enzymol 130:132–181
19. Abramoff MD, Magalhaes PJ, Ram SJ (2004) Image processing with ImageJ. Biophotonics Int 11:36–42
20. Das R, Laederach A, Pearlman SM, Herschlag D, Altman RB (2005) SAFA: semi-automated footprinting analysis software for high-throughput quantification of nucleic acid footprinting experiments. RNA 11:344–354
21. Soper T, Mandin P, Majdalani N, Gottesman S, Woodson SA (2010) Positive regulation by small RNAs and the role of Hfq. Proc Natl Acad Sci U S A 107:9602–9607
22. Hopkins JF, Panja S, Woodson SA (2011) Rapid binding and release of Hfq from ternary complexes during RNA annealing. Nucleic Acids Res 39:5193–5202
23. Zaug AJ, Grosshans CA, Cech TR (1988) Sequence-specific endoribonuclease activity of the *Tetrahymena* ribozyme: enhanced cleavage of certain oligonucleotide substrates that form mismatched ribozyme-substrate complexes. Biochemistry 27:8924–8931

Part V

RNA–RNA Interactions

Chapter 14

Computational Identification of sRNA Targets

Brian Tjaden

Abstract

Many small noncoding RNAs (sRNAs) in bacteria act as posttranscriptional regulators by base pairing to their message targets. `TargetRNA` is a program that predicts the targets of a sRNA by identifying messages with significant potential to base pair with the sRNA. Since base pairing potential alone is insufficient to accurately identify sRNA targets, `TargetRNA` integrates several additional features of RNA interactions when predicting regulatory targets of a sRNA. In this chapter, we provide a detailed guide on how to use `TargetRNA` to identify targets of sRNA regulation.

Key words: sRNA target prediction, sRNA regulatory targets, `TargetRNA`

1. Introduction

1.1. Base Pairing Between sRNAs and Their Targets

The best studied group of sRNAs in bacteria corresponds to *trans*-acting regulators that base pair to their message targets. In contrast to *cis*-acting sRNAs, the *trans*-acting sRNAs typically have limited complementarity to their targets, often pairing over a region of one to two dozen nucleotides. Within the region of interaction between a sRNA and one of its message targets, a few mutational studies have suggested that only about 4–9 nucleotides are essential for regulatory effect (1).

As in the case of microRNAs in eukaryotes, *trans*-acting sRNA regulators that bind via base pairing to messages typically affect the translation and stability of their targets. Commonly, these sRNAs inhibit the translation of their mRNA target, e.g., by binding in the neighborhood of the translation initiation site and blocking ribosome binding. Either as an alternative or complement to ribosome occlusion, sRNAs may decrease the stability of the message and target it for degradation by RNase E (2, 3). Thus, sRNAs often act stoichiometrically where they are degraded along with their targets. Less commonly in cases studied to date, sRNAs can

Kenneth C. Keiler (ed.), *Bacterial Regulatory RNA: Methods and Protocols*, Methods in Molecular Biology, vol. 905,
DOI 10.1007/978-1-61779-949-5_14, © Springer Science+Business Media, LLC 2012

activate translation, e.g., by freeing translation initiation sites that would otherwise be occluded by an inhibitory secondary structure (4, 5). One sRNA can interact with multiple mRNAs (reviewed in ref. (6)), enabling sRNAs to be effective participants in bacterial global responses to specific, often stress related, conditions. In *Escherichia coli*, where they are perhaps best understood, there are approximately 100 such *trans*-acting sRNAs. Because pairing between a sRNA and its target is imperfect and over a short region, computational methods for predicting sRNA regulatory targets have met with only moderate success, often generating large numbers of false positive predictions.

1.2. The `TargetRNA` Approach for Predicting sRNA Targets

`TargetRNA` operates under the assumption that the base pairing potential of two nucleotide sequences, corresponding to a sRNA and a mRNA target, can serve as a predictor, albeit imperfect, for the interaction of the two RNAs (7). Base pairing potential of two nucleotide sequences is determined using one of two hybridization scoring methods. The *individual* base pair model of hybridization scoring is analogous to the Smith–Waterman dynamic program (8), except that instead of assessing homology potential, base pairing potential is assessed. The *stacked* base pair model of hybridization scoring is based on stacking and destabilizing energies of interacting nucleotides (9, 10), specifically stacked base pairs, bulge loops, and internal loops of hybridizing RNAs. The stacked base pair model determines the minimum free energy of hybridization for two RNAs without allowing intramolecular base pairing (11).

Once `TargetRNA` computes the hybridization scores for two RNAs, the statistical significance of the interaction is assessed (11). Ten thousand random RNA sequences are generated such that the nucleotides in the random sequences are drawn from the distribution of nucleotides in the candidate target sequences. The hybridization score is calculated for each of these random sequences and the sRNA sequence. The resulting distribution of 10,000 hybridization scores is used to estimate a *p*-value for a sRNA:mRNA hybridization score by determining the probability of observing a score, by chance, equal to or less than the given sRNA:mRNA hybridization score.

In addition to the base pairing potential of a sRNA and a candidate target, other features may be suggestive of sRNA:mRNA interactions. Many interactions contain a seed, representing the initial interaction between a sRNA and its target, corresponding to a stretch of consecutive unpaired nucleotides in a hairpin loop of a sRNA that first base pairs with the target message. sRNAs often base pair with their targets in the 5′ UTR of the message. Interactions are often conserved in closely related species, and RNA secondary structures can play important roles in facilitating interactions. By considering these and other features when predicting sRNA:mRNA interactions, TargetRNA aims to distinguish true targets from messages that have base pairing potential to a sRNA but are not a target of the sRNA.

1.3. Chapter Overview

We begin below with a detailed explanation of using TargetRNA to identify regulatory targets of a sRNA. We discuss all of the various options and advanced features of the program, and we provide guidance on interpreting the results. Our goal is to provide a step-by-step tutorial on using the program as well as an understanding of the results so that the utility and limitations of the application may be fully fathomed.

2. Materials

TargetRNA is available for use via a web interface (http://snowwhite.wellesley.edu/targetRNA/). The companion program, RNATarget, is similarly available via a web interface (http://snowwhite.wellesley.edu/targetRNA/index_2.html).

3. Methods

3.1. Basic Usage of TargetRNA

3.1.1. Program Input

TargetRNA takes as input the name of a genome and the nucleotide sequence of a sRNA gene. The sRNA sequence may be either DNA or RNA. The input sequence need not correspond precisely to the gene sequence, for example if the exact transcription initiation and termination sites are not known, but input sequences that correspond well with the extent of a sRNA gene are more likely to yield accurate results, i.e., higher sensitivity and specificity. When a TargetRNA user submits a sRNA sequence, a message appears in the web page indicating the estimated amount of time required for TargetRNA to analyze the sequence and process the results. Given the name of a genome and an input sRNA sequence, TargetRNA considers the base pairing potential of the sRNA with each mRNA from the genome. mRNAs for a genome are determined from a RefSeq .ptt file retrieved from that National Center for Biotechnology Information (12).

3.1.2. Program Output

Once TargetRNA processes an input sRNA sequence, four sets of information are output. First, the input sRNA sequence is output along with the minimum free energy secondary structure of the sequence as computed by RNAfold from the Vienna RNA package (13). Second, the parameter settings are output so that the user has a record of the particular parameter values that generated the output results. Third, a summary of mRNA targets predicted by TargetRNA is output. Predicted targets are listed in order from the most significant to the least significant. The summary list of predicted targets includes the name of the predicted gene target, any synonym name for the gene, the raw hybridization score computed by TargetRNA for the sRNA and mRNA, the *p*-value of the sRNA:mRNA interaction, and a graphical representation of where the predicted interaction occurs within the sRNA sequence. While the third set of information output by TargetRNA contains a summary of the predicted targets, the fourth set of information output contains detailed information about each predicted target and its interaction with the input sRNA. The detailed information includes an annotation of the predicted target, the location of the interacting regions within the sRNA and the mRNA, a graphical representation of the interacting base pairs, and a link to further information about the target at the Entrez Gene database (14). Finally, a link at the bottom of the output web page enables output in text rather than HTML format. Figure 1 illustrates TargetRNA's output when executed with default parameter values on the sequence for the sRNA GcvB in *E. coli*. Four of the targets predicted by TargetRNA in Fig. 1, dppA, cycA, oppA, and ilvC, have been shown to be regulated by GcvB (15–17).

3.2. Advanced Usage of TargetRNA

TargetRNA offers a number of advanced options for specialized analyses. The advanced options can be grouped into one of three categories: those relating to the sRNA, those relating to the target, those relating to the scoring system used by TargetRNA.

3.2.1. sRNA Parameters

Many sRNA sequences contain Rho-independent hairpin terminators. TargetRNA offers an option whereby it attempts to identify such a terminator and remove it from the input sRNA sequence. Since most but not all sRNA terminators do not participate in an interaction with a mRNA target, excluding these regions may reduce false positive predictions from TargetRNA.

3.2.2. Target Parameters

By default, TargetRNA searches for interactions around the ribosome binding site of a mRNA, specifically from 30 nucleotides upstream of the start of translation to 20 nucleotides downstream of the start of translation. However, this region can be modified by the user as desired, including focusing the search to the entire mRNA or to the 3′ UTR of the message. TargetRNA can require the existence of a seed region in all predicted interactions. A seed region is meant to represent a stretch of consecutive unpaired nucleotides in a hairpin loop of a sRNA that first base pairs with the target message. By default, a seed of nine consecutive base pairs, disallowing G:U wobble pairs, is required of all predicted interactions, but the length of the seed and the inclusion of wobble pairs may be adjusted by the user. Also, rather than considering all

A tool for identifying targets of small (non-coding) RNA action

TargetRNA
||||||||||
TargetRNA

Home Advanced Search Options Contact

Predicted targets for sRNA: GcvB

ACUUCCUGAGCCGGAACGAAAAGUUUUAUCGGAAUGCGUG
UUCUGGUGAACUUUUGGCUUACGGUUGUGAUGUUGUGUUG
UUGUGUUUGCAAUUGGUCUGCGAUUCAGACCAUGGUAGCA
AAGCUACCUUUUUUCACUUCCUGUACAUUUACCCUGUCUG
UCCAUAGUGAUUAAUGUAGCACCGCCUAAUUGCGGUGCUU
UUUUU

Species	Remove terminator	Before Start Codon	After Start Codon	Hybridization Seed	G:U Pairs in Seed	Single Target	P-value	Thermodynamic	Orthologs
e_coli	yes	-30	20	9	no	-	0.01	no	-

Rank	Gene	Synonym	Score	Pvalue
1	dppA	b3544	-84	0.000393413
2	yhfS	b3376	-81	0.000646868
3	cycA	b4208	-79	0.000901103
4	ybdH	b0599	-73	0.00243491
5	yeiG	b2154	-73	0.00243491
6	yeiJ	b2161	-72	0.00287339
7	htrC	b3989	-71	0.0033907
8	oppA	b1243	-70	0.00400096
9	ndk	b2518	-70	0.00400096
10	trpE	b1264	-69	0.00472078
11	tynA	b1386	-69	0.00472078
12	aroC	b2329	-69	0.00472078
13	eutP	b2461	-69	0.00472078
14	mltC	b2963	-69	0.00472078
15	yfcJ	b2322	-66	0.00775129
16	yqcE	b2775	-66	0.00775129
17	yhiP	b3496	-65	0.00914275
18	ilvC	b3774	-65	0.00914275
19	yifK	b3795	-65	0.00914275
20	ysgA	b3830	-65	0.00914275

1 dppA dipeptide transport protein

Score: -84 Pvalue: 0.000393413

```
sRNA (GcvB)   24   UUUUAUCGGAAUGCGUGUUCUGGUGAACUUUUGGCUUACGGUUGUGAUGUUGUGUUGUU   82
                   |||||  ||||||||||||         ||||| ||| |||||||||||| |||||||
mRNA (dppA)   18   AAAGUUCCUUUAUGCGUAAU--------AAGAC-GAG-GUUAACACUACAA-ACAACAA  -30
```

Fig. 1. The figure shows output from the TargetRNA web server when the sRNA GcvB in *Escherichia coli* is input. The sequence for GcvB is shown followed by a link to GcvB's predicted secondary structure. The search parameters are then indicated in *yellow* followed by a summary of the 20 target predictions. At the *bottom* of the figure, details for the first of the 20 predictions are shown.

messages from a genome, `TargetRNA` can analyze a single message for its propensity to interact with a sRNA. Analyzing a single message rather than all messages may be useful when exploring the parameter space for a particular high-confidence sRNA:mRNA interaction.

3.2.3. Scoring Parameters

`TargetRNA` outputs candidate targets whose hybridization score with the sRNA has a *p*-value at or below a threshold of 0.01, by default (see Note 1). The *p*-value threshold can be set by the user. By default, `TargetRNA` uses an individual model for hybridization scoring, however a stacked base pair model for hybridization scoring may be selected (see Note 2). The stacked base pair model for hybridization determines the minimum free energy of hybridization for two RNAs. Finally, beyond the reference genome being searched, the user can select any set of other bacterial genomes in which to search for orthologous sRNA:mRNA interactions. For each predicted sRNA:mRNA interaction, `TargetRNA` will search for similar sRNA sequences and similar mRNA sequences in other genomes and if it finds putative orthologs then it will compute and output a predicted sRNA:mRNA interaction for the orthologous pair of RNAs.

3.3. RNATarget

`RNATarget` is a companion program to `TargetRNA`. While `TargetRNA` predicts mRNA targets of a given sRNA, `RNATarget` predicts sRNAs that regulate a given mRNA target (see Note 3). `RNATarget` takes as input the name of a genome and a mRNA. It then screens all intergenic regions of the genome for sequences that evince significant hybridization with the mRNA. Such intergenic sequences are output as possible regions containing regulating sRNAs. In Fig. 2, output from `RNATarget` is shown after using the program to search for a possible sRNA that regulates the known mRNA target dppA.

`RNATarget` offers a number of advanced options that can be specified by the user. Either the name of the mRNA or any subsequence of a mRNA can be input to the program. A specific region within the mRNA sequence can be specified either around the 5′ UTR or the 3′ UTR. By default, `RNATarget` uses the region of the mRNA from 30 nucleotides upstream of the start of translation to 20 nucleotides downstream from the start of translation. As in the case of `TargetRNA`, a seed region can be specified for `RNATarget`. The seed corresponds to an initial interaction between the two RNAs represented by a stretch of consecutive base pairs. By default, `RNATarget` only identifies interactions with a seed of at least 9 base pairs. Further, the seed may be permitted to contain G:U wobble pairs or not at the user's discretion. Finally, a *p*-value can be specified such that only interactions with significance at or below the *p*-value will be reported. By default, the *p*-value threshold is set to 0.01.

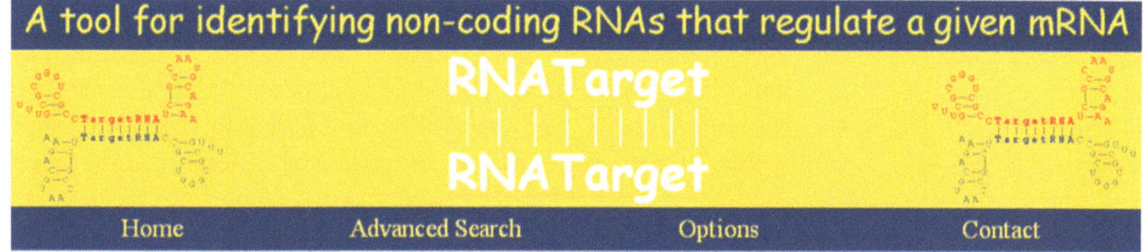

Candidate regulators for mRNA: dppA

Species	Before Start Codon	After Start Codon	Hybridization Seed	G:U Pairs in Seed	P-value
e_coli	-30	20	9	no	0.01

Rank	IG	mRNA	Score	Pvalue	IG_start	IG_stop	mRNA_start	mRNA_stop
1	dppA...yhjW	dppA	-150	1.48783e-06	1	30	-30	-1
2	osmY...yjjU	dppA	-86	0.00220347	241	266	-27	2
3	gcvA...ygdI	dppA	-84	0.00245038	152	210	-30	18
4	ydjL...yeaC	dppA	-76	0.00817033	108	137	-29	3
5	yjiJ...yjiK	dppA	-63	0.00915752	13	29	-30	-14

3 gcvA...ygdI 350 nucleotides

Score: -84 Pvalue: 0.00245038

```
IG (gcvA...ygdI)   152   UUUUAUCGGAAUGCGUGUUCUGGUGAACUUUUGGCUUACGGUUGUGAUGUUGUGUUGUU   210
                         |||:|  |:||||:||::||          ||:|| ||: |::|||||||||| |||||||
mRNA (dppA)        18    AAAGUUCCUUUAUGCGUAAU--------AAGAC-GAG-GUUAACACUACAA-ACAACAA   -30
```

Fig. 2. The figure illustrates output from RNATarget after searching for intergenic sequences with significant potential to hybridize with the message target dppA in *E. coli*. The search parameters are shown in *yellow* followed by a summary of the five intergenic regions demonstrating base pairing potential with the target dppA. At the *bottom* of the figure, details of the predicted hybridization between dppA and the third predicted intergenic sequence are shown (only the third of the five is shown for brevity). The third predicted intergenic sequence corresponds to GcvB, a known regulator of dppA, and the predicted hybridization corresponds exactly with that predicted by TargetRNA in Fig. 1 when searching for targets of GcvB.

4. Notes

1. With its default parameter settings including a *p*-value threshold of 0.01, TargetRNA predicts on average ten mRNA targets for a sRNA. At this level of significance, TargetRNA correctly identifies 21% of 73 documented sRNA:mRNA interactions involving 24 sRNAs in *E. coli* (18).
2. TargetRNA offers two models for hybridization scoring. The individual base pair model is the default and it is the faster of the two, typically requiring only seconds to screen a genome.

The stacked base pair model is computationally more intensive and, as a result, increases the time required for calculating base pair binding potential by a factor of approximately five for most sequence inputs. Since the individual base pair model generally yields better results in less time, it is the default and recommended model for most applications.

3. `RNATarget`, the converse program of `TargetRNA`, is not a general purpose sRNA gene finder. It identifies intergenic sequences that demonstrate significant base pairing potential to a mRNA regulatory target. It does not employ information such as transcription signals, RNA secondary structure, or conservation to identify candidate sRNA sequences.

References

1. Kawamoto H, Koide Y, Morita T, Aiba H (2006) Base-pairing requirement for RNA silencing by a bacterial small RNA and acceleration of duplex formation by Hfq. Mol Microbiol 61:1013–1022
2. Pfeiffer V, Papenfort K, Lucchini S, Hinton JCD, Vogel J (2009) Coding sequence targeting by MicC RNA reveals bacterial mRNA silencing downstream of translational initiation. Nat Struct Mol Biol 16:840–846
3. Desnoyers G, Morissette A, Prevost K, Masse E (2009) Small RNA-induced differential degradation of the polycistronic mRNA *iscRSUA*. EMBO J 28:1551–1561
4. Prevost K, Salvail H, Desnoyers G, Jacques JF, Phaneuf E, Masse E (2007) The small RNA RyhB activates the translation of *shiA* mRNA encoding a permease of shikimate, a compound involved in siderophore synthesis. Mol Microbiol 64:1260–1273
5. Urban JH, Vogel J (2008) Two seemingly homologous noncoding RNAs act hierarchically to activate *glmS* mRNA translation. PLoS Biol 6:e64
6. Papenfort K, Vogel J (2009) Multiple target regulation by small noncoding RNAs rewires gene expression at the post-transcriptional level. Res Microbiol 160:278–287
7. Tjaden B, Goodwin SS, Opdyke JA, Guillier M, Fu DX, Gottesman S, Storz G (2006) Target prediction for small, noncoding RNAs in bacteria. Nucleic Acids Res 34:2791–2802
8. Smith TF, Waterman MS (1981) Identification of common molecular subsequences. J Mol Biol 147:195–197
9. Xia T, SantaLucia J, Burkhard ME, Kierzek R, Schroeder SJ, Jiao X, Cox C, Turner DH (1998) Thermodynamic parameters for an expanded nearest-neighbor model for formation of RNA duplexes with Watson-Crick base pairs. Biochemistry 37:14719–14735
10. Mathews DH, Sabina J, Zuker M, Turner DH (1999) Expanded sequence dependence of thermodynamic parameters provides robust prediction of RNA secondary structure. J Mol Biol 288:910–940
11. Rehmsmeier M, Steffen P, Hochsmann M, Giegerich R (2004) Fast and effective prediction of microRNA/target duplexes. RNA 10:1507–1517
12. Pruitt KD, Tatusova T, Maglott DR (2005) NCBI reference sequence (RefSeq): a curated non-redundant sequence database of genomics, transcripts and proteins. Nucleic Acids Res 33:D501–D504
13. Hofacker IL, Fontana W, Stadler PF, Bonhoeffer LS, Tacker M, Schuster P (1994) Fast folding and comparison of RNA secondary structure. Monatsh Chem 125:167–188
14. Maglott D, Ostell J, Pruitt KD, Tatusova T (2007) Entrez gene: gene-centered information at NCBI. Nucleic Acids Res 35:D26–D31
15. Urbanowski ML, Stauffer LT, Stauffer GV (2000) The *gcvB* gene encodes a small untranslated RNA involved in expression of the dipeptide and oligopeptide transport systems in *Escherichia coli*. Mol Microbiol 37:856–868
16. Vogel J (2009) A rough guide to the non-coding RNA world of *Salmonella*. Mol Microbiol 71:1–11
17. Pulvermacher SC, Stauffer LT, Stauffer GV (2009) Role of the sRNA GcvB in regulation of *cycA* in *Escherichia coli*. Microbiology 155: 106–114
18. Peer A, Margalit H (2011) Accessibility and evolutionary conservation mark bacterial small-RNA target-binding regions. J Bacteriol 193: 1690–1701

Chapter 15

Detection of sRNA–mRNA Interactions by Electrophoretic Mobility Shift Assay

Teppei Morita, Kimika Maki, and Hiroji Aiba

Summary

Electrophoretic mobility shift assay is a simple, rapid, and sensitive technique to analyze the RNA–RNA interaction. A ^{32}P-labeled RNA is incubated with another unlabeled RNA and subjected to electrophoresis on a native polyacrylamide gel. If two RNA molecules base pair stably, the movement of the probe RNA through the gel is retarded resulting in a characteristic band corresponding to the RNA duplex. Here, we describe the methods to study the interaction of an Hfq-binding small RNA (sRNA) and its target mRNA. Although we focus on the interaction of SgrS and its target *ptsG* mRNA, the methods can be applied to the analysis of base pairing between any sRNAs and their targets.

Key words: Electrophoretic mobility shift assay, RNA–RNA interaction, Base pairing, sRNA, Hfq, Native polyacrylamide gel, In vitro transcription

1. Introduction

This chapter deals with the methods to detect the interaction between base-pairing small RNAs (sRNAs) and their target mRNAs in vitro by using the electrophoretic mobility shift assay (EMSA). EMSA was first used to detect the binding of the purified in *Escherichia coli* Lac repressor protein to the *lac* operator DNA (1, 2). Since then, it had been developed widely as a powerful technique to study DNA–protein interaction. The method also has been successfully used to analyze RNA–protein interaction and RNA–RNA (DNA) base pairing. The assay is a simple, rapid, and sensitive to detect the interaction of a nucleic acid with other macromolecules. It permits quantitative and kinetic analysis of complex formation between two or more molecules. The basis of EMSA is the change

Kenneth C. Keiler (ed.), *Bacterial Regulatory RNA: Methods and Protocols*, Methods in Molecular Biology, vol. 905, DOI 10.1007/978-1-61779-949-5_15,

in the electrophoretic mobility of an RNA (DNA) molecule on a non-denaturing or native gel upon binding with other molecules.

We focus on the interaction of an Hfq-binding sRNA SgrS and its target *ptsG* mRNA. However, the methods can be applied to the analysis of base pairing between any sRNAs and their targets. Briefly, both ^{32}P-labeled 5′ portion of *ptsG* mRNA and unlabeled SgrS RNA are prepared. The labeled RNA probe is incubated with the unlabeled RNA and subjected to electrophoresis on a native polyacrylamide gel. If two RNA molecules base pair stably, the movement of the probe RNA through the gel is retarded resulting in a characteristic band corresponding the RNA duplex. First, we describe a basic protocol to detect the duplex formation between two RNA molecules. By using this protocol, we defined the SgrS region required for base pairing with *ptsG* mRNA. Then, we describe the protocols to measure the kinetics of the duplex formation, and to examine the effect of Hfq.

The Hfq-binding sRNAs act by imperfect base-pairing to regulate the translation and stability of target mRNAs under specific physiological conditions (3, 4). SgrS, an example for this class of sRNAs, is induced in response to accumulation of glucose phosphates and down-regulates the *ptsG* mRNA encoding the glucose transporter IICB^{Glc} in *E. coli* (5). Hfq, originally identified as a host factor required for the in vitro replication of the RNA phage Qβ in *E. coli* (6), is an RNA binding protein extensively involved in sRNA action (7, 8). Hfq has been shown to stimulate the base pairing between a given sRNA and its target mRNA in vitro either by acting as an RNA chaperone. We showed that Hfq dramatically enhances the rate of duplex formation between *ptsG* RNA and SgrS (9).

2. Materials

2.1. Electrophoresis Components

1. 0.7% Agarose gel: Dissolve 0.7 g agarose by heating in 100 mL of 0.5× TBE (45 mM Tris base, 45 mM boric acid, 1 mM EDTA) and add 5 μL of 5 mg/mL ethidium bromide. We used 52 mm × 60 mm × 10 mm gels.
2. 30% Acrylamide solution: 29% acrylamide, 1% *N, N'*-methylene bisacrylamide.
3. Ammonium persulfate (APS): 10% solution in H_2O.
4. *N,N,N′,N′*-tetramethylene diamine (TEMED).
5. Loading buffer: 50% glycerol, 0.1% bromophenol blue.
6. 2.5× TBE electrophoresis buffer: 225 mM Tris base, 225 mM boric acid, 5 mM EDTA.

7. 6% Polyacrylamide native gel: Mix 2 mL 30% acrylamide solution, 2 mL 2.5× TBE, and 6 mL H_2O. Add 50 μL of 10% APS and 12 μL of TEMED (see Note 1). Gel dimensions ($W \times H \times D$): 85 mm × 70 mm × 1 mm.
8. 4% Polyacrylamide native gel for EMSA: Mix 4 mL 30% acrylamide solution, 6 mL 2.5× TBE, 3 mL 50% glycerol, and 17 mL H_2O. Add 150 μL of 10% APS and 36 μL of TEMED (see Note 1). We used 150 mm × 150 mm × 1 mm gels.

2.2. RNA Preparation Components

1. Vector containing T7 promoter: We used pBluescript SK(-) (Stratagene) to clone the DNA for the RNA sequence of interest (see Note 2).
2. Plasmids containing the DNA for the RNA sequence of interest: We used pTH111 harboring the *ptsG* gene (10) and pQES-grS1 harboring *sgrS* gene (9).
3. PCR primers: We used an upstream primer containing *Kpn*I site and a downstream primer containing *Sac*I site to clone the DNA template into pBluescript SK(-). An upstream primer containing T7 promoter sequence and a downstream primer containing Flag tag sequence were used to amplify the DNA template for in vitro transcription.
4. In vitro transcription kit. We used CUGA®7 in vitro transcription kit (Nippon Genetech).
5. [α-^{32}P]-UTP.
6. Gene Elute Minus EtBr Spin Columns (Sigma).
7. Phenol/chloroform: Mix 200 mL phenol saturated with TE and 200 mL of chloroform.
8. TLC plate PE SIL-G/UV254 (GE Healthcare).
9. Gilbert buffer: 0.5 M ammonium acetate, 10 mM magnesium acetate, 1 mM EDTA, 0.1% SDS.
10. Other reagents: deionized water; 99.9% ethanol; 70% ethanol; 3 M sodium acetate (pH 5.5); TE (10 mM Tris–HCl pH 8.0, 1 mM EDTA).

2.3. EMSA Components

1. sRNA of interest: We used SgrS RNA and synthetic RNA oligomers corresponding the base-pairing region of SgrS (see Note 3).
2. Target RNA: We used ^{32}P-labeled *ptsG'* RNA (see Note 4) and ^{32}P-labeled *ptsG'-flag* RNA (see Note 5).
3. Binding buffer 1: 20 mM Tris–HCl, pH 8.0, 1 mM DTT, 1 mM $MgCl_2$, 20 mM KCl, 10 mM Na_2HPO_4-NaH_2PO_4, pH 8.0.
4. Binding buffer 2: 10 mM Tris–HCl, pH 8.0, 50 mM NaCl, 50 mM KCl, 10 mM $MgCl_2$, 200 ng/mL yeast tRNA.
5. Hfq-His_6 (see Note 6).
6. Whatman 3MM filter paper.

3. Methods

We use CUGA®7 in vitro transcription kit for preparation of RNAs. In brief, the DNA template for the RNA sequence of interest (SgrS or 5′ portion of *ptsG* mRNA) is cloned downstream of the T7 promoter of pBluescript SK(-). The resulting plasmid is linearized with a restriction enzyme. An in vitro transcription results in a run-off transcript containing the RNA sequence of interest. The DNA template can also be generated by PCR using an upstream primer containing the T7 promoter sequence. Chemically synthesized RNA oligomers can also be used for base-pairing reaction.

3.1. Preparation of DNA Template for RNA from Plasmid

1. Clone the DNA template for SgrS or 5′ portion of *ptsG* RNA (*ptsG*′) between *Kpn*I and *Sac*I sites of pBluescript SK(-) (see Notes 7 and 8).
2. Purify the resulting plasmid.
3. Digest 10 μg of the purified plasmid with *Sac*I in 50 μL of reaction mixture. Add an equal volume of phenol/chloroform. Mix by vortexing. After centrifugation at 13,000×*g* for 5 min at room temperature, transfer the aqueous phase to a fresh tube.
4. Add 0.1 volume of 3 M sodium acetate (pH 5.5). Mix well and add 2.5 volumes of ethanol. Mix well and store at –20°C for 15 min. Recover the DNA by centrifugation at 13,000×*g* for 10 min at 4°C. Wash the pellet with 70% ethanol and recentrifuge at 4°C. Dissolve the precipitated DNA in 50 μL of TE.
5. Add 0.1 volume of loading buffer. Fractionate the DNA sample by electrophoresis on 0.7% agarose gel containing ethidium bromide.
6. Cut out the linearized DNA band with a razor blade under a long-wave UV irradiation.
7. Extract DNA from the gel slice using Gene Elute Minus EtBr Spin Columns. Add an equal volume of phenol/chloroform. Mix by vortexing. After centrifugation at 13,000×*g* for 5 min at room temperature, transfer the aqueous phase to a fresh tube.
8. Add 0.1 volume of 3 M sodium acetate (pH 5.5). Mix well and add 2.5 volumes of ethanol. Mix well and store at –20°C for 15 min. Recover the DNA by centrifugation at 13,000×*g* for 10 min at 4°C. Wash the pellet with 70% ethanol and recentrifuge at 4°C. Dissolve the precipitated DNA in 10–20 μL of TE.
9. Check the quality and amount of the DNA by gel electrophoresis and by measuring the optical density at 260 nm (see Note 9).

3.2. Preparation of DNA Template for RNA by PCR

1. Amplify a DNA template containing the *ptsG′-flag* RNA by PCR using an upstream primer possessing T7 promoter sequence and a downstream primer (see Note 5). Check the PCR product by gel electrophoresis.
2. Add 0.1 volume of 3 M sodium acetate (pH 5.5). Mix well and add 2.5 volumes of ethanol. Mix well and store at −20°C for 15 min. Recover the DNA by centrifugation at 13,000 × *g* for 10 min at 4°C. Wash the pellet with 70% ethanol and recentrifuge at 4°C. Dissolve the precipitated PCR product in 20 μL of TE.
3. Add 0.1 volume of loading buffer. Fractionate the PCR product by electrophoresis on 6% polyacrylamide native gel.
4. Remove one glass plate and cover the exposed gel surface with plastic wrap. Place a TLC plate containing florescent material against the wrapped surface of the gel, turn upside down, and remove another glass plate. Locate and cut out the DNA band with a razor blade under UV irradiation.
5. Place the gel slice on a glass plate and chop it into fine pieces with a razor blade. Transfer the pieces to a tube and add 800 μL of Gilbert buffer and incubate at 37°C on a rotating wheel overnight.
6. Centrifuge at 13,000 × *g* for 5 min at room temperature and transfer the supernatant to a fresh tube. Add an equal volume of phenol/chloroform. Mix by vortexing. After centrifugation at 13,000 × *g* for 5 min at room temperature, transfer the aqueous phase to a fresh tube.
7. Add 0.1 volume of 3 M sodium acetate (pH 5.5). Mix well and add 2.5 volumes of ethanol. Mix well and store at −20°C for 15 min. Recover the DNA by centrifugation at 13,000 × *g* for 10 min at 4°C. Wash the pellet with 70% ethanol and recentrifuge at 4°C. Dissolve the precipitated DNA in 10–20 μL of TE.
8. Check the quality and amount of the DNA by gel electrophoresis and by measuring the optical density at 260 nm (see Note 10).

3.3. In vitro Transcription and Purification of RNA

1. Transcribe 0.05–1 pmol of DNA template (either linearized plasmid or PCR product) by using CUGA®7 in vitro transcription kit according to manufacturer's instruction (see Note 11). When needed, add [α-^{32}P]-UTP to the reaction mixture to obtain uniformly ^{32}P-labeled transcripts (see Note 12).
2. Add 0.1 volume of 3 M sodium acetate (pH 5.5). Mix well and add 2.5 volumes of ethanol. Mix well and store at −20°C for 15 min. Centrifuge at 13,000 × *g* for 10 min at 4°C. Wash the pellet with 70% ethanol and recentrifuge at 4°C. Dissolve the precipitate in 20 μL of TE.
3. Add 0.1 volume of loading buffer. Fractionate the sample by electrophoresis on 6% polyacrylamide native gel.

4. Remove one glass plate and cover the exposed gel surface with plastic wrap. Place a TLC plate containing florescent material against the wrapped surface of the gel, turn upside down, and remove another glass plate. Locate and cut out the RNA band with a razor blade under UV irradiation.
5. Place the gel slice on a glass plate and chop it into fine pieces with a razor blade. Transfer the pieces to a tube and add 800 μL of Gilbert buffer and incubate at 37°C on a rotating wheel overnight.
6. Centrifuge at 13,000 × *g* for 5 min at room temperature and transfer the supernatant to a fresh tube. Add an equal volume of phenol/chloroform. Mix by vortexing. After centrifugation at 13,000 × *g* for 5 min at room temperature, transfer the aqueous phase to a fresh tube.
7. Add 0.1 volume of 3 M sodium acetate (pH 5.5). Mix well and add 2.5 volumes of ethanol. Mix well and store at –20°C for 15 min. Recover the RNA by centrifugation at 13,000 × *g* for 10 min at 4°C. Wash the pellet with 70% ethanol and recentrifuge at 4°C. Dissolve the precipitated RNA in 50 μL of H_2O.
8. Check the quality and concentration of RNA by gel electrophoresis and by measuring the optical density at 260 nm.

3.4. Analysis of Base Pairing Reaction with or Without Preheating

Examples of the experiment are shown in Figs. 1 and 2. The full-length SgrS RNA can form quickly an RNA duplex with ^{32}P-labeled *ptsG′-flag* RNA when the mixture is preheated (Fig. 1b, lane 2) but it takes longer times to form the duplex without preheating (Fig. 2). The preheating is able to bypass the Hfq-mediated step by melting RNA secondary structure. RNA Olig14 is sufficient to form a stable duplex while Oligo8 is unable to form the stable duplex although it contains the critical bases for pairing with *ptsG* mRNA in vivo (Fig. 1).

1. Mix 0.2 pmol of ^{32}P-labeled *ptsG′-flag* RNA with 0.4 pmol of SgrS RNA or synthetic RNA oligomers in 3–10 μL of binding buffer 1 on ice (see Notes 13 and 14).
2. Incubate the reaction mixture at 70°C for 5 min and then cool down to 30°C. Skip this step to analyze base pairing without preheating.
3. Incubate the reaction mixture at 30 or 37°C for various times.
4. Add 0.2 volume of loading buffer, load the mixture onto a 4% polyacrylamide native gel in 0.5× TBE containing of 5% glycerol, and run the gel at 20 mA until the bromophenol blue migrates to 2/3 of the gel at 4°C.
5. Vacuum dry the gel on a Whatman 3MM filter paper for autoradiography.

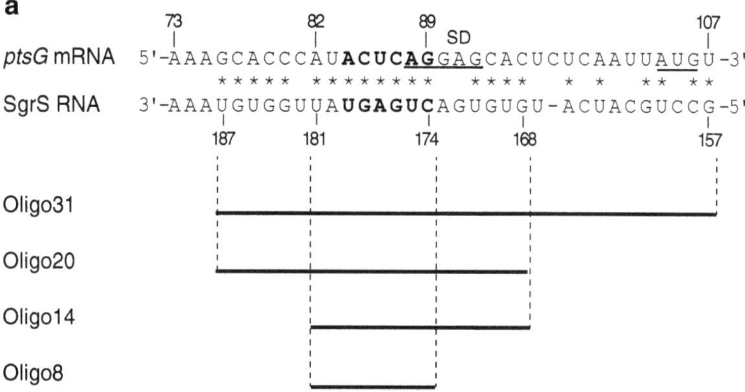

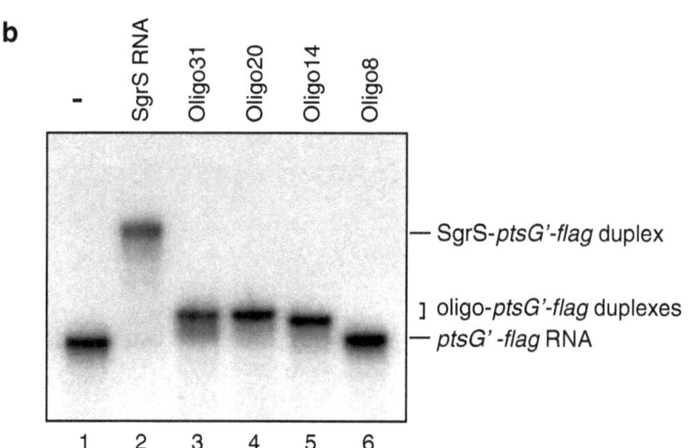

Fig. 1. Detection of SgrS–*ptsG* interaction by EMSA. (**a**) The nucleotide sequences of *ptsG* mRNA and SgrS RNA around the base-pairing region. The SD sequence and the initiation codon of *ptsG* mRNA are underlined. The *asterisks* indicate the predicted base pairs between SgrS and *ptsG* RNAs (5). The *bold letters* represent the crucial base pairs for SgrS action (9). Oligo31, 20, 14, and 8, representing the whole or part of the predicted base-pairing region of SgrS, are chemically synthesized. (**b**) Duplex formation between *ptsG'*-*flag* RNA and SgrS RNA or RNA oligomers. The ^{32}P-labeled *ptsG'-flag* mRNA (0. 2 pmol) was incubated with 0.4 pmol (**a**) of each of the indicated RNA in 3.5 μL of binding buffer at 70°C for 5 min and then cooled down to 30°C. Samples were analyzed by EMSA on a 4% native polyacrylamide gel after the addition of 0.7 μL of RNA loading buffer. The figures have been redrawn based on the data published previously (12).

3.5. Kinetic Analysis of Base Pairing

An example of kinetic analysis is shown in Fig. 3 (9). The duplex formation between SgrS and *ptsG'* RNA proceeds very slowly in the absence of Hfq (see also Fig. 2) while it occurs rapidly in the presence of Hfq, indicating that Hfq stimulates the rate of base pairing reaction (compare odd lanes and even lanes).

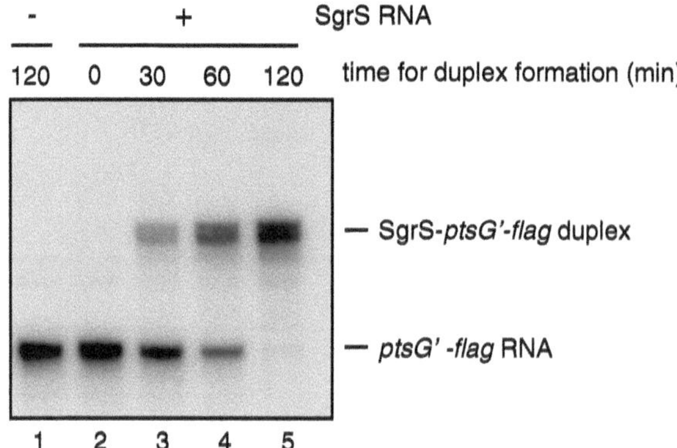

Fig. 2. Formation of SgrS–*ptsG* duplex without preheating. ^{32}P-labeled *ptsG'-flag* RNA (0. 2 pmol) was incubated with SgrS RNA (0.4 pmol) in 3.5 μL of binding buffer at 37°C for indicated times. Samples were analyzed by EMSA on a 4% native polyacrylamide gel after adding 1 μL of RNA loading buffer. The figure has been redrawn based on data published previously (11).

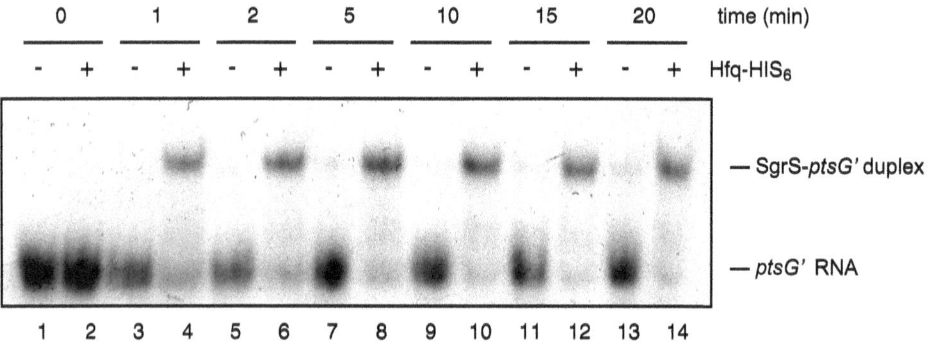

Fig. 3. Time-course for duplex formation. ^{32}P-labeled *ptsG'* RNA (0.2 pmol) was incubated with SgrS RNA (0.4 pmol) and/or Hfq-His$_6$ (1 pmol) in 20 μL of binding buffer 2 at 37°C. At indicated times, the samples were treated with phenol and were analyzed by EMSA on a 4% native polyacrylamide gel. The figure has been redrawn based on data published previously (9).

1. Mix 0.2 pmol of ^{32}P-labeled *ptsG'* RNA with 0.4 pmol of SgrS RNA with and without 1 pmol of Hfq-His$_6$ in 20 μL of binding buffer 2 on ice.
2. Incubate the reaction mixture at 37°C for 0–20 min.
3. Add an equal volume of phenol (see Note 15). Mix by vortexing. Centrifuge at 13,000×*g* for 5 min at room temperature. Transfer 15 μL of the aqueous phase to a fresh tube and add 2 μL of loading buffer.
4. Load each sample onto a 4% polyacrylamide gel in 0.5× TBE containing 5% glycerol (see Note 16).
5. Run the gel at 20 mA until the bromophenol blue migrates to 2/3 of the gel at 4°C Vacuum dry the gel on a Whatman 3MM filter paper for autoradiography.

4. Notes

1. Deaerate the acrylamide solution under reduced pressure prior to addition of TEMED.
2. Any other plasmid vectors containing SP6, T7, or T3 promoters can be used.
3. The SgrS RNA prepared by in vitro transcription consists of the whole region of SgrS (+1 to +227) and additional sequences derived from the vector and/or primers at both ends.
4. The *ptsG'* RNA prepared by in vitro transcription consists of the 5′ portion of *ptsG* mRNA (from +1 to +120) and additional sequences derived from the vector and/or primers at both ends.
5. The *ptsG'-flag* RNA prepared by in vitro transcription consists of the *ptsG* region (from +55 to +455) and the sequence for Flag tag at 3′ end.
6. Hfq-His$_6$ was purified from cells harboring pQE80L-Hfq-His as described (11).
7. The DNA templates for the 5′ portion of *ptsG* mRNA (from +1 to +120) and for SgrS RNA (+1 to +227) were amplified by PCR from the plasmid pTH111 harboring the *ptsG* gene and pQESgrS1 harboring *sgrS* gene, respectively, using an upstream primer containing *Kpn*I site and a downstream primer containing *Sac*I site. The amplified DNAs were digested with *Kpn*I and *Sac*I. The linearized DNAs were inserted between *Kpn*I and *Sac*I sites of pBluescript SK(-).
8. Other restriction sites within the multicloning sites of pBluescript SK(-) also can be used for cloning.
9. The linearized DNAs were used as DNA templates for in vitro transcription to generate the *ptsG'* RNA and SgrS RNA.
10. The *ptsG'-flag* DNA was used as a DNA template to generate the *ptsG'-flag* RNA.
11. We added 1 μL (20 unit) of RNase inhibitor (SIN-101, TOYOBO LIFE SCIENCE) in the reaction mixture to prevent RNA degradation.
12. The specific activity of the RNA probe can be adjusted by changing the ratio of labeled and unlabeled UTP in the reaction. Other [α-^{32}P]-nucleoside triphosphates (NTPs) can also be used during the transcription to achieve uniform labeling of RNA.
13. The amounts of RNAs used in a reaction mixture can be increased or decreased depending on the specific activity of ^{32}P-labeled probe RNA.

14. The *ptsG'-flag* RNA can be used also as an mRNA template for in vitro translation reaction by PURESYSTEM (11). The presence of Flag tag allows one to detect easily the translation product by using anti-Flag antibody.
15. Add phenol together with Hfq-His_6 for time 0.
16. Skip the steps 3 and 4 to observe the ternary complex containing SgrS, *ptsG'*, and Hfq on the gel. The relative amount of Hfq-His_6 to SgrS or *ptsG'* RNA should be reduced to detect a clear band corresponding to the ternary complex.

Acknowledgement

This work was supported by Grants-in-Aid from the Ministry of Education, Culture, Sports, Science and Technology of Japan.

References

1. Fried M, Crothers DM (1981) Equilibria and kinetics of lac repressor-operator interactions by polyacrylamide gel electrophoresis. Nucleic Acids Res 9:6505–6525
2. Garner MM, Revzin A (1981) A gel electrophoresis method for quantifying the binding of proteins to specific DNA regions: application to components of the *Escherichia coli* lactose operon regulatory system. Nucleic Acids Res 9:3047–3060
3. Gottesman S, Storz G (2010) Bacterial small RNA regulators: versatile roles and rapidly evolving variations. Cold Spring Harb Perspect Biol 1:1–16
4. Waters LS, Storz G (2009) Regulatory RNAs in bacteria. Cell 136:615–628
5. Vanderpool CK, Gottesman S (2004) Involvement of a novel transcriptional activator and small RNA in post-transcriptional regulation of the glucose phosphoenolpyruvate phosphotransferase system. Mol Microbiol 54: 1076–1089
6. Franze de Fernandez MT, Eoyang L, August JT (1968) Factor fraction required for the synthesis of bacteriophage Qbeta-RNA. Nature 219:588–590
7. Valentin-Hansen P, Eriksen M, Udesen C (2004) The bacterial Sm-like protein Hfq: a key player in RNA transactions. Mol Microbiol 51:1525–1533
8. Vogel J, Luisi BF (2011) Hfq and its constellation of RNA. Nat Rev Microbiol 9:578–589
9. Kawamoto H, Koide Y, Morita T, Aiba H (2006) Base-pairing requirement for RNA silencing by a bacterial small RNA and acceleration of duplex formation by Hfq. Mol Microbiol 61:1013–1022
10. Takahashi H, Inada T, Postma P, Aiba H (1998) CRP down-regulates adenylate cyclase activity by reducing the level of phosphorylated IIA(Glc), the glucose-specific phosphotransferase protein, in *Escherichia coli*. Mol Gen Genet 259:317–326
11. Maki K, Morita T, Otaka H, Aiba H (2010) A minimal base-pairing region of a bacterial small RNA SgrS required for translational repression of ptsG mRNA. Mol Microbiol 76:782–792
12. Maki K, Uno K, Morita T, Aiba H (2008) RNA, but not protein partners, is directly responsible for translational silencing by a bacterial Hfq-binding small RNA. Proc Natl Acad Sci U S A 105:10332–10337

Chapter 16

Activity of Small RNAs on the Stability of Targeted mRNAs In Vivo

Guillaume Desnoyers and Eric Massé

Abstract

Northern blots are extremely useful to monitor the steady state level of small regulatory RNAs (sRNAs) as well as their target mRNAs. In combination with the drug rifampicin, which blocks cellular transcription, Northern blots can be used to determine the stability of sRNAs and mRNAs. Here we describe a protocol to assess the activity of the sRNA RyhB on the stability of targeted mRNAs *sodB*, *fumA*, and *iscRSUA*. We also describe how to identify a sRNA-induced initial cleavage site on a target mRNA. This protocol can be used for any sRNAs and their target mRNAs.

Key words: Small RNA, RNAse E, RNA degradosome, mRNA decay, Translation initiation

1. Introduction

Bacterial small RNAs (sRNAs) typically function by pairing to sequences located in the 5′-end of their target mRNAs at or near the translation initiation region (1). In most cases, the result of a sRNA-mRNA pairing is either activation or repression of translation initiation, which in turn affects mRNA stability. In cases where the sRNA activates the initiation of translation, the protective effect of elongating ribosomes against cellular ribonucleases will likely increase the mRNA stability (2, 3). In contrast, when the sRNA represses the initiation of translation, there is frequently the formation of a ribonucleoprotein (RNP) complex whose function is to destroy both the sRNA and mRNA (4–7). This RNP complex is composed of the sRNA and its target mRNA, the RNA chaperone Hfq and the endoribonuclease RNase E (8, 9). Thus, the turnover of a targeted mRNA can vary dramatically, depending on the expression level of the sRNA and its function.

Kenneth C. Keiler (ed.), *Bacterial Regulatory RNA: Methods and Protocols*, Methods in Molecular Biology, vol. 905, DOI 10.1007/978-1-61779-949-5_16, © Springer Science+Business Media, LLC 2012

A relatively straightforward approach to address the effect of a sRNA on its targeted mRNAs is to compare the level of a given mRNA in the presence or absence of the regulatory sRNA. Genomic methods such as microarray analysis and RNA sequencing have been used in the past to successfully identify new targets for sRNAs (10–13). Although they are very informative, these techniques often require validation to confirm the expression levels of sRNA and mRNA. A technique of choice for the validation of cellular RNA levels is to perform Northern blottings, which allow a highly specific and direct visualization of a sRNA or mRNA using an RNA or DNA probe. Northern blotting also gives additional information on the size and alternative processing products of a given RNA. In addition, the use of the antibiotic rifampicin, which specifically blocks RNA polymerase-dependent transcription initiation, helps to monitor the effect of a sRNA on the stability of a given mRNA.

Recently, we have combined the use of Northern blotting and construction of the reporter gene *lacZ* (β-galactosidase) fused to mRNAs to determine the minimal sequence necessary for sRNA-induced degradation of a mRNA (7). By constructing various lengths of *sodB*, *iscRSUA*, and *fumA* target mRNAs fused to the *lacZ* gene, we showed that the sRNA-induced initial cleavage of a target mRNA was frequently located far downstream, sometimes hundreds of nucleotides away from the sRNA pairing site.

Here, we describe a protocol to assess the effect of any given sRNA on the stability of a target mRNA. The approach described here is based on a previously published result from our laboratory (Fig. 2 of (14)). Briefly, rifampicin is used to stop transcription and total RNA is extracted with hot phenol. RNA is then resolved on a denaturing polyacrylamide gel and analyzed by Northern blotting. Using this technique, we have determined that the iron responsive sRNA RyhB destabilized the 3′ region of the *iscRSUA* polycistronic mRNA (*iscSUA*) but that it had no effect on the upstream 5′ region (*iscR*) (Fig. 1). Moreover, the protocol described here can be used to study the effect of any sRNA on the stability of any mRNA. We also give a brief description of how to determine the sRNA-dependant initial cleavage site on a target mRNA as we have previously published (Fig. 1 of (7)).

2. Materials

RNA is highly sensitive to degradation. It is therefore of great importance to take special precautions to avoid any RNases contamination. In this respect, all solutions should be prepared with sterile ultrafiltered water. RNase-free chemicals should be used and clean laboratory gloves should be worn during all the operations.

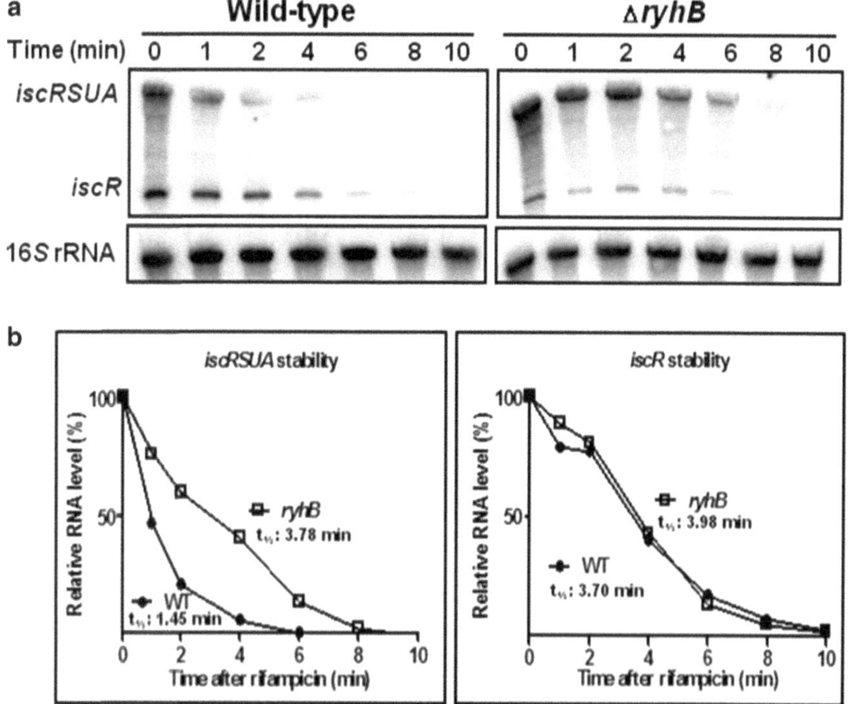

Fig. 1. RyhB decreases the stability of the *iscRSUA* transcript, not the *iscR* transcript. (**a**) Northern blot on total RNA extracted from wild-type and Δ*ryhB* cells and hybridized with an *iscR*-specific probe. Cells were grown in LB media to an OD_{600} of 0.5, at which point 2.2'-dipyridyl was added (200 μM final) for 10 min to induce the expression of RyhB and *iscRSUA*. Rifampicin was added (250 μg/mL final) at time 0 to block transcription. Total RNA was extracted at indicated times. 16S rRNA was used as a loading control. (**b**) Densitometry analysis of three Northern blots performed as in (**a**). (Reproduced from (14)).

2.1. RNA Extraction Components

1. LB media: Dissolve 10 g of tryptone, 5 g of yeast extract and 10 g of NaCl in 1 L of water. Sterilize by autoclaving.
2. 8× Lysis solution: 320 mM sodium acetate (pH 5.2), 16 mM ethylenediaminetetraacetic acid (EDTA), 8%w/v sodium dodecyl sulfate (SDS).
3. Phenol-water: Thaw phenol crystals at 65°C and preheat an equal volume of ultrafiltered water to the same temperature. Carefully mix equal parts of hot phenol and hot water in a glass cylinder or beaker. Add 0.1% (final concentration) 8-hydroxy-quinoline and mix with care (phenol is highly corrosive and can cause severe skin burns!). Incubate for 5 min at 65°C. Aliquot the water-saturated phenol in 50 mL conical tubes. Keep at –20°C until further use.
4. Phenol-chloroform-isoamyl alcohol: Mix water-saturated phenol, chloroform, and isoamyl alcohol at a 25:24:1 ratio.

5. Rifampicin: 12.5 mg/mL solution in methanol (see Note 1).
6. Agitating heating block that accommodates 1.5 mL tubes.

2.2. Acrylamide Gel Electrophoresis Components

1. 10× TBE: 0.89 M Tris, 0.89 M boric acid, 0.02 M EDTA (pH 8.0): Dissolve 108 g of Tris base and 55 g of boric acid in 700 mL of water. Add 40 mL of EDTA 0.5 M (pH 8.0). Adjust the volume to 1 L with water.
2. Acrylamide: 40% solution in water.
3. Ammonium persulfate (APS): 10% w/v solution in water.
4. *N*, *N*, *N*, *N*′-tetramethyl-ethylenediamine (TEMED).
5. Urea.
6. Formamide loading dye: 95% formamide, 0.5× TBE, 0.025% w/v of xylene cyanol and 0.025% w/v of Bromophenol blue.

2.3. Acrylamide Gel Blotting Components

1. Positively charged nylon membrane designed for nucleic acid transfer.
2. 3 mm Blotting paper.
3. Electroblot transfer device.
4. Transfer buffer: 0.5× TBE.
5. UV cross-linker.

2.4. Membrane Hybridization Components

1. Formamide.
2. 20× SSC: 3 M NaCl, 0.3 M trisodium citrate.
3. Denhardt's reagent: 1% Ficoll 400, 1% PVP (polyvinylpyrrolidone), 1% BSA.
4. Salmon sperm DNA.
5. PCR product serving as a template for the RNA probe transcription (see Note 2).
6. 10× Transcription buffer for T7 RNA polymerase (Ambion). Supplied with T7 RNA polymerase.
7. 100 mM ATP, CTP, GTP and UTP nucleotides.
8. RNase Out (Invitrogen).
9. $\alpha[^{32}P]$UTP (0.37 MBq/μL).
10. T7 RNA polymerase 20 U/μL.
11. Turbo DNase (Fermentas).
12. G50 Sephadex resin (Sigma-Aldrich).
13. Wash solution #1: 1× SSC, 0.1% SDS.
14. Wash solution #2: 0.1× SSC, 0.1% SDS.
15. Phosphorimager screen (GE Healthcare).
16. Phosphorimager and ImageQuant software.

3. Methods

3.1. RNA Extraction

1. Dilute an overnight bacterial culture 1/1,000 in 50 mL of LB medium containing the appropriate antibiotics (if necessary). Typically, we use a wild-type *E. coli* K-12 strain containing the endogenous copy of the RyhB sRNA and a strain harboring a chromosome deletion of the sRNA gene as control.
2. Mix 85 μL of 8× lysis solution with 685 μL of water-saturated phenol-water in 1.5 mL tubes. At least 15 min before the RNA extraction, preheat the phenol-lysis solution mix to 65°C in an agitating heating block (see Note 3).
3. Once the cells are at an OD_{600} of 0.5, induce the expression of the sRNA by the addition of 2,2'-dipyridyl (iron chelator) to a final concentration of 200 μM (see Note 4).
4. Incubate for 10 min and add rifampicin to a final concentration of 250 μg/mL (see Note 5).
5. At the designated time, take out 600 μL of the bacterial culture and mix with the preheated phenol-lysis solution mix. Initial time (Time 0) is typically selected immediately before the addition of rifampicin. Given the relatively short half-lives of bacterial mRNAs, a short time course is suggested (0, 1, 2, 4, 6, 8 and 10 min).
6. Using the agitating heating block, incubate with shaking (1,400 rpm) for 5 min at 65°C (see Note 3).
7. Centrifuge the tubes for 15 min at 16,000 ×*g*.
8. Perform a phenol-chloroform-isoamyl alcohol extraction of the aqueous phase by mixing an equal volume of PCI with the aqueous phase (~550 μL). Mix the phases by vortexing vigorously and spin 15 min at 16,000 ×*g*. Remove the aqueous phase (top layer) to a new tube and repeat the phenol-chloroform-isoamyl alcohol extraction once.
9. After the second phenol-chloroform-isoamyl alcohol extractions, remove the aqueous phase (top layer) to a new tube and add 3 V of 95% ethanol and allow precipitation to proceed for 1 h at –80°C (see Note 6).
10. Centrifuge at 16,000 ×*g* at 4°C for 20 min.
11. Discard the supernatant and wash the pellet with 500 μL of ice-cold 75% ethanol.
12. Centrifuge at 16,000 ×*g* at 4°C for 5 min.
13. Discard the supernatant and air-dry the pellet for about 10 min.
14. Resuspend the pellet in 15 μL of 10 mM Tris buffer.
15. Quantitate the RNA using a spectrophotometer (see Note 7)

3.2. Acrylamide Gel Electrophoresis

1. Prepare a 4% acrylamide (see Note 8), 8 M urea gel by mixing 1 mL of 40% acrylamide, 1 mL of 10× TBE and 5 g of urea. Add water to 10 mL final volume.
2. Slightly heat the mix to completely dissolve urea.
3. Add 100 μL of APS and 10 μL of TEMED, and pour the mix inside a gel cassette.
4. Insert a 10-well comb immediately avoiding introduction of any air bubbles.
5. Prepare the RNA samples by mixing 5 μg of RNA (complete to 2.4 μL with water) and 12.6 μL of formamide loading dye.
6. Assemble the migration apparatus and fill the upper and lower reservoir with 1× TBE.
7. Heat the samples for 2 min at 90°C before loading and flush all the wells with a syringe to remove leached urea.
8. Load the sample and run migration at 100 V. Use the tracking dye to determine the right amount of time to run your samples (Fig. 2).

3.3. Electro-Transfer

1. Set up the transfer system as shown in Fig. 3.
2. Prepare 1 L of transfer buffer (0.5× TBE).

% polyacrylamide	Xylene cyanol	Bromophenol blue
4	155 nt	30 nt
6	110 nt	25 nt
8	75 nt	20 nt
10	55 nt	10 nt

Fig. 2. Relative mobility of tracking dyes on a denaturing polyacrylamide gel. Adapted from ref. (15).

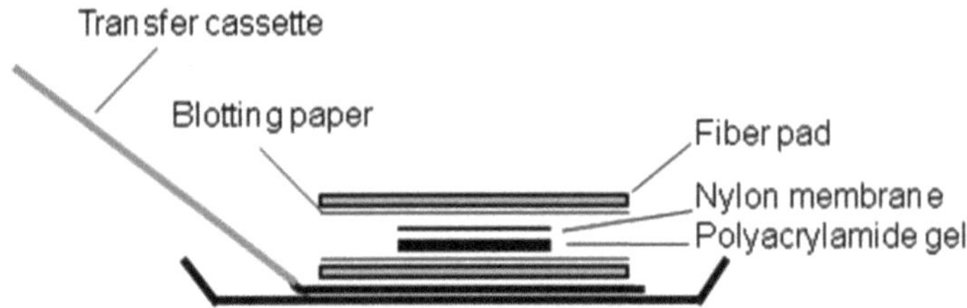

Fig. 3. Schematic representation of the transfer system described in this paper.

3. Cut out a nylon membrane slightly larger than the size of the acrylamide gel. Pre-wet in water for 2 min and then equilibrate for 5 min in the transfer buffer.
4. Cut out two pieces of 3 mm blotting paper of the size of the fiber pads and pre-wet in transfer buffer.
5. Open the transfer cassette in a glass dish containing 1–2 cm deep of transfer buffer. Place one soaked fiber pad onto the transfer cassette (black side of the Biorad transfer cassette at the bottom).
6. Place one pre-wetted blotting paper onto the fiber pad.
7. Place the gel onto the blotting paper.
8. Lay the nylon membrane over the gel. It is important to remove any air bubbles by pressing in a rolling motion with a pipette over the membrane.
9. Lay the second pre-wetted blotting paper over the nylon membrane and again press by rolling with a pipette to remove any air bubbles and to make sure that the gel and the membrane are in close contact.
10. Cover with the second pre-wet fiber pad and close the cassette.
11. Place the cassette in the buffer tank and fill to the maximum with transfer buffer.
12. Transfer at 4°C for 2 h at 100 mA or overnight at 30 mA.

3.4. Probe Hybridization

1. After the transfer, remove the membrane and place on a tray. Gently wash with 0.5× TBE. Cross-link (1,200 J) using a UV crosslinker.
2. Prepare the pre-hybridization solution (20 mL total): Mix 10 mL of formamide, 5 mL of 20× SSC, 2 mL of 10% SDS, 2 mL of 50× Denhardt's solution, 800 μL of water and 200 μL of salmon sperm DNA (incubate the salmon sperm DNA for 5 min at 95°C before using). Pre-heat the pre-hybridization solution to 60°C for 10 min.
3. Place the cross-linked nylon membrane in a hybridization bottle, RNA facing the inside. Add the pre-hybridization solution and incubate at 60°C for a minimum of 4 h in a rotating hybridization oven (see Note 9).
4. Transcribe an RNA probe by mixing the following components: 2 μL of 10× transcription buffer, 2 μL of 4 mM ATP/CTP/GTP mix, 2 μL of 0.1 mM UTP, 7.5 μL of water, 0.5 μL of RNase Out, 2 μL of a PCR product which contains the T7 RNA polymerase promoter (~0.5 μg of DNA), 3 μL of UTP-α-P^{32}, and 1 μL (20 U) of T7 RNA polymerase.

5. Incubate for 2 h at 37°C.
6. Add 1 μL of DNase Turbo.
7. Incubate for 15 min at 37°C.
8. Complete the volume to 100 μL and perform a phenol-chloroform extraction.
9. Purify the probe on a G-50 Sephadex column.
10. Add the radiolabeled RNA probe into hybridization bottle and hybridize for a minimum of 4 h (see Note 9).
11. After the hybridization, discard the pre-hybridization solution containing the RNA probe. Wash three times for 15 min with the wash solution #1 and once for 15 min with the wash solution #2 (see Note 10). Perform all the washes at 65°C.
12. Wrap the membrane in a plastic wrap and expose onto a phosphorscreen cassette (see Note 11).
13. Scan the phosphorscreen with the Typhoon scanner.

3.5. Data Analysis

1. Using an image analysis software (e.g., ImageQuant), quantify the intensity of each band of interest.
2. Subtract the background value (the intensity of an area of the same size than your bands of interest) to the value of each band of interest.
3. Relativize the value of each band to the value of the band at time 0, which correspond to 100% of RNA level.
4. Using scientific graphing software (e.g., GraphPad Prism), plot a graph (relative RNA level (%) vs. time after rifampicin (min)) and calculate the half-lives of each RNA species of interest (value at which there is 50% of RNA left compared to time 0).
5. Compare the half-lives values obtained in the presence and absence of sRNA expression to determine the effect of a given sRNA on the stability of a target mRNA.

3.6. Method for Mapping a sRNA-Induced Initial Cleavage Sites on a mRNA

1. Construct translational and transcriptional (see Note 12) fusions of a gene of interest with the *lacZ* gene (see Note 13). Vary the length of the gene of interest included in the fusion as done in Fig. 4.
2. Extract the RNA as described above from strains harboring the different *lacZ* fusions and expressing or not the sRNA of interest.
3. Perform a Northern blot analysis as described above. Use a probe specific to the *lacZ* gene.
4. The initial cleavage site is located in the region included between the longest sRNA-resistant fusion and the shortest sRNA-sensitive fusion.

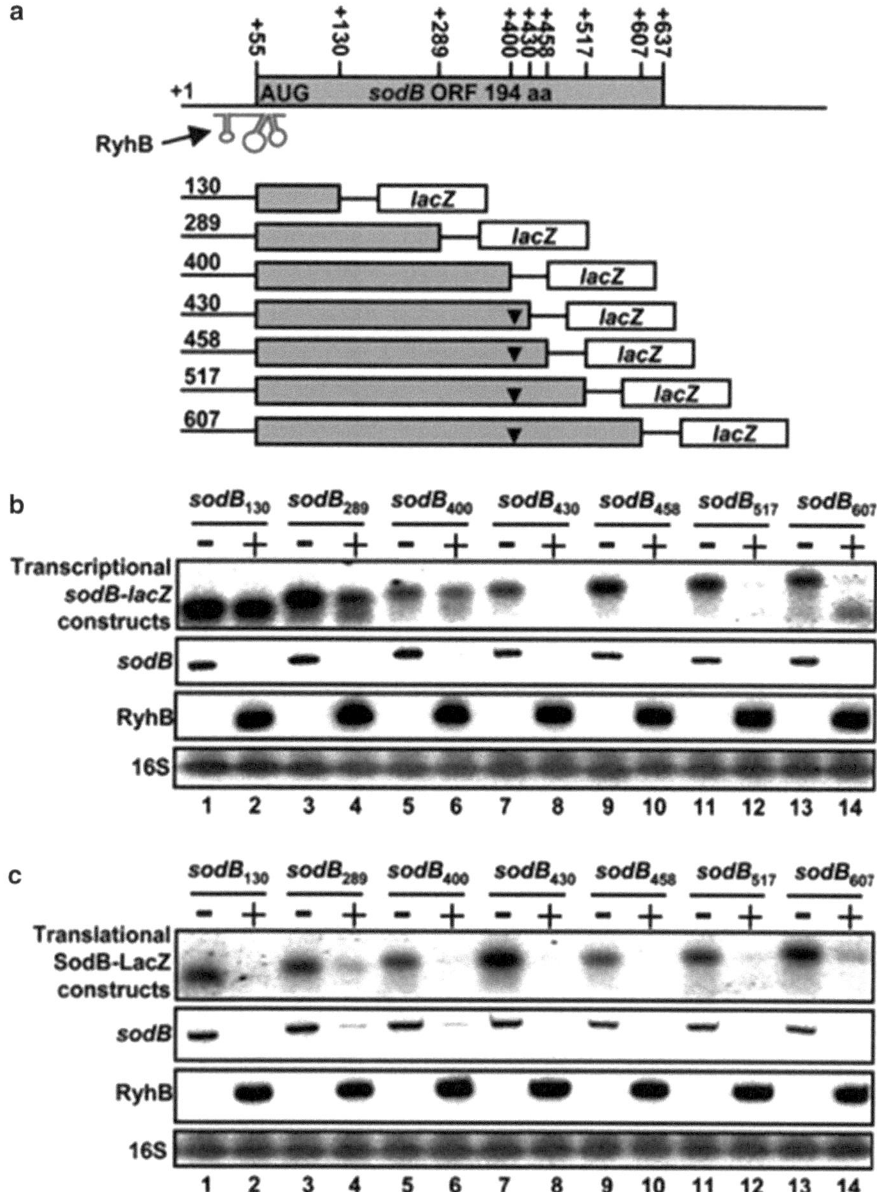

Fig. 4. Minimal sequence requirement for sRNA-induced mRNA degradation in vivo. (**a**) Description of the different constructs used to determine the minimal *sodB* sequence to generate RyhB-induced mRNA degradation. (**b**) Northern blots with a *lacZ*-specific probe showing the effect of RyhB expression (after 10 min) on various transcriptional *sodB-lacZ* constructs. The expression of endogenous *sodB* transcript and RyhB is also shown. 16S ribosomal RNA was used as a loading control. (**c**) Northern blots using a *lacZ*-specific probe showing the effect of RyhB expression (after 10 min) on various translational SodB-LacZ constructs. The expression of endogenous *sodB* transcript and RyhB is also shown. 16S ribosomal RNA was used as a loading control. (Reproduced from (7)).

4. Notes

1. Rifampicin is toxic. Always wear protective gloves when working with it. In addition, always use a fresh solution (maximum shelf life is 2 weeks).
2. This PCR product should contain a T7 polymerase promoter sequence (taatacgactcactatag) and produce an anti-sense RNA of approximately 300 nts.
3. If an agitating heating block is not available, a regular heating block maintained at 65°C can be used and the sample-phenol mix should be mixed every 20 s.
4. When working with a sRNA other than RyhB, use the appropriate physiological conditions that allow expression of the sRNA (different growth phase or inducer). It is also possible to work with plasmids (e.g., pBAD), which allow the expression of a given sRNA following the addition of arabinose.
5. The time of sRNA expression before the addition of rifampicin needs to be carefully selected. A significant level of sRNA expression is preferred. Furthermore, the level of target mRNA must be enough to allow visualization by Northern blotting.
6. It is also possible to precipitate the RNA overnight at −80°C.
7. Using a spectrophotometer, measure the absorbance of your sample at 260 nm (A_{260}). A solution of RNA at 40 μg/mL has an A_{260} of 1 for a path length of 1 cm. It is also possible to assess the purity of your RNA sample. For purified RNA, the A_{260}/A_{280} ratio should be approximately of 2. A ratio <2 suggests protein contamination while a ratio >2 suggests phenol contamination.
8. Depending on the size of the RNA of interest, a final concentration of acrylamide between 4 and 10% is possible. Both the sRNA and its mRNA target can be detected on the same acrylamide gel by preparing a two acrylamide concentration gel. We typically use 10% (bottom)/5% (top) acrylamide gels.
9. The pre-hybridization and hybridization steps can be performed from 4 h up to 48 h.
10. Waste containing radioactive material should be discarded in the radioactive waste.
11. Exposure time of the membrane can be from a few seconds to a few days, depending on the expression level of the RNA of interest as well as the strength of the probe.
12. In transcriptional fusions, both the gene of interest and the β-galactosidase gene harbor a translation initiation region, which allows independent translation from each other.

However, in translational fusions both genes are in-frame and produce a single fused peptide.

13. For more details on the cloning procedures to construct transcriptional and translational β-galactosidase fusions, please refer to (7).

Acknowledgments

We thank Gilles Dupuis for comments on the manuscript. This work was funded by an operating grant MOP69005 to E.M. from the Canadian Institutes for Health Research (CIHR). G.D. is a Ph.D. Scholar from the FQRNT (*Fonds Québécois de la Recherche sur la Nature et les Technologies*). E.M. is a FRSQ (*Fonds de la Recherche en Santé du Québec*) Junior II Scholar.

References

1. Waters LS, Storz G (2009) Regulatory RNAs in bacteria. Cell 136:615–628
2. Deana A, Belasco JG (2005) Lost in translation: the influence of ribosomes on bacterial mRNA decay. Genes Dev 19:2526–2533
3. Prévost K, Salvail H, Desnoyers G, Jacques JF, Phaneuf E, Massé E (2007) The small RNA RyhB activates the translation of shiA mRNA encoding a permease of shikimate, a compound involved in siderophore synthesis. Mol Microbiol 64:1260–1273
4. Massé E, Escorcia FE, Gottesman S (2003) Coupled degradation of a small regulatory RNA and its mRNA targets in *Escherichia coli*. Genes Dev 17:2374–2383
5. Morita T, Maki K, Aiba H (2005) RNase E-based ribonucleoprotein complexes: mechanical basis of mRNA destabilization mediated by bacterial noncoding RNAs. Genes Dev 19:2176–2186
6. Ikeda Y, Yagi M, Morita T, Aiba H (2011) Hfq binding at RhlB-recognition region of RNase E is crucial for the rapid degradation of target mRNAs mediated by sRNAs in *Escherichia coli*. Mol Microbiol 79:419–432
7. Prévost K, Desnoyers G, Jacques JF, Lavoie F, Massé E (2011) Small RNA-induced mRNA degradation achieved through both translation block and activated cleavage. Genes Dev 25:385–396
8. Caron MP, Lafontaine DA, Massé E (2010) Small RNA-mediated regulation at the level of transcript stability. RNA Biol 7:140–144
9. Vogel J, Luisi BF (2011) Hfq and its constellation of RNA. Nat Rev Microbiol 9:578–589
10. Massé E, Vanderpool CK, Gottesman S (2005) Effect of RyhB small RNA on global iron use in *Escherichia coli*. J Bacteriol 187:6962–6971
11. Papenfort K, Pfeiffer V, Mika F, Lucchini S, Hinton JC, Vogel J (2006) SigmaE-dependent small RNAs of *Salmonella* respond to membrane stress by accelerating global omp mRNA decay. Mol Microbiol 62:1674–1688
12. Guillier M, Gottesman S (2006) Remodelling of the *Escherichia coli* outer membrane by two small regulatory RNAs. Mol Microbiol 59:231–247
13. Beisel CL, Storz G (2011) The base-pairing RNA spot 42 participates in a multioutput feedforward loop to help enact catabolite repression in *Escherichia coli*. Mol Cell 41:286–297
14. Desnoyers G, Morissette A, Prévost K, Massé E (2009) Small RNA-induced differential degradation of the polycistronic mRNA iscRSUA. EMBO J 28:1551–1561
15. Sambrook J, Russell DW (2001) Molecular cloning: a laboratory manual, vol 2. Cold Spring Harbour, New York

Part VI

Interactions of Regulatory RNA with RNA Polymerase and the Ribosome

Chapter 17

Native Gel Electrophoresis to Study the Binding and Release of RNA Polymerase by 6S RNA

Karen M. Wassarman

Abstract

RNA–protein interactions are critical in diverse aspects of gene expression and often serve to mediate regulatory events. Many procedures are available to gain information about RNA–protein interactions. They span from initial identification of an interaction, such as through co-immunoprecipitation studies, to highly detailed atomic resolution definition of the interaction gained from crystallographic and NMR studies. One of the most versatile techniques uses native gel electrophoresis to study RNA–protein complexes, which is often called band shift, gel retardation, or electrophoretic mobility shift assays. Gel shift assays have been used to study a plethora of RNA–protein interactions in all organisms, but here we will use the 6S RNA:RNA polymerase interaction from *Escherichia coli* as an example to direct discussion of questions that can be addressed, including the ability to follow the dynamics of complexes over time.

Key words: 6S RNA, RNA polymerase, pRNA synthesis, RNA–protein interaction, RNA complex, EMSA

1. Introduction

In the simplest terms, gel shift assays rely on the fact that an RNA–protein complex is larger than the RNA alone, and the "shift" refers to the change in apparent size of the free RNA to a larger species. However, factors in addition to mass of the RNA–protein complex influence migration distance, most notably charge and shape of the complex, and therefore the distance of the "shift" is not a true indication of size (see ref. (1, 2)). However, the binding of a protein and the altered migration allow for the fraction of RNA in a bound or free state to be easily analyzed. One of the big advantages of gel shift assays is the ease with which they can be carried out in any lab with basic molecular biology equipment, and the

Kenneth C. Keiler (ed.), *Bacterial Regulatory RNA: Methods and Protocols*, Methods in Molecular Biology, vol. 905, DOI 10.1007/978-1-61779-949-5_17,

high sensitivity that requires low quantities of RNA and protein. In addition, depending on the nature of questions being addressed, protein purity is often not a factor and assays can even be carried out using crude cell lysates. In all cases, appropriate controls to monitor specificity of detected complexes are necessary for interpretation of complexes detected. Gel shift assays do require fairly high affinity interactions between RNA and protein to maintain complexes in the gel, but there is an apparent increase in the stability of RNA–protein complexes in gels compared to solution (3), making this assay amenable to study a wide range of interactions.

The most straightforward question that can be addressed by a gel shift assay is whether a given RNA and a given protein form a complex. Typically a purified, labeled RNA of interest is incubated with a purified protein of interest and then free RNA is separated from bound RNA on a native gel, such as shown for 6S RNA binding to the σ^{70} containing form of RNA polymerase ($E\sigma^{70}$) (Fig. 1). Once a complex is detected, the specificity of that complex can be

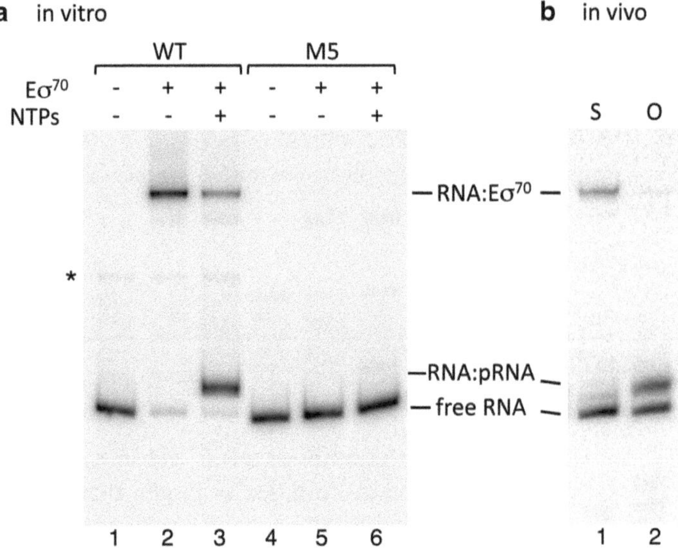

Fig. 1. Native gel separation of the 6S RNA:RNA polymerase complex from free 6S RNA and the 6S RNA:pRNA duplex. (**a**) In vitro assay using 20 nM labeled wild-type 6S RNA (WT) or an inactive mutant RNA (M5) was incubated in the absence of RNA polymerase (*lanes 1* and *4*) or in the presence of 20 nM active $E\sigma^{70}$ RNA polymerase followed by incubation in the absence (*lanes 2* and *5*) or presence (*lanes 3* and *6*) of 100 μM NTPs to promote pRNA synthesis. Reactions were incubated with 0.1 mg/mL heparin prior to loading onto a 5% polyacrylamide, 0.5× TBE, 5% glycerol gel, and the gel was run at room temperature for 2 h. Note the presence of a low level of alternatively folded 6S RNA present in the input RNA, readily identified as it is present in the free RNA lane (marked with *asterisk*). (**b**) Analysis of in vivo generated complexes. Cell lysate from stationary phase cells (S) or cells in outgrowth (O; 2 min after dilution of stationary phase cells into LB) was incubated with 0.1 mg/mL heparin and loaded onto a native gel as above. 6S RNA containing complexes were detected by Northern analysis using an RNA probe specific for 6S RNA.

tested by making mutations in the presumed RNA binding site, in the RNA binding region of the protein, or by competition with other specific and/or nonspecific RNAs. Alternatively, for an RNA–protein complex for which there is little information about the requirements for binding, examining RNA mutants and protein variants can be very helpful for defining the region or nature of the interaction. For instance, in the case of 6S RNA, which is a mostly double-stranded RNA with a large central single-stranded bulge, mutant RNAs that decrease the single-stranded bulge abolish binding activity (e.g., M5 RNA in Fig. 1a) (4). Testing protein variants also identified residues within the specificity subunit of RNA polymerase, σ^{70}, which are required for 6S RNA binding (5).

In addition to determining qualitatively whether an RNA and protein interact, gel shift assays can be used to quantitatively determine binding constants and kinetic parameters dictating the interaction. For example, the binding constant for CsrA binding to *hfq* mRNA and association rates of 6S RNA for RNA polymerase have been determined by gel shift assays (5, 6).

Finally, gel shift assays can be quite facile at studying changes in complexes over time or as a result of the influence of additional factors. For example, the *Escherichia coli* 6S RNA has been shown to be a template for RNA polymerase to generate a product RNA (pRNA) (7). Denaturing gels easily demonstrated the de novo generation of pRNA upon incubation of 6S RNA:RNA polymerase complexes with substrate nucleotide tri-phosphates (NTPs). However, the real biological significance of this reaction is likely to be the release of RNA polymerase from 6S RNA through the process of pRNA synthesis. Gel shift assays readily demonstrated this release (Fig. 1a), as well as allowing the identification of various intermediate complexes through differential labeling techniques (7). Specifically, 6S RNA-containing complexes were identified by following labeled 6S RNAs under different conditions (e.g., ±RNA polymerase, NTPs), and complexes that contained the pRNA were identified by performing reactions with labeled nucleotides that were incorporated into pRNA to allow detection of pRNA containing complexes.

Interestingly, the 6S RNA is released from RNA polymerase as a 6S RNA:pRNA duplex that migrates differently from free 6S RNA (see Fig. 1a). The ability to follow generation of 6S RNA:pRNA duplexes was instrumental in demonstrating that the pRNA synthesis reaction occurs in vivo during outgrowth from stationary phase (Fig. 1b; (7)). In this case, neither purified RNA nor protein was used, but instead native gel electrophoresis was used to examine in vivo generated complexes by fractionating a crude cell lysate and the complexes of interest were identified by Northern analysis using a 6S RNA-specific probe.

Here we outline the method for a typical gel shift assay that can be used to analyze an RNA protein interaction such as 6S RNA and RNA polymerase. Labeled RNA is generated in vitro and gel purified prior to binding assays. Purified RNA is incubated with purified protein or protein complexes. The extent of complex formation can then be visualized after separation of free and bound RNA by native gel electrophoresis.

2. Materials

2.1. In Vitro RNA Synthesis

1. Phage RNA polymerases (available commercially).
2. 10× transcription buffer: 400 mM Tris (pH 7.8), 200 mM NaCl, 60 mM $MgCl_2$, 20 mM Spermidine, 100 mM DTT.
3. Nucleotide tri-phosphates: α-[^{32}P]-CTP, 3,000 Ci/mmol, 2 mM ATP, 2 mM GTP, 2 mM UTP, and 0.5 mM CTP (see Note 1).
4. Linear template DNA, 1 mg/mL.
5. TE pH 7.5: 10 mM Tris (pH 7.5), 1 mM EDTA.
6. 5 M ammonium acetate.
7. 3 M sodium acetate.
8. Ethanol.
9. Megascript kit (Ambion, Applied Biosystems).
10. DNase I, RNase free.
11. 5× TBE: 0.445 M Tris base, 0.445 M boric acid, 10 mM Na_2EDTA, pH 8.3.
12. Formamide load buffer: 80% formamide, 1× TBE, ~0.1% each xylene cyanol and bromophenol blue.
13. Gel elution buffer: 10 mM Tris (pH 7.5), 0.3 M sodium acetate, 5 mM EDTA, 0.1% SDS.

2.2. In Vitro Complex Formation

1. 10× HM buffer: 200 mM Hepes (pH 7.9), 50 mM $MgCl_2$.
2. 10× HK buffer: 200 mM Hepes (pH 7.9), 1.2 M KCl.
3. 50% Glycerol.
4. 0.1 M DTT.
5. Purified protein (e.g., RNA polymerase available from Epicentre).
6. Heparin, 1 mg/mL.
7. Total yeast RNA, 10 mg/mL.

2.3. Cell Lysates

1. Lysis buffer: 20 mM Tris (pH 7.5), 150 mM KCl, 1 mM $MgCl_2$, 1 mM DTT.
2. Glass beads, 0.1 mm diameter, RNase free (see Note 2).
3. Heparin, 1 mg/mL.

2.4. Native Gel Electrophoresis

1. 30% Polyacrylamide (37.5:1) such as Protogel (National Diagnostics).
2. 50% Glycerol.
3. 10% Ammonium persulfate.
4. TEMED.
5. Native load dyes: 50% glycerol, 0.5× TBE, ~0.1% each xylene cyanol and bromophenol blue.
6. Heparin, 1 mg/mL.

2.5. Detection

1. Whatman 3MM paper and plastic wrap.
2. Phosphor screen/Phosphorimager.
3. Uncharged nylon membrane, such as Hybond N (GE Healthcare).
4. 0.5× TBE: 0.0445 M Tris base, 0.0445 M boric acid, 1 mM Na_2EDTA, pH 8.3.

3. Methods

For all methods here, care must be taken to work in an RNase-free environment and with RNase-free materials (see Note 3). In addition, safety precautions for use and disposal of radiolabeled material must be followed.

3.1. Preparation of Purified RNA for In Vitro Studies

To generate RNA in vitro, phage RNA polymerase (e.g., T3, SP6, or T7) is used to transcribe off a linearized DNA template (see Note 4). Both labeled and unlabeled RNAs are done similarly, but the specific activity of the RNA desired depends on the concentration needed in the reaction (see Note 5).

1. For [^{32}P]-labeled RNA, mix 5 μL of 10× transcription buffer, 1 μL of α-[^{32}P]-CTP (5 μCi), 0.5 μL of 0.5 mM CTP, 1 μL each of 2 mM ATP, 2 mM GTP, and 2 mM UTP, 1 μL of DNA template at 1 mg/mL (see Note 6), 1 μL of phage RNA polymerase, and 38.5 μL of water, and incubate for 1 h at 37°C. If higher yields are desired, after the first incubation add an additional 1 μL of phage RNA polymerase and incubate for 1 more hour at 37°C.
2. Ethanol precipitate the RNA taking safety precautions because the samples contain a high level of ^{32}P (see Note 7). Add 50 μL of TE pH 7.5, 100 μL of 5 M ammonium acetate and 500 μL of ethanol, incubate on ice for at least 20 min, centrifuge for 15 min, remove the supernatant, and wash the pellet with ~1 mL of 75% ethanol (see Note 8).

3. Resuspend the RNA pellet in 7.5 μL of formamide load buffer.
4. RNAs must be purified on denaturing polyacrylamide gels; typically 6% polyacrylamide with 0.4 mm spacers and wells ~1.5 cm wide are used for 6S RNA. RNAs are visualized after exposure to a phosphor screen or film, and RNA bands are excised from the gel and eluted overnight at 37°C in 500 μL gel elution buffer.
5. RNA is then ethanol precipitated (add 1/10 volume 3 M sodium acetate, 2.5 volumes ethanol, incubate on ice >20 min, spin 15 min, remove supernatant), resuspended into water, and quantified by counting in a scintillation counter.
6. For unlabeled RNA, repeat step 1 but omit radiolabeled CTP and include unlabeled CTP at the same concentration as the other NTPs. Alternatively, Megascript kits are considerably more efficient at generating high quantities of unlabeled RNA (see Note 9).
7. Following the reaction, add 1 μL of DNase I (RNase free) and incubate at 37°C for 20 min.
8. Ethanol precipitate the RNA as in step 2.
9. Resuspend the RNA in 10–15 μL of formamide load buffer (see Note 10).
10. Unlabeled RNA should be gel purified similarly to step 4 except 1 mm thick spacers and 10 mm wide wells are used, RNA is visualized by UV shadowing (8) (see Note 11), RNA bands are eluted overnight at 37°C in ~2 mL gel elution buffer, and RNA is quantified by absorbance at 260 nm.

3.2. Preparation of Samples to Examine In Vitro Generated Complexes

1. To prepare RNA for binding assays, labeled RNA (10^5 cpm) and unlabeled RNA (195 nM) are mixed in 1× HM buffer to allow for folding (see Note 12–14).
2. RNA and protein are mixed together under conditions appropriate for binding, such as 50 mM Hepes pH 7.9, 150 mM KCl, 5% glycerol, 1 mM DTT, 1 mM $MgCl_2$ with 6S RNA at 20 nM and RNA polymerase at 20 nM (4) (see Notes 15 and 16).
3. The specificity of an interaction should be monitored by adding a competitor (e.g., heparin, nonspecific RNA, or specific RNA "sink") followed by further incubation to allow nonspecific or short-lived complexes to dissociate. For 6S RNA, 1.1 μL of 1 mg/mL heparin is added to a 10 μL reaction, followed by incubation for 2 min at room temperature (see Note 17).
4. Samples should be loaded onto the gel immediately (see Subheading 3.4).

3.3. Preparation of Samples to Examine In Vivo Generated Complexes

To study in vivo generated complexes, a cell lysate needs to be used. There are many methods to make cell lysates including sonication, French press, and bead beater. The following is one simple method that is particularly useful for generation of small volume lysates which can be readily applied to preparation of multiple samples in parallel.

1. Collect ~10^{10} cells by centrifugation and resuspend in 200 μL of lysis buffer (see Note 18).
2. Add 200 μL of glass beads, vortex for 30 s followed by 15 s on ice, and repeat nine times more.
3. Add 400 μL additional lysis buffer and spin in a microcentrifuge tube for 10 min at 4°C. The supernatant is the cell lysate and should be used immediately (see Note 19).
4. Typically, 10 μL of cell lysate is separated by Native gel electrophoresis (see Subheading 3.4).

3.4. Preparation of a Native Polyacrylamide Gel

1. To make a gel, mix 8.3 mL of 30% polyacrylamide, 5 mL of 5× TBE, 5 mL of 50% glycerol, and 31.7 mL of water. Add 500 μL of 10% ammonium persulfate and 50 μL of TEMED to the gel mix, pour into taped plates (17 cm × 15 cm × 1 mm), and insert a comb (10 well comb, each well 10 mm width). Let polymerize completely (≥20 min) (see Note 20).
2. Set up the gel on a gel box with 0.5× TBE for running buffer, and gently rinse out wells with 0.5× TBE. Pre-run for 30 min at 200 V at room temperature (see Note 21).
3. To load each sample, add an equal volume of native load dyes, quickly but gently mix by pipetting up and down, and load one-half of the sample onto the running gel before moving on to the next sample (see Note 22). For time-sensitive samples (such as those treated with heparin to reduce nonspecific interactions) the samples should be loaded on a precise time course that matches the time course of heparin addition (see Note 23).
4. Continue running the gel at 200 V at room temperature for 2 h (see Note 24).

3.5. Detection of Complexes

1. If complexes are labeled to allow direct detection, open the gel plates, place two sheets of dry 3MM paper larger than the gel onto the gel. If the complexes are not labeled, start at step 4.
2. Peel off the paper and the gel should stick well to the paper (see Note 25). Cover the gel with plastic wrap and dry the gel >1 h at 80°C on a gel drier.
3. Expose the gel to a phosphorimager screen and visualize on a phosphorimager. Analyze results as described in Subheading 3.6.

4. If complexes are not labeled, Northern analysis can be done (see Note 26) after transfer of RNA to nylon membrane by electroblotting as below (see Note 27).
5. For Northern analysis, soak the gel in 0.5× TBE for 5–10 min. Also wet a nylon membrane about the same size as the gel in a separate dish of 0.5× TBE.
6. Remove the gel from the soaking dish while still on the glass plate (e.g., use the glass plate to support the gel as it is lifted out of the solution) (see Note 28).
7. Wet two sheets of 3MM paper slightly larger than the gel and place on the gel.
8. Flip over the glass plate/gel/paper stack and gently peel the paper with the gel stuck to it off the glass plate.
9. Place the gel and paper (paper side down) onto a sponge from the transfer device.
10. Place the nylon membrane about the same size as the gel onto the gel.
11. Roll out any bubbles between the gel and membrane with a glass tube or pipet.
12. Wet two more sheets of 3MM paper slightly larger than the gel and place on top of the membrane.
13. Roll out any bubbles with glass tube or pipet.
14. Put the gel/membrane sandwich into the tank filled with 0.5× TBE. The membrane should be towards the positive pole.
15. Transfer overnight at 10 V at 4°C.
16. Remove the membrane, keeping track of which side was against the gel (i.e., the side with the RNA on it) and bake between two sheets of 3MM paper at 80°C, such as on a gel drier.
17. Probe the membrane according to your favorite method. We prefer to use RNA probes generated by the same method as in Subheading 3.1, steps 1 and 2. Gel purification of the RNA probe is not necessary (see Note 29).

3.6. Analysis of Results

The presence of a shifted band compared to free RNA (e.g., the location of RNA on the gel without added protein) indicates an RNA–protein complex has formed. More complicated questions, and deserving of some consideration, are whether the complex(es) detected is specific and biologically relevant. Specificity can be addressed by examination of mutant RNAs and proteins, which also provide more information about the nature of the interaction. However, they often require much additional work. At a minimum, binding of an unrelated RNA to the protein of interest can be assessed by inclusion of nonspecific RNA (total RNA, e.g.) or chemicals such as heparin to test if they disrupt the complex. In the case of RNA polymerase, which has a fairly high nonspecific RNA binding activity, differentiation between specific interaction with 6S RNA and nonspecific interactions was readily distinguished by challenge with either heparin (as in Fig. 1a) or by inclusion of nonspecific RNA in binding reactions (4). Biological relevance is a harder question, but looking at interactions in more complex situations (a cell lysate, e.g.) and by other techniques is likely to help ascertain when and why the complex of interest might be important. For 6S RNA, genetic and phenotypic studies have characterized roles for 6S RNA in regulation of transcription and cell survival (reviewed in ref. (9)).

In some cases, a specific shifted species is not observed, but instead the RNA smears throughout the lane on native gels. Such results may indicate the lack of a specific complex, but might also result if the gel conditions are not as well suited to the RNA–protein complex of interest. For example, the 6S RNA:RNAP interaction was not well resolved in gels lacking glycerol, which may serve to stabilize RNA polymerase. Experimentation with binding and gel conditions often can resolve these difficulties to allow native gel electrophoresis to be applied to many diverse systems.

4. Notes

1. We typically use labeled CTP, but any α-labeled NTP can be used with adjustments to unlabeled nucleotide concentrations. Nonradioactive labeling methods also could be used (10, 11), but we prefer ^{32}P to avoid potential interference of the addition of bulky chemical groups to our RNA.
2. To ensure the beads are RNase-free, bake them at 190°C for 2 h and store in clean bottles. Optimal bead diameter depends on cell size; 0.1 mm beads work well for *E. coli* and *Bacillus subtilis*. Some commercial glass beads are more variable in size than others; Thomas Scientific glass beads are recommended.
3. Gloves are used at all times in the lab. Glassware is baked at 190°C for 2 h prior to use, or sterile plastics are used. Equipment and reagents are kept for RNA work only. Water is autoclaved in baked glassware, but no further treatment is required or recommended.
4. We typically use a linearized plasmid containing a phage promoter to generate a run-off RNA. Care should be taken to generate an RNA as close to the endogenous RNA as possible. Phage RNA polymerases prefer to initiate with GGG, but we have found initiation with GG does not reduce efficiency of 6S RNA transcription significantly and decreases the number of changed nucleotides at the 5′ end of the RNA. In addition, design of a restriction site to generate a run-off RNA with an

appropriate 3′ end is necessary. Alternatively, PCR products can be used for in vitro reactions.

5. The specific activity desired depends on whether one chooses to spike hot RNA into a reaction where most of the RNA concentration comes from unlabeled RNA, or if the labeled RNA is at the concentration needed in the reaction. The method here describes how to make a high specific activity RNA that would be supplemented with unlabeled RNA to use at concentrations >1 nM in binding reactions.

6. After restriction digestion of plasmid DNA to make linear template, it is best to avoid phenol extraction as carryover of phenol can be inhibitory in the transcription reaction. Reactions should be directly ethanol precipitated after restriction digestion and resuspended in water.

7. If desired, free label or excess NTPs can be removed prior to ethanol precipitation using G-50 microspin columns. Centrifuge the columns at 3,000 × *g* for 1 min to remove storage buffer, place the column in a clean microcentrifuge tube, load 50 μL of the transcription reaction onto the column, and centrifuge the column for 2 min at 3,000 × *g*. The RNA will flow through the column and is collected in the clean tube.

8. Using ammonium acetate as the counter ion in ethanol precipitation removes a higher percentage of free label in ethanol precipitation.

9. Follow the Megascript protocol for generation of small RNAs, which includes an incubation time of 4–6 h at 37°C.

10. Unlabeled RNA can be stored at −80°C indefinitely. Gel purified labeled RNA should be used within a couple of days for best results.

11. For UV shadowing, place gel between two sheets of plastic wrap and put against a white background (e.g., 3MM paper). In the darkroom using a handheld UV lamp, shine 254 nm light onto the gel and you should see a dark band ("shadow") appear. Circle with a marker to facilitate cutting out the band. Minimize time of UV exposure of gel to prevent RNA degradation and crosslinking.

12. We find that 10^4 cpm of labeled RNA per 10 μL reaction is optimal to obtain a reasonable signal (e.g., high enough radioactivity to detect easily, low enough to work with comfortably). For 6S RNA prepared as stated, the final concentration will be 20 nM in the protein binding reaction. If a higher or lower concentration is desired, unlabeled RNA can be increased or decreased in the folding mixture, or the ratio of hot to cold input CTP in the in vitro transcription reaction generating the RNA can be altered to make the desired concentration of RNA in 10^4 cpm/10 μL reaction.

13. Some RNAs require a more active folding regimen, such as heating to 90°C followed by slow cooling, but 6S RNA does not require further steps for folding beyond being present in 1× HM buffer for >2 min.
14. In some cases, it might be desired to generate RNA–protein complexes using unlabeled RNA. For example, complexes containing the 6S RNA-templated pRNA can be detected by incubating unlabeled 6S RNA:Eσ^{70} complexes with labeled nucleotides to generate labeled pRNA to follow by native gel electrophoresis.
15. Typically, binding is assessed with near equimolar ratios of 6S RNA and active RNA polymerase at 20–40 nM, at room temperature for 5–15 min. However, the concentrations, timing, and temperature of incubation will depend on the specific experiment. For example, binding may be slowed by decreasing temperature and concentrations to allow measurement of association rates (see ref. (5)). Conditions and timing of binding will vary depending on RNA:protein complex studied, and optimal conditions should be determined experimentally.
16. Alternatively, cell lysate can be used as a source of protein, in which case incubation is typically done in lysis buffer. If the lysate is fairly concentrated (e.g., ~2 mg/mL protein as prepared here), heparin treatment is required to simplify higher order complexes and allow samples to run into the gel.
17. The time of incubation with and choice of competitor will depend on the dissociation rate of the complex of interest. We typically use a 2 min incubation at room temperature with 0.1 mg/mL heparin to remove most nonspecific RNA:RNA polymerase interactions with no effect on specific 6S RNA:RNAP complexes. Alternatively, competitor can be added into the binding reaction; total yeast RNA at 100 μL/mL abolishes most nonspecific RNA binding to RNA polymerase while not effecting 6S RNA:RNAP complex formation or stability.
18. DTT should be added to lysis buffer immediately before use. Buffer should be chilled to 4°C before use.
19. The lysis buffer volume to bead ratio is critical for efficient lysis. Vortexing should be done in 2 mL flat bottomed microcentrifuge tube to maximize bead mixing. Additional lysis buffer is added after vortexing to facilitate recovery of the extract. These volumes and cell concentrations work well for *E. coli* and *B. subtilis*.
20. Gel dimensions can be altered, but this type of gel works well for our studies. Reducing the gel thickness or well width decreases the capacity of the gel system to allow separation of cell lysate effectively and to run with minimal band distortion

across wells. In addition, running time and voltage may need to be altered for different gel dimensions to prevent warming of gel during the run time.

21. Samples must be timed reasonably well to allow a 30 min pre-run, although ±10 min does not appear to alter gel performance significantly. If binding reactions are done at temperatures below room temperature, such as for kinetic studies, it is recommended that gels are run at 4°C rather than room temperature. In this case, gels and running buffer should be chilled to 4°C before use, and migration distances into gel will be altered relative to gels run at room temperature.
22. Loading a running gel results in a slight variation in migration distances of the same RNA across the gel. It is recommended to load the samples in order (e.g., left to right or right to left) to avoid apparent discontinuity in gel migration distances.
23. I find 10 s per sample is the limit for how fast an experienced researcher can load samples and remain precisely on time; >15–20 s is more reasonable for newer researchers.
24. Gels can be run for shorter or longer times depending on optimal distances for the complexes of interest, and the size/structure of free RNAs examined.
25. Native gels are very soft and easy to distort, so care should be taken to prevent stretching or pulling on the gel. Two sheets of 3MM paper are used to help prevent bleed through of radioactivity onto the gel drier.
26. Technically western analyses also could be done to examine proteins, but this type of native gel is best at differentiating free and bound nucleic acid and is not as useful for proteins. For example, RNA polymerase generally smears throughout the gel when not complexed with nucleic acid so it is difficult to quantify "free" protein.
27. Uncharged nylon membrane is preferred for use with RNA probes. Semi-dry blotting is difficult to make uniform across polyacrylamide gels forRNA:protein complexes and is not recommended.
28. Native gels are very soft and easily distorted, so care should be taken to move the gel around on a hard surface such as a glass plate to support the gel shape.
29. For RNA probes 100–300 nucleotides in length, Northern blots can be done by hybridization in 50% formamide, 1.5× SSPE, 1% SDS, 0.5% Blotto (powdered milk) at 55°C. Membranes are equilibrated with hybridization buffer without probe (prehybridization, 12.5 mL per blot in hybridization bottles) for >30 min at 55°C. The RNA probe (>10^6 cpm recommended) is heated to 90°C with 100 μL 10 mg/mL cRNA

for 2 min. The prehybridization buffer is exchanged for prewarmed hybridization buffer (12.5 mL) and the heated probe is added to the hybridization buffer in the hybridization tube. After >12 h at 55°C, blots are rinsed twice quickly with room temperature 4× SSC, 0.1% SDS (~50 mL each), and then three times 20 min with 0.1× SSC, 0.1% SDS (~75 mL each). Membranes are placed in plastic wrap, exposed on a phosphor screen, and RNAs detected with a phosphorimager. 20× SSPE: 3 M NaCl, 0.3 M sodium acetate at pH 7.0. 20× SSC: 3 M NaCl, 0.2 M sodium phosphate monobasic pH 7.4, 0.02 M EDTA

References

1. Fried MG (1989) Measurement of protein-DNA interaction parameters by electrophoresis mobility shift assay. Electrophoresis 10: 366–376
2. Lane D, Prentki P, Chandler M (1992) Use of gel retardation to analyze protein-nucleic acid interactions. Microbiol Rev 56:509–528
3. Cann JR (1989) Phenomenological theory of gel electrophoresis of protein-nucleic acid complexes. J Biol Chem 264:17032–17040
4. Trotochaud AE, Wassarman KM (2005) A highly conserved 6S RNA structure is required for regulation of transcription. Nat Struct Mol Biol 12:313–319
5. Klocko AD, Wassarman KM (2009) 6S RNA binding to Eσ^{70} requires a positively charged surface of σ^{70} region 4.2. Mol Microbiol 73:152–164
6. Baker CS, Eöry LA, Yakhnin H, Mercante J, Romeo T, Babitzke P (2007) CsrA inhibits translation initiation of *Escherichia coli hfq* by binding to a single site overlapping the Shine-Dalgarno sequence. J Bacteriol 189:5472–5481
7. Wassarman KM, Saecker RM (2006) Synthesis-mediated release of a small RNA inhibitor of RNA polymerase. Science 314:1601–1603
8. Hassur SM, Whitlock HW Jr (1974) UV shadowing–a new and convenient method for the location of ultraviolet-absorbing species in polyacrylamide gels. Anal Biochem 59: 162–164
9. Wassarman KM (2007) 6S RNA: a regulator of transcription. Mol Microbiol 65:1425–1431
10. Mansfield ES, Worley JM, McKenzie SE, Surrey S, Rappaport E, Fortina P (1995) Nucleic acid detection using non-radioactive labeling methods. Mol Cell Probes 9: 145–156
11. Pagano JM, Clingman CC, Ryder SP (2011) Quantitative approaches to monitor protein-nucleic acid interactions using fluorescent probes. RNA 17:14–20

Chapter 18

Ribosome Purification Approaches for Studying Interactions of Regulatory Proteins and RNAs with the Ribosome

Preeti Mehta*, Perry Woo*, Krithika Venkataraman*, and A. Wali Karzai

Abstract

Ribosomes are large complexes of RNA and protein that perform the essential task of protein synthesis in the cell. Ribosomes also serve as the initiation point for several translation-associated functions. To perform these tasks efficiently, ribosomes interact with a myriad of nonribosomal proteins and RNAs. Given that most of these interactions are transient, purification of the interacting factors in complex with the ribosome can be a challenging undertaking. Here, we review methods commonly used to isolate ribosomes and study ribosome-associated factors. We also discuss crucial parameters for designing and executing ribosome association studies. Finally, we present a detailed protocol for reporter based enrichment assays that are employed to selectively isolate ribosomes translating a particular message of interest. These protocols can be used to study a wide range of ribosome-associated functions.

Key words: tmRNA, SmpB, *trans*-translation, Ribosome, Translation, Sucrose gradient

1. Introduction

The bacterial 70S ribosome is a 2.7 MDa particle composed of two subunits, the large or 50S subunit and the small or 30S subunit. Basic techniques to isolate bacterial ribosomes were developed in the 1960s and 1970s (1–4), and are still widely used with relatively minor modifications. Most of these techniques require differential or density gradient ultra-centrifugation of cell lysates to yield ribosomes or ribosomal subunits of varying purity (Fig. 1). Crude ribosomes can be obtained by ultra-centrifugation of clarified cell lysate at 100,000×g. Tight-coupled ribosomes are compact and highly active ribosomal particles prepared by sucrose cushion centrifugation (5, 6). Linear sucrose gradients are the predominant

*The authors "Preeti Mehta, Perry Woo, and Krithika Venkataraman" have contributed equally to this work.

Kenneth C. Keiler (ed.), *Bacterial Regulatory RNA: Methods and Protocols*, Methods in Molecular Biology, vol. 905, DOI 10.1007/978-1-61779-949-5_18, © Springer Science+Business Media, LLC 2012

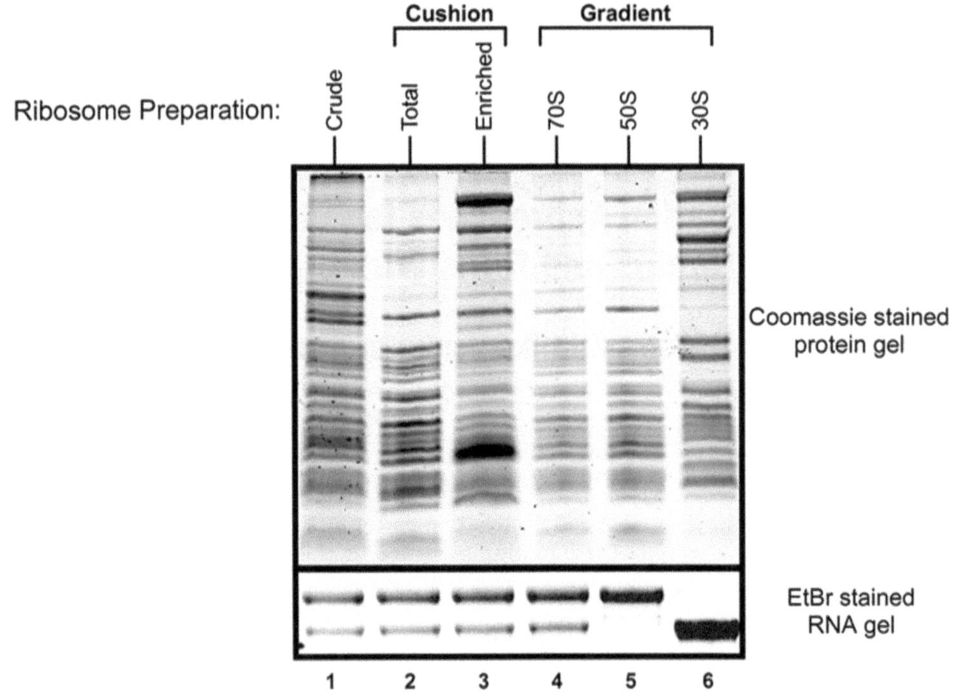

Fig. 1. Protein and RNA profile of ribosomes purified by various isolation techniques. From *left* to *right*, *lane 1*: crude ribosomes, *lane 2*: tight-coupled ribosomes from sucrose cushion, *lane 3*: ribosomes enriched post sucrose cushion, *lanes 4*, *5*, and *6* are 70S, 50S, and 30S ribosomal profiles, respectively, from an analytical sucrose gradient. *Top*: Coomassie stained protein gel. *Bottom*: ethidium bromide (EtBr) stained RNA gel showing the 23S and 16S rRNA bands.

option for obtaining ribosome profiles, separating polyribosomes and intact 70S ribosomes from the 30S and 50S subunits. Selection of the optimal method depends on the application, the protein or RNA of interest, and the specific requirements of the experiment.

In addition to decoding genetic information for protein synthesis, ribosomes also serve as a platform for a number of co-translational processing events such as protein folding (7, 8), enzymatic processing of the nascent polypeptide chain (9), and degradation of defective or nonstop mRNAs (10, 11). These processes require the ribosome to associate with a number of nonribosomal proteins and RNAs during the course of its function. Furthermore, ribosomes are known to associate with several regulatory and translation quality control factors. These include the SmpB-tmRNA ribosome rescue complex (12) and stress/starvation sensors such as RelA (13) and RelE (14). Non-ribosomal factors can interact with actively translating ribosomes or with one of the two ribosomal subunits. Knowing where these ribosome-binding factors lie on the functional ribosome landscape can provide significant information about their cellular roles. For instance, ribosome maturation factors usually associate with individual subunits and may not be present on mature 70S particles (15). Enzymes involved in post-translational modifications of the nascent polypeptide generally associate with the large subunit of the intact ribosome. Initial information on the binding site and subunit preference of a protein or RNA of interest can be obtained by Western or Northern blot analysis of a linear sucrose gradient profile of ribosomes. More directed experiments should then be designed to further characterize these interactions.

Factors that interact with functional ribosomes, particularly those that interact transiently or in a stage-specific manner with translating ribosomes, are best studied by isolating translationally active ribosomes. Such active ribosomes can be obtained either by polyribosome preparations, which can be arduous, or by specially designed affinity purification protocols that enable the enrichment of actively translating ribosomes using marked nascent polypeptide or RNA tags (16–18). Reporter based ribosome enrichment provides a powerful method for isolating ribosomes actively translating a particular mRNA. A protocol for enriching ribosomes translating a nonstop mRNA that lacks in-frame stop codons was developed in our laboratory to study *trans*-translation factors. This protocol can be easily tailored to study other translation-associated functions. In this approach, a reporter mRNA encoding an N-terminally His6-tagged protein is expressed in the desired strain. Total ribosomes are isolated and passed over a Ni^{2+}-NTA column to specifically capture/enrich ribosomes translating the reporter transcript. Eluted ribosomes can then be used for Western or Northern blot analysis to determine if the factor of interest associates with the captured ribosomes (Fig. 2a). We typically use an N-terminally His6-tagged λ-cI-N nonstop reporter mRNA to study *trans*-translation factors ((16, 19, 20) and Fig. 2b). Quantitative estimation of the differential enrichment of *trans*-translation factors (SmpB, tmRNA, or RNase R) on ribosomes can be obtained by using a nonstop reporter and a control "normal" reporter that contains an in-frame stop codon (Fig. 3, and (19)). Modifications of this method can afford a simple and powerful means of isolating active ribosomes, which can then be analyzed for the relevant interactions. Any affinity tag or epitope can be used to capture the protein, RNA, ribosome, and/or a combination of these factors. However, care must be exercised to ensure that the placement of the epitope does not interfere with the biological function of these factors. For instance, we have demonstrated that the SmpB protein is essential for recognition of stalled ribosomes by tmRNA (12), and that the C-terminal domain of SmpB plays a critical role in accommodation of tmRNA into the ribosomal A-site (21). Therefore, appending a His6 epitope, or any other epitope, to the C-terminus of SmpB would be a poor choice, as it severely affects the biological function of the protein. In contrast, appending an N-terminal His6 tag to SmpB does not interfere with its biological activity and has been successfully used to identify SmpB interacting partners (22).

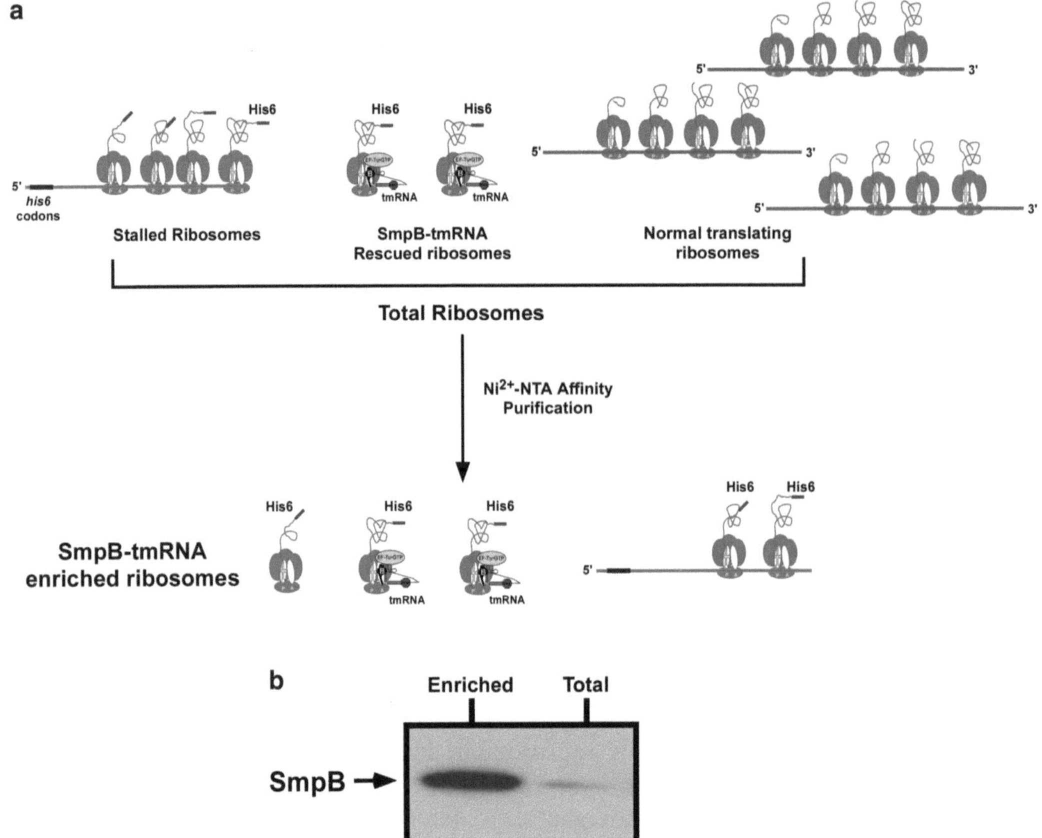

Fig. 2. Reporter based enrichment assay for ribosome-associated *trans*-translation factors. (**a**) Schematic representation of a ribosome enrichment experiment. Total ribosomes can be obtained via any of the methods described in the main text. Isolated total ribosomes comprise a mixture of those translating normal cellular mRNA, or the reporter nonstop mRNAs encoding a His6 epitope tag. Stalled and rescued *trans*-translating ribosomes can be separated from the normal ribosome pool by using a suitable affinity column. The figure depicts enrichment of ribosomes translating a nonstop mRNA encoding an N-terminal His6-tagged reporter protein using a Ni^{2+}-NTA affinity column. Enriched ribosomes are subjected to Western blot analysis, using antibodies specific to the protein of interest. (**b**) Representative Western blot showing enrichment of SmpB on ribosomes translating λ-cI-nonstop mRNA. Cushion purified total ribosomes were segregated into normal and stalled or *trans*-translating ribosomes using Ni^{2+}-NTA column chromatography. Total and eluted enriched ribosomes were normalized by A_{260} and resolved by electrophoresis on a 10% SDS-acrylamide gel. The gel was used for electrophoretic transfer and Western blot analysis using anti-SmpB antibodies. SmpB was enriched on captured ribosomes translating the reporter nonstop mRNA.

While designing co-purification experiments, careful consideration should be given to buffer composition. Protein–ribosome interactions exhibit a range of salt sensitivities. For example, ribosomal protein S1 and trigger factor are at the two extremes of the salt sensitivity spectrum. Interactions of S1 with the ribosome are known to be highly salt sensitive, with S1 falling off the ribosome at as low as 100 mM NH_4Cl (23). In contrast, interactions of trigger factor with the ribosome are highly resistant to salt washes, with trigger factor remaining bound to the ribosome at >500 mM NH_4Cl (24). In general, proteins with binding sites within the

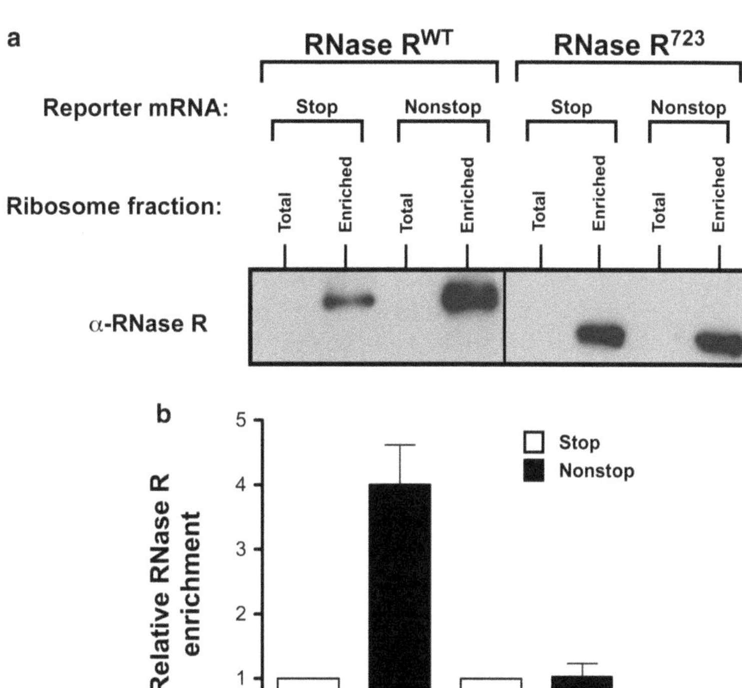

Fig. 3. Selective enrichment of RNase R on ribosomes, translating a nonstop mRNA, is dependent on its C-terminal tail. (**a**) Ribosomes were enriched for fractions translating the λ-cI nonstop and stop reporter mRNAs and tested for presence of RNase R using Western blot analysis. A C-terminal truncation variant of RNase R (RNase R^{723}) was also tested in this assay. (**b**) Quantitation of the fold-increase in levels of RNase R on ribosomes translating nonstop message compared to the control stop message. The level of RNase R is higher on enriched ribosomes that are stalled on the defective nonstop reporter mRNA. RNase R^{WT} enriches more on stalled ribosomes than the C-terminally truncated RNase R^{723}. This result illustrates that the ribosome enrichment assay is specific and quantitative.

interior of intact 70S ribosome are more resistant to dissociation as compared to surface bound proteins.

Most ribosome-binding proteins will pellet along with crude ribosomes under low stringency salt conditions (<100 mM NH_4Cl). A caveat to be mindful of is that use of low stringency condition could result in higher background signal, due largely to association of proteins that do not normally bind ribosomes or have any translation related function. Therefore, it is preferable to isolate ribosomes using either higher stringency conditions or the sucrose cushion approach (Fig. 1). However, it should be kept in mind that the higher stringency approaches might also cause the dissociation of some ribosome-surface associated factors. Consequently, the optimal method for one's favorite ribosome-associated factor often needs to be empirically determined. Use of non-physiological salt concentration can result in artifactual interactions and erroneous conclusions. For instance, SmpB and tmRNA are essential components of the ribosome rescue system that function as a complex during all stages of the *trans*-translation process (20, 25, 26). SmpB had been suggested to interact with the ribosome in the absence of tmRNA. However, some of these interactions were most likely due to nonspecific and off-pathway binding of SmpB to stalled ribosomes. Indeed, work done by Sundermeier et al. (20) convincingly demonstrated that the reported SmpB-ribosome interactions in the absence of tmRNA were non-physiological and attributable to the use of low stringency salt conditions during ribosome purification and binding studies.

Another important aspect while designing ribosome association studies is the scale of the experiment. At any given time, only a fraction of ribosomes in the cell may be associated with the protein or RNA of interest. These interactions can also be very transient and labile. This makes detection of these factors in small-scale cultures very difficult. The amount of starting material should depend on the sensitivity of the detection method and the nature of the interaction. Plasmid-borne expression of the desired protein or RNA is sometimes used to improve the signal, as it shifts the equilibrium towards greater interaction. However, over-expression might lead to increased nonspecific interactions that can be misleading. It is therefore important to keep the expression of the factor of interest as close to its endogenous level as possible.

As discussed above, isolation of auxiliary factors in context of the ribosome needs robust and physiologically relevant ribosome isolation techniques. Here, we provide a compilation of methods routinely used in our laboratory to obtain ribosomes, with special emphasis on ribosome–protein interactions. Finally, we provide details of specialized protocols developed in our laboratory to study interactions of ribosomes with RNase R and other *trans*-translation factors. This protocol can be readily adapted to studying specialized factors that transiently associate with ribosomes in other functional contexts. Particular care has been taken to maintain, as much as possible, physiologically relevant buffer and salt conditions during the entire processing and purification process.

2. Materials

All solutions should be prepared using ultrapure Milli-Q or DEPC treated water (see Notes 1 and 2). All plasticware and glassware should be new or treated with RNase ZAP (Ambion) (see Note 3). All reagents are analytical grade or higher. All buffers should be filter sterilized and stored at 4°C (see Note 4). Change gloves frequently to minimize RNase contamination (see Note 1).

2.1. Stocks

1. 2 M Tris pH 7.5: Weigh 242.27 g of Tris and transfer to a glass beaker. Add 750 mL of Milli-Q water. Mix using a magnetic stir-bar and adjust the pH to 7.5 using HCl (see Note 5). Adjust the volume to 1 L with Milli-Q water.
2. 1 M $MgCl_2$: Add 95.21 g of $MgCl_2$ to 700 mL of water. Adjust the volume to 1 L with Milli-Q water. Use caution while mixing $MgCl_2$ in water, as the reaction is exothermic.
3. 500 mM EDTA: Add 186.12 g of EDTA to 800 mL of water. Adjust the pH to 8.0 with NaOH. Adjust the volume to 1 L with Milli-Q water, filter, and store at room temperature.
4. 100 mM PMSF: Dissolve 0.87 g of PMSF (phenylmethylsulfonyl fluoride) in 50 mL of ethanol and store at –20°C (see Note 6).

2.2. Buffers

1. Buffer A: 20 mM Tris (pH 7.5), 300 mM NH_4Cl, 10 mM $MgCl_2$, 0.5 mM EDTA, 6 mM β-mercaptoethanol (β-ME) (see Note 7), 10 U/mL SuperASE-In (Ambion).
2. Buffer B: 20 mM Tris (pH 7.5), 300 mM NH_4Cl, 10 mM $MgCl_2$, 2 mM β-ME.
3. Buffer C: 20 mM Tris (pH 7.5), 300 mM NH_4Cl, 10 mM $MgCl_2$, 2 mM β-ME, 250 mM Imidazole. Store away from light at 4°C.
4. Storage Buffer: Buffer A containing 10% glycerol.
5. Wash Buffer: 20 mM Tris pH 7.5, 100 mM NaCl, 10 mM $MgCl_2$.

2.3. Cell Growth and Lysis

LB broth: Dissolve 25 g of LB broth Miller mixture per 1 L of water. Sterilize by autoclaving.

1. Wash buffer (see Subheading 2.2)

2.4. French Press

1. Prechilled 32 mL French press cell.
2. Buffer A (see Subheading 2.2).
3. PMSF and DNase I (RNase free).

2.5. Chemical Lysis

1. Lysozyme.
2. Bacterial protein extraction reagent: B-PER (Pierce # 78243).
3. DNase I (RNase free).

2.6. Crude Ribosome Preparation

1. Buffer A (see Subheading 2.2).
2. Beckman Ultra-Clear centrifuge tubes (# 344058).
3. Beckman SW28 swinging bucket rotor and ultracentrifuge (see Note 8).

2.7. Tight-Coupled Ribosomes

1. Buffer A (see Subheading 2.2).
2. Storage buffer (see Subheading 2.2).
3. 32% Sucrose solution in Buffer A.
4. Beckman polycarbonate bottles (# 355649).
5. Beckman 50.2 Ti rotor and ultracentrifuge.

2.8. Gradient Purified Ribosomes

1. Buffer A (see Subheading 2.2).
2. 10% and 40% Sucrose solutions in Buffer A.
3. Storage buffer (see Subheading 2.2).
4. 20 mL syringes.
5. Stainless steel blunt-end large bore needle.
6. 18 gauge or 22 gauge needles.
7. Beckman Ultra-Clear centrifuge tubes (# 344058).
8. Beckman SW28 swinging bucket rotor and ultracentrifuge.
9. BioComp Gradient Master 107ip.
10. UV-transparent 96 well plates.
11. Plate reader (Molecular devices SpectraMax M5E, or equivalent).

2.9. Reporter Based Enrichments Assays

1. Buffer B (see Subheading 2.2). Depending on the Ni^{2+}-NTA slurry used for purification, Buffer B can contain 10 mM Imidazole.
2. Ni^{2+}-NTA resin (GE Healthcare # 17-5318-02 or Sigma-Aldrich # P6611).
3. Buffer C (see Subheading 2.2).
4. Bio-Rad micro-Biospin columns.
5. Collection tubes.
6. Suitable antibodies for detection of protein of interest.
7. Materials for SDS-PAGE and Western blot analysis.

3. Methods

3.1. Cell Culture and Preparation of S30 Extracts

1. Grow cells to mid-log phase, Optical Density at 600 nm of ≈0.5 (OD_{600}≈0.5), in LB using baffled flasks at 37°C with vigorous shaking (250 rpm). Culture volumes and growth conditions can be varied depending on the nature of the experiment (see Note 9). Generally, the N-terminally His6-tagged reporter of interest is induced at OD_{600}≈0.5 for 1 h.

2. Harvest and wash cells with Wash Buffer (see Subheading 2.2). The cell pellet may be stored at –80°C. Perform all subsequent manipulations at 4°C.
3. Although a myriad of cell lysis methods exist, we routinely use either the French press or chemical lysis method. For isolating intact and functional ribosomes, we use the French press method for cell lysis. The French press cell should be prechilled at 4°C. Resuspend cell pellets in Buffer A (15 mL Buffer A containing 0.1 mM PMSF/L of culture) and transfer to a pre-chilled French press cell (see Note 10). Lyse cells by passage through a French press at 10,000 psi. Collect the lysed cells at a rate of ~15 drops per minute. Add DNase I to a final concentration of 5 U/mL and incubate for 15 min at 4°C. Spin the cell lysate at 30,000 × *g* for 30 min. To avoid transferring unwanted cellular debris, recover the top 75–85% of the supernatant and spin again at 30,000 × *g* for 30 min. The supernatant obtained in this step is referred to as the S30 fraction.

For analytical gradients and culture volumes of less than 200 mL, we use the chemical lysis approach. Resuspend cell pellets in Buffer A (100 μL/100 mL of culture) containing 0.4 μg/μL of lysozyme. Incubate at RT for 1 min. Freeze cells at –80°C. Add 500 μL of B-PER cell lysis reagent containing 10 U/mL of DNase I and 10 mM $MgCl_2$. Incubate on ice for 5 min. Subject cells to two additional cycles of freezing at –80°C and thawing on ice (see Note 11). Adjust the volume up to 1 mL with Buffer A and spin at 30,000 × *g* for 30 min at 4°C. To avoid transferring unwanted cellular debris, use the top 75–85% of the supernatant for ribosome preparation.

3.2. Crude Ribosome Preparations

This method separates ribosomes from the majority of other lower molecular weight cellular components.

1. Spin the S30 fraction from a 750 mL culture at 100,000 × *g* for 1 h in Ultra-Clear centrifuge tubes, using a Beckman SW28 rotor. Culture volumes can vary depending on the experiment.
2. Rinse the resulting crude ribosome pellet with Buffer A, with a gentle swirling motion. It is important to note that a salt wash procedure will remove some ribosomal proteins thus generating a heterogeneous ribosome population (see Note 12).
3. The ribosome pellet can be resuspended in any buffer of choice for further analysis. If intact ribosomes are required for downstream processing, use a buffer containing 10 mM $MgCl_2$. For subunit preparations, resuspend the ribosome pellet in low $MgCl_2$ Buffer A (containing 1 mM $MgCl_2$) before layering on a sucrose gradient. The quality of the ribosome preparation should be checked at this stage (see Note 13).

3.3. Tight-Coupled Ribosomes

Tight-coupled ribosomes refer to compact and highly active ribosomal particles purified via a sucrose cushion.

1. Prepare 32% sucrose solution in Buffer A.
2. Transfer 12.5 mL of 32% sucrose solution to an ultracentrifuge tube and carefully layer an equal volume of the S30 supernatant on top by pipetting along the walls of the tube (see Note 14). The tube must be filled to the brim to prevent cracking during ultra-centrifugation. The exact volumes can vary depending on the tube used.
3. Normalize weight across the tubes using Buffer A (see Subheading 2.2).
4. Spin in a Beckman 50.2 Ti rotor at 100,000 × g for 16 h at 4°C.
5. Discard the supernatant. The resulting ribosomes form a clear pellet. Occasionally, the clear ribosome pellet may be covered with a brown film. Wash the pellet in Buffer A to remove the brownish material. Resuspend the tight-coupled ribosomes in a buffer of choice for downstream processing. Ribosomes can be flash frozen in storage buffer and kept at −80°C for long-term storage.

3.4. Gradient Purified Ribosomes

Sucrose gradients are used to separate intact ribosomes from the individual subunits. Analytical gradients are used to determine the profile of ribosomes and subunits in cells. Gradients can be used to separate polyribosomes and 70S ribosomes from the 50S and 30S subunits based on their density. Fractions from analytical gradients can be used to determine the binding sites of non-ribosomal proteins on the ribosome. Small cultures (100–200 mL) are sufficient for this analysis. For this purpose, S30 fractions from chemically lysed cells (maximum 1 mL, 100–200 A_{260} units) can be directly layered on a gradient (see below). Preparative gradients from larger cultures may be needed to observe more transient interactions. For preparative gradients, crude ribosomes (see Subheading 3.2), or tight-coupled ribosomes (see Subheading 3.3) are completely resuspended in Buffer A and used. For ribosomal subunit preparations, it is important to limit the magnesium concentration in the buffer to 1.0 mM. Below, we provide a detailed protocol for preparative gradients:

1. Resuspend tight-coupled ribosomes (see Subheading 3.3), or crude ribosomes (see Subheading 3.2), in 1 mL Buffer A for isolating 70S ribosomes and in low $MgCl_2$ Buffer A (containing 1 mM $MgCl_2$) for subunit preparations. Typically, we use ribosomes isolated from 750 mL cultures.

2. Prepare a 10–40% linear sucrose gradient. Linear sucrose gradients can be prepared in several different ways. We generate linear sucrose gradients using the BioComp Gradient Master 107ip (see Note 15).
3. Load resuspended ribosomes (250–500 μL volume) onto a 40 mL 10–40% sucrose gradient. The percent range of the gradient can be varied when better resolution of the 70S ribosomes, or one of the subunits, is required. Overloading the gradient can impair the ability of the gradient to sufficiently resolve ribosomal subunits.
4. Spin the gradients at 82,705 ×*g* for 16 h in a Beckman SW28 rotor.
5. Fractionate the gradient either by using a fraction collector (see Note 16) or as follows: Mount the gradient tube onto a ring stand. Carefully pierce the bottom of the tube with an 18- or 22-gauge needle and remove the needle such that the flow is drop-wise (see Note 17). Fast flow rates can reduce resolution and make it more difficult to collect precise fractions. Collect the fractions in a 96 well plate, approximately 300 μL per well.
6. Determine the A_{260} values for each fraction by using a plate reader. Fractions might need to be diluted to obtain readings within the accuracy range of the spectrophotometer (see Note 18). Remember to use UV-transparent plates and mix the fractions well before measurement.
7. Plot the A_{260} values against the fraction number to determine the profile of the 70S, 50S and 30S fractions (Fig. 4).
8. The concentration of intact 70S ribosome and its subunits can be calculated based on the observation that 1 A_{260} unit of 50S is equivalent to 36 pmol/mL, 1 A_{260} unit of 30S is equivalent to 72 pmol/mL, and 1 A_{260} unit of 70S is equivalent to 24 pmol/mL (27).
9. Pool and flash freeze the required fractions in liquid nitrogen and store at −80°C. Glycerol can be added to a final concentration of 10% to serve as a cryo-protectant.

3.5. Reporter Based Ribosome Enrichment

Crude ribosome pellets (see Subheading 3.2) or tight-coupled ribosomes prepared by the sucrose cushion method (see Subheading 3.3) can be used to perform ribosome enrichment assays. Generally, a 750 mL starting culture is required per enrichment experiment.

1. Resuspend ribosome pellets in 5 mL of Buffer B (enrichment buffer) by gentle rocking at 4°C. The buffer normally contains 10 mM Imidazole, unless a different source of Ni^{2+}-NTA is used for downstream processing (see Note 19).

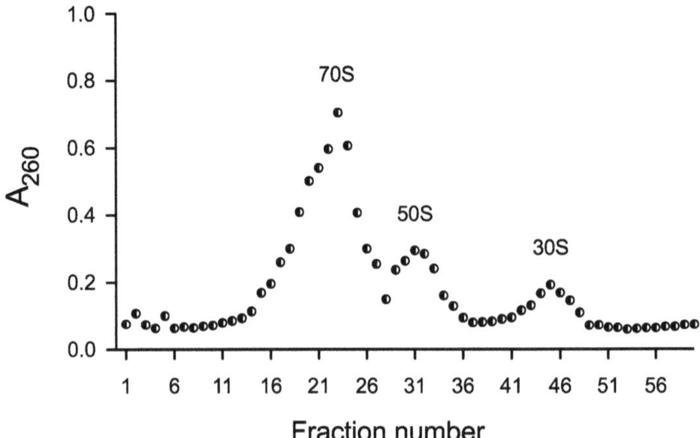

Fig. 4. A_{260} profile of S30 extract separated on a sucrose gradient. S30 extract from 100 mL of cell culture was layered onto a 10–40% sucrose gradient and spun at 82,705 × *g* for 16 h in a Beckman SW28 rotor. Fractions were collected in a UV-transparent 96-well plate and normalized for volume. A_{260} readings were obtained using a plate reader and plotted vs. fraction number.

2. Centrifuge the resuspended sample at 30,000 × *g* for 30 min. Carefully transfer the supernatant to a fresh tube.
3. Add 40 μL of Buffer B equilibrated Ni^{2+}-NTA resin (GE Healthcare) and incubate for 2 h at 4°C.
4. Mount a 1 mL Bio-Rad spin column onto a collection tube. Load the ribosome-Ni^{2+}-NTA slurry into the column. Allow the liquid to drain by gravity flow. The column should not be spun at this stage, and care must be exercised to prevent the resin from drying out.
5. Wash the column by addition of 1 mL of Buffer B. Allow the resin to settle for 1 min before opening the stopcock to permit even flow of the buffer through the settled resin. Repeat this procedure four times.
6. To elute bound ribosomes, wipe the bottom of the column with Kimwipe paper and transfer it to a collection tube. Add 200 μL of Buffer C, allow it to stand for 1 min, and then spin in a microcentrifuge at 6,000 × *g* for 1 min. Repeat the elution step by addition of another 200 μL of Buffer C.
7. Quantify the eluted samples by measuring A_{260} values. Assess enrichment of the factor of interest using Western or Northern blot analysis (see Note 20, and Fig. 5).

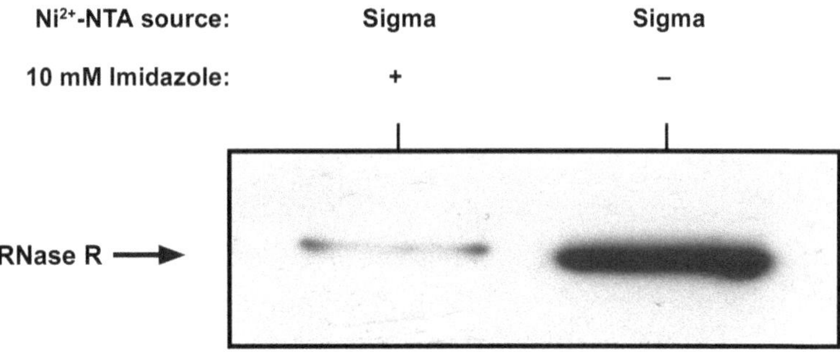

Fig. 5. Ribosome enrichment and the effect of Imidazole on the Ni^{2+}-NTA resin. Ni^{2+}-NTA resins from various manufacturers have different binding capacities and exhibit distinct sensitivities to the presence of Imidazole in enrichment buffer (Buffer B). We typically use Ni^{2+}-NTA resin from GE Healthcare with enrichment Buffer B that contains 10 mM Imidazole. We have also used Ni^{2+}-NTA slurry from Sigma-Aldrich. However, Ni^{2+}-NTA slurry from Sigma-Aldrich has lower binding capacity and does not work well with buffers containing Imidazole. To illustrate this point, we chose Ni^{2+}-NTA slurry from Sigma-Aldrich and performed ribosome enrichment assays in the presence or absence of 10 mM Imidazole in the Buffer B. Shown is a Western blot of enriched ribosomes probed with anti-RNase R antibodies.

4. Notes

Ribosome isolation protocols are generally robust and, with a little care, can yield reproducible and clean preparations. Certain important considerations and precautions are listed below:

1. To prevent RNase contamination, several important guidelines must be followed.
   (a) Use RNase-free water.
   (b) Wear disposable gloves to prevent contamination by ribonucleases present on your hands.
   (c) Use prewrapped disposable plasticware when possible.
   (d) Use reagents of the highest quality available.
   (e) Filter-sterilize all buffers for long-term storage.
2. DEPC treatment is commonly used to inactivate RNases. DEPC inactivates RNases by covalent modification. It is a strong but not an absolute inhibitor of RNases. Direct treatment of buffers with DEPC is not recommended because DEPC reacts with primary amines. As an alternative, high quality Milli-Q water can be used. DEPC treatment of water can be accomplished as follows:
   (a) Add DEPC to a final concentration of 0.1% to water.
   (b) Incubate at 37°C overnight.
   (c) Autoclave to degrade DEPC.

3. Glassware can be made RNase free by incubating in 0.1% DEPC overnight at 37°C followed by autoclaving for 30 min. Alternatively, use commercially available products such as RNase ZAP (Ambion).
4. All manipulations must be carried out at 4°C. We recommend that all solutions be prepared 1 day in advance to permit sufficient time for cooling.
5. The pH adjustment of Tris-containing solutions should be done at the final working temperature.
6. PMSF is toxic and should be handled with care. A stock solution of PMSF can be prepared in isopropanol, methanol, or ethanol at a concentration of 100 mM. It can be stored at -20°C. However, long-term storage is not recommended. PMSF should be added to solutions and buffers immediately before use.
7. Reducing agents (β-ME/DTT) should always be added to buffers immediately before use.
8. To help minimize potential RNase activity, we recommend prechilling rotors and centrifuges.
9. Broad specificity periplasmic RNases can be avoided by using an RNase I deficient strains, such as MRE600 or the Keio RNase I knockout strain. *E. coli* K12 strains A19, D10, or CAN20-19E can also be used. For functional studies, it is preferable to use RNase I deficient strains. However, with a little care, intact and fully functional ribosomes can easily be isolated from wild type *E. coli* strains.
10. It is advisable to maintain low concentrations of EDTA in buffers during the initial stages of cell lysis and S30 preparation, even in the presence of $MgCl_2$, to prevent metal-induced RNA cleavage. For isolation of intact 70S ribosomes the $MgCl_2$ concentration should be much higher (tenfold or more) than the EDTA concentration.
11. Additional freeze–thaw cycles can be used if required. If the lysate is very viscous add an additional 2 U/mL of DNase I and incubate for 1–5 min at room temperature
12. It is important to keep in mind that a salt wash procedure on any ribosome pellet will partially remove some ribosomal proteins, and can thus generate a heterogeneous ribosome population. A number of ribosomal proteins have been observed to dissociate with a buffer containing 0.5 M NH_4Cl (28).
13. The quality of a ribosome preparation should be determined by standard RNA gel analysis. The integrity of 16S, 23S, and 5S rRNA can be directly visualized using 1.0–1.5% formaldehyde agarose gels (see Fig. 1).

14. Care must be taken while layering the cleared cellular lysate on top of 32% sucrose solution. Two distinct layers should be observed in order to achieve consistency between experiments.
15. There are several protocols available for making sucrose gradients. These alternative protocols can be used if a gradient maker is not available. For example, Luthe (29) provides a simple and reproducible protocol that does not require any special equipment. Here, we describe a method using a BioComp Gradient Master 107ip.
    (a) Prepare a 10 and a 40% sucrose solution in Buffer A. For subunit preparation, use Buffer A containing 1 mM $MgCl_2$.
    (b) To layer 10% sucrose solution on top of the 40% sucrose solution, we use a BioComp marker block to designate the half-full point in the centrifuge tube. This point can vary depending on the type of tube-cap that is used. The long cap is designed to leave a 10 mm gap above the finished gradient and a short cap is designed to leave a 4 mm gap above the finished gradient. We prefer using the long cap. Using the marker block, make a half-full mark on the tube with a fine-tip permanent marker. Gently pipette the 10% sucrose solution to the half-full mark. Place an equivalent volume of the 40% sucrose solution in a syringe equipped with a blunt-end stainless steel needle. The stainless steel needle must be sufficiently long to span the length of the centrifuge tube. Insert the tip of the needle to the bottom of the tube and slowly dispense the 40% sucrose solution to upwardly displace the 10% solution. It is important to maintain the distinct interface between the two solutions.
    (c) Add approximately 1 mL of the 10% sucrose solution to the top, gently close the tube with the long BioComp cap, and place the tubes in the BioComp Gradient Master. When inserting the cap, it is important to ensure that there are no air bubbles trapped inside the tube. The presence of any air bubbles can interfere with gradient formation. Select the parameters of the desired gradient (10–40% w/v sucrose in this case) according to the manufacturer's instructions. Once the gradient is ready, carefully remove the cap and load the desired volume (250–500 μL) of your ribosome sample.
16. A density gradient fractionator available from Teledyne Isco can be used when fractioning the sucrose gradient instead of the described method. This system allows the A_{260} of a fractionated sample to be monitored throughout the fractionation process.

17. Use caution when piercing the bottom of the tube with a needle. Do not push the needle all the way into the gradient. Holding the centrifuge tube firmly in one hand and the needle in the other, carefully push the needle into the bottom of the tube with a slight twisting motion. The needle should be barely visible on the interior of the tube. Once the tube is pierced the needle should be removed before collection. Alternatively, a generic flow regulator (two-way stopcock) can be attached to the needle prior to piercing the tube to ensure an even flow through the needle and stopcock.
18. For preparative gradients, we usually dilute the sample 1:20 in water. It is important to mix the samples in the original plate before taking aliquots for dilution.
19. Not all Ni^{2+}-NTA agarose resins work similarly in enrichment experiments. Our enrichment protocol is optimized with the GE Healthcare Ni^{2+}-NTA agarose resin (#17-5138-02). We have also used Ni^{2+}-NTA resin from Sigma-Aldrich (# P6611) with good success. It is essential to be cognizant of the differences in the protein binding capacity (mg protein/mL of resin) and sensitivity to Imidazole of the various Ni^{2+}-NTA resins. Choosing the optimal buffer conditions, during the binding and washing steps, are critical for obtaining valid and reproducible ribosome enrichment results. As an example, we have examined the difference in enrichment of RNase R on ribosomes using the Sigma-Aldrich slurry under two different buffer conditions. We performed enrichment assays to test the efficiency of the Sigma slurry in Buffer B with and without 10 mM Imidazole. The samples were eluted in enrichment Buffer C. Normalized samples were evaluated by Western blot analysis, using anti-RNase R antibodies, to probe for the presence of RNase R (Fig. 5).
20. To obtain an accurate quantification of enrichment of a factor of interest on the captured ribosomes, it is important to normalize samples based on their A_{260} value.

References

1. Britten RJ, Roberts RB (1960) High-resolution density gradient sedimentation analysis. Science 131:32–33
2. McQuillen K, Roberts RB, Britten RJ (1959) Synthesis of nascent protein by ribosomes in *Escherichia coli*. Proc Natl Acad Sci U S A 45:1437–1447
3. Nomura M, Tissières A, Lengyel P (1974) Ribosomes. Cold Spring Harbor Laboratory, Cold Spring Harbor
4. Ron EZ, Kohler RE, Davis BD (1968) Magnesium ion dependence of free and polysomal ribosomes from *Escherichia coli*. J Mol Biol 36:83–89
5. Bonincontro A, Nierhaus KH, Onori G, Risuleo G (2001) Intrinsic structural differences between “tight couples” and Kaltschmidt-Wittmann ribosomes evidenced by dielectric spectroscopy and scanning microcalorimetry. FEBS Lett 490:93–96
6. Risuleo G, Gualerzi C, Pon C (1976) Specificity and properties of the destabilization, induced by initiation factor IF-3, of ternary complexes of the 30-S ribosomal subunit, aminoacyl-tRNA and polynucleotides. Eur J Biochem 67:603–613
7. Kramer G et al (2002) L23 protein functions as a chaperone docking site on the ribosome. Nature 419:171–174

8. Wegrzyn RD, Deuerling E (2005) Molecular guardians for newborn proteins: ribosome-associated chaperones and their role in protein folding. Cell Mol Life Sci 62:2727–2738
9. Bingel-Erlenmeyer R et al (2008) A peptide deformylase-ribosome complex reveals mechanism of nascent chain processing. Nature 452:108–111
10. Ge Z, Karzai AW (2009) Co-evolution of multipartite interactions between an extended tmRNA tag and a robust Lon protease in *Mycoplasma*. Mol Microbiol 74:1083–1099
11. Richards J, Mehta P, Karzai AW (2006) RNase R degrades non-stop mRNAs selectively in an SmpB-tmRNA-dependent manner. Mol Microbiol 62:1700–1712
12. Karzai AW, Susskind MM, Sauer RT (1999) SmpB, a unique RNA-binding protein essential for the peptide-tagging activity of SsrA (tmRNA). EMBO J 18:3793–3799
13. Block R, Haseltine AW (1975) Purification and properties of stringent factor. J Biol Chem 250:1212–1217
14. Hurley JM, Cruz JW, Ouyang M, Woychik NA (2011) Bacterial toxin RelE mediates frequent codon-independent mRNA cleavage from the 5′ end of coding regions in vivo. J Biol Chem 286:14770–14778
15. Shajani Z, Sykes MT, Williamson JR (2010) Assembly of bacterial ribosomes. Annu Rev Biochem 80:501–526
16. Sundermeier T et al (2008) Studying tmRNA-mediated surveillance and nonstop mRNA decay. Methods Enzymol 447:329–358
17. Youngman EM, Green R (2005) Affinity purification of in vivo-assembled ribosomes for in vitro biochemical analysis. Methods 36: 305–312
18. Zhou Z, Reed R (2003) Purification of functional RNA-protein complexes using MS2-MBP. Curr Protoc Mol Biol chap 27:unit 27.3
19. Ge Z, Mehta P, Richards J, Karzai AW (2010) Non-stop mRNA decay initiates at the ribosome. Mol Microbiol 78:1159–1170
20. Sundermeier TR, Karzai AW (2007) Functional SmpB-ribosome interactions require tmRNA. J Biol Chem 282:34779–34786
21. Sundermeier TR, Dulebohn DP, Cho HJ, Karzai AW (2005) A previously uncharacterized role for small protein B (SmpB) in transfer messenger RNA-mediated trans-translation. Proc Natl Acad Sci U S A 102: 2316–2321
22. Karzai AW, Sauer RT (2001) Protein factors associated with the SsrA.SmpB tagging and ribosome rescue complex. Proc Natl Acad Sci U S A 98:3040–3044
23. Szer W, Hermoso JM, Leffler S (1975) Ribosomal protein S1 and polypeptide chain initiation in bacteria. Proc Natl Acad Sci U S A 72:2325–2329
24. Hesterkamp T, Hauser S, Lutcke H, Bukau B (1996) *Escherichia coli* trigger factor is a prolyl isomerase that associates with nascent polypeptide chains. Proc Natl Acad Sci U S A 93: 4437–4441
25. Dulebohn D et al (2007) Trans-translation: the tmRNA-mediated surveillance mechanism for ribosome rescue, directed protein degradation, and nonstop mRNA decay. Biochemistry 46:4681–4693
26. Karzai AW, Roche ED, Sauer RT (2000) The SsrA-SmpB system for protein tagging, directed degradation and ribosome rescue. Nat Struct Biol 7:449–455
27. Christodoulou J et al (2004) Heteronuclear NMR investigations of dynamic regions of intact *Escherichia coli* ribosomes. Proc Natl Acad Sci U S A 101:10949–10954
28. Gnirke A, Geigenmuller U, Rheinberger HJ, Nierhaus LH (1989) The allosteric three-site model for the ribosomal elongation cycle. Analysis with a heteropolymeric mRNA. J Biol Chem 264:7291–7301
29. Luthe DS (1983) A simple technique for the preparation and storage of sucrose gradients. Anal Biochem 135:230–232

Chapter 19

Analysis of Aminoacyl- and Peptidyl-tRNAs by Gel Electrophoresis

Brian D. Janssen, Elie J. Diner, and Christopher S. Hayes

Abstract

During protein synthesis, ribosomes translate the genetic information encoded within messenger RNAs into defined amino acid sequences. Transfer RNAs (tRNAs) are crucial adaptor molecules in this process, delivering amino acid residues to the ribosome and holding the nascent peptide chain as it is assembled. Here, we present methods for the analysis of aminoacyl- and peptidyl-tRNA species isolated from *Escherichia coli*. These approaches utilize denaturing gel electrophoresis at acidic pH to preserve the labile ester bonds that link amino acids to tRNA. Specific aminoacyl- and peptidyl-tRNAs are detected by Northern blot hybridization using probes for tRNA isoacceptors. Small peptidyl-tRNAs can be differentiated from aminoacyl-tRNA through selective deacylation of the latter with copper sulfate. Additionally, peptidyl-tRNAs can be detected through metabolic labeling of the nascent peptide. This approach is amenable to pulse-chase analysis to examine peptidyl-tRNA turnover in vivo. We have applied these methods to study programmed translational arrests and the kinetics of paused ribosome turnover.

Key words: Aminoacyl-tRNA, Gel electrophoresis, Northern blot hybridization, Peptidyl-tRNA, Peptidyl-tRNA hydrolase, Pulse-chase analysis, Ribosome pausing, Translation

1. Introduction

Ribosomes use aminoacyl-tRNAs as precursors for the assembly of protein chains during translation. This strategy serves at least two purposes. First, activation of amino acids as acyl esters helps to promote peptide bond formation on the ribosome. Second, the covalently linked tRNAs are essential nucleic acid adaptors that decode the genetic information within mRNAs. Apart from translation initiation, all aminoacyl-tRNAs are delivered to the ribosome A site, where base-pairing interactions between codon and anticodon are monitored. Cognate codon–anticodon interactions promote

Kenneth C. Keiler (ed.), *Bacterial Regulatory RNA: Methods and Protocols*, Methods in Molecular Biology, vol. 905, DOI 10.1007/978-1-61779-949-5_19,

full accommodation of aminoacyl-tRNA into the A site for transpeptidation with the nascent peptidyl-tRNA, which is bound within the ribosome P site. As a result, the nascent peptide chain transferred onto the A-site tRNA, and the P-site tRNA is deacylated. Subsequent ribosome translocation repositions the deacylated and peptidyl-tRNAs into the E and P sites (respectively) and presents a new codon in the A site for decoding. Thus, the ribosome iteratively adds new amino acid residues to the C-terminus of the nascent polypeptide, which remains linked to tRNA continuously. This elongation cycle is repeated until a stop codon is presented in the A site. Protein release factors bind to A-site stop codons and promote hydrolysis of P-site peptidyl-tRNA to release newly synthesized proteins from the ribosome. Although aminoacyl-tRNAs play a central role in protein synthesis, they also perform other extra-ribosomal functions. In some bacteria, aminoacyl-tRNAs are used to modify membrane lipids and the peptidoglycan cell wall, and can also be involved in the synthesis of antibiotics (1).

Cellular aminoacyl-tRNA levels are largely determined by the relative rates of synthesis by aminoacyl-tRNA synthetases and consumption by the ribosome. In *Escherichia coli* and other bacteria, aminoacyl-tRNA synthesis is often limited by amino acid availability. Acute amino acid starvation leads to an abrupt decrease in tRNA aminoacylation and an attendant halt to protein synthesis. The resulting deacylated tRNA is a potent signal of nutritional stress in bacteria and plays a critical role in redirecting gene expression to adapt to amino acid starvation (2, 3). Deacylation of specific tRNAs can also occur during the overproduction of heterologous proteins in *E. coli*, resulting in translational pausing and poor protein yields. This phenomenon is often associated with non-preferred or "rare" codons, which are typically decoded by low-abundance tRNA isoacceptors. Although protein production can usually be increased through over-expression of the limiting isoacceptor, the underlying deficit is due to selective aminoacylation rather than the absolute level of tRNA (4–6). The over-expressed tRNAs compete more effectively with synonymous isoacceptors for aminoacylation, thereby providing sufficient aminoacyl-tRNA to support translation.

The first method to analyze aminoacyl-tRNAs in complex biological samples was described by Ho & Kan (7). Their method utilizes acid-urea polyacrylamide gel electrophoresis followed by Northern blot hybridization to detect specific aminoacyl-tRNAs. This approach exploits a small gel mobility shift to resolve aminoacylated tRNA from deacylated tRNA. Although the mass difference between aminoacyl- and deacylated tRNAs is generally small (<1%), acid-urea gel electrophoresis has been used successfully with several tRNAs and is capable of differentiating *N*-formyl-methionyl-$tRNA_i^{Met}$ from deformylated methionyl-$tRNA_i^{Met}$ (5, 8–10). More recently, Tao Pan, Måns Ehrenberg, and their colleagues have developed a microarray approach to quantitatively monitor the aminoacylation of all tRNA isoacceptors (4). This elegant approach gives a comprehensive view of tRNA aminoacylation, but requires specialized reagents and equipment that may not be available to many researchers.

Acid-urea gel electrophoresis can also be used to detect and quantify peptidyl-tRNAs in biological samples. As outlined above, peptidyl-tRNAs are intermediates of protein synthesis and are generally bound stably to the ribosome P site. However, in some instances peptidyl-tRNA dissociates from the ribosome in process termed "drop-off" (11–15). Drop-off occurs frequently with small peptidyl-tRNAs (three to seven amino acid residues), but these abortive translation products are rapidly broken down into constituent peptide and tRNA by peptidyl-tRNA hydrolase (Pth). Pth is an essential enzyme in bacteria, but peptidyl-tRNA drop-off products can be detected and studied in conditional *pth* mutants (12, 16, 17). Pth is unable to hydrolyze ribosome-bound peptidyl-tRNA (18, 19), and therefore peptidyl-tRNAs isolated from *pth*$^+$ cells most likely occupy the ribosome P site at the time of extraction. Thus, in principle, it is possible to infer the positions of ribosomes on a given mRNA by identifying the associated peptidyl-tRNAs. In practice, it is difficult to detect individual peptidyl-tRNA species even during gratuitous protein overproduction (20). Presumably, this reflects the asynchrony of protein synthesis, in which each of the many nascent peptidyl-tRNA intermediates is present at a low level. However, specific peptidyl-tRNAs accumulate to relatively high levels during translational pauses (20–24). In these instances, the position of the arrested ribosome can be determined by identifying the tRNA isoacceptor that is linked to the nascent chain.

Here, we apply acid-urea gel electrophoresis and Northern blot hybridization to detect and differentiate aminoacyl- and peptidyl-tRNA isolated from *E. coli* cells. Peptidyl-tRNAs carrying nascent peptides of ten or more amino acid residues are readily resolved from aminoacyl-tRNA on acid-urea gels, but it is often difficult to unambiguously identify smaller peptidyl-tRNAs. In these instances, aminoacyl-tRNAs can be specifically deacylated using copper sulfate treatment. We employ this strategy to detect small peptidyl-tRNAs that dissociate from the ribosome in response to erythromycin treatment. Finally, we present a pulse-chase protocol to examine the turnover of specific peptidyl-tRNAs in vivo.

2. Materials

All solutions should be prepared using NANOpure water (Thermo-Barnstead) or equivalent source of 18 MΩ-cm resistivity water. All chemical and radioactive wastes should be disposed of according to the appropriate institutional safety protocols.

2.1. Alkali Treatment

1. AcE storage buffer: 10 mM sodium acetate (pH 5.0), 1 mM EDTA.
2. 1 M Tris–HCl (pH 8.9).
3. 10 mM ethylenediaminetetraacetic acid (EDTA) (pH 8.0).
4. 3 M sodium acetate (pH 5.0).
5. Ice bucket.
6. Microcentrifuge.
7. 95% Ethanol.
8. 75% Ethanol.

2.2. Copper Sulfate Treatment

1. AcE storage buffer: 10 mM sodium acetate (pH 5.0), 1 mM EDTA.
2. 100 mM $CuSO_4$.
3. Microcentrifuge.
4. 95% Ethanol.
5. 75% Ethanol.
6. AcE loading buffer: 8 M urea, 10 mM sodium acetate (pH 5.0), 1 mM EDTA, 0.01% xylene cyanol, 0.01% bromophenol blue.

2.3. Acid-Urea Polyacrylamide Gel Electrophoresis

1. 3 M sodium acetate (pH 5.0).
2. 500 mM ethylenediaminetetraacetic acid (EDTA) (pH 8.0).
3. AcE gel-loading buffer: 8 M urea, 10 mM sodium acetate pH 5.0, 1 mM EDTA, 0.01% xylene cyanol, 0.01% bromophenol blue.
4. Urea (98% or greater purity).
5. 30% Acrylamide: bisacrylamide (29:1) solution.
6. 10% Ammonium persulfate (APS), prepare every few weeks and store at 4°C.
7. *N*,*N*,*N*′,*N*′-tetramethylethane-1,2-diamine (TEMED).
8. Prepare gel solution in 15 mL conical tube: add 3 g of urea, 0.6 mL of 1 M sodium acetate (pH 5.0), 12.5 μL of 500 mM EDTA (pH 8.0) and 1.2 mL of 30% acrylamide solution (29:1 acrylamide:bisacrylamide). For 12% gels, add 2.4 mL of 30% acrylamide solution. Add water to a final volume of 6 mL and mix solution on a rotisserie at room temperature until the urea is completely dissolved.
9. Sodium acetate electrophoresis running buffer: 100 mM sodium acetate (pH 5.0), 1 mM EDTA. Prepare in prechilled (4°C) water.
10. Mini-Protean 3 (Bio-Rad) or equivalent mini-gel electrophoresis system.

2.4. Northern Blot

1. 10× TBE: 0.89 M Tris–HCl (108 g/L), 0.87 M boric acid (54 g/L), 6.4 mM EDTA (1.86 g/L).
2. 20× SSC: 3 M sodium chloride (175.32 g/L), 0.3 M sodium citrate (88.23 g/L), adjust to pH 7.0 with 1 N HCl and filter.
3. 10% Sodium dodecyl sulfate (SDS).
4. 100× Denhardt's solution (250 mL): 5 g Ficoll 400, 5 g polyvinylpyrrolidone, 5 g bovine serum albumin (fraction V), freeze 50 mL aliquots at –20°C for long-term storage.
5. Hybridization solution (50 mL): to 29 mL of water, add 5 mL 100× Denhardt's solution, 15 mL 20× SSC and 0.5 mL of 10% SDS.
6. Whatman 3MM Chr filter paper: cut four pieces (70 mm × 90 mm) for each gel to be transferred.
7. Nylon membrane (0.45 μm pore, Nytran® SPC, Whatman).
8. Semi-dry electrotransfer apparatus.
9. Power supply.
10. Adenosine 5′-triphosphate [γ-^{32}P]-labeled (3,000 Ci/mmol).
11. Bacteriophage T4 polynucleotide kinase (PNK) and 10× reaction buffer.
12. Oligonucleotide probes: *E. coli* $tRNA_2^{Pro}$ probe, (5′-CAC CCC ATG ACG GTG CG); *E. coli* $tRNA_3^{Gly}$ probe, (5′-CTT GGC AAG GTC GTG CT); and *E. coli* $tRNA_2^{Arg}$ probe, (5′-CCT CCG ACC GCT CGG TTC G).
13. Sephadex G-25 spin columns equilibrated in 10 mM Tris–HCl (pH 8.0), 1 mM EDTA (see Note 1).
14. Bambino II rotisserie hybridization oven.
15. Plastic food wrap.
16. 8 × 10-in. Phosphorimaging screen.
17. Phosphorimager.

2.5. Pulse-Chase

1. MOPS-tricine-NaCl solution: 0.5 M 3-morpholinopropane-1-sulfonic acid (MOPS), 0.05 M tricine, 0.625 M NaCl (pH 7.4).
2. $FeCl_2$/trace metal solution (for 100 mL): 5 g $FeCl_2 \cdot 4H_2O$, 184 mg $CaCl_2 \cdot 2H_2O$, 64 mg H_3BO_3, 40 mg $MnCl_2 \cdot 4H_2O$, 18 mg $CoCl_2 \cdot 6H_2O$, 4 mg $CuCl_2 \cdot 2H_2O$, 340 mg $ZnCl_2$, 605 mg $Na_2MoO_4 \cdot 2H_2O$, 8 mL concentrated HCl, q/s to 100 mL.
3. O solution (500 mL): 28.8 g $MgCl_2 \cdot 6H_2O$, 10 mL of $FeCl_2$/trace metal solution, q/s to 500 mL and filter to sterilize.
4. S solution: 275 mM K_2SO_4 (4.8 g/100 mL).

5. N solution: 3.74 M NH_4Cl (20 g/100 mL).
6. 0.5 M K_2HPO_4 (dibasic).
7. 1 mg/mL thiamine-HCl.
8. 40% D-glucose.
9. MOPS defined media (10 mL): To 8.97 mL of sterile water add 20 μL of O solution, 800 μL of MOPS-tricine-NaCl solution, 10 μL of S solution, 50 μL of N solution, 26.4 μL of 0.5 M K_2HPO_4, 100 μL of 40% D-glucose, 10 μL of 1 mg/mL thiamine-HCl. Supplement with appropriate antibiotics and all amino acids (final concentration 20 μg/mL) except L-methionine and L-cysteine. MOPS defined media is based on the protocol of Neidhardt et al. (25).
10. 125 mL Erlenmeyer flask with cap.
11. Ice bucket.
12. 1 M isopropyl β-D-1-thiogalactopyranoside (IPTG).
13. Radiolabeling solution: [^{35}S]-L-methionine and L-cysteine (1,175 Ci/mmol, MP Biomedicals).
14. Chase solution: unlabeled L-methionine and L-cysteine, 25 mg/mL of each.
15. 50% Trichloroacetic acid (TCA).
16. Microcentrifuge.
17. Cell wash buffer: 50 mM Tris-acetate (pH 7.0), 1 mM EDTA.
18. Cell lysis buffer: 1% SDS, 50 mM Tris-acetate (pH 7.0), and 1 mM EDTA.
19. 2% Cetyltriethylammonium bromide (CTABr).
20. 0.5 M sodium acetate (pH 5.0).
21. 100% Acetone.
22. AcE storage buffer: 8 M urea, 10 mM sodium acetate (pH 5.0), 1 mM EDTA.
23. AcE gel-loading buffer: 8 M urea, 10 mM sodium acetate (pH 5.0), 1 mM EDTA, 0.01% xylene cyanol, 0.01% bromophenol blue.
24. 30% Methanol.
25. 15 × 15 cm Tupperware® or equivalent plastic container.
26. 3MM Chr filter paper (Whatman).
27. Plastic food wrap.
28. Gel dryer and vacuum pump.
29. 8 × 10-in. Phosphorimaging screen.
30. Phosphorimager.

3. Methods

The methods outlined in Subheadings 3.1, 3.2 and 3.3 are designed to analyze total cellular RNA prepared by the one-step guanidinium isothiocyanate (GITC)-phenol extraction method of Chomczynski & Sacchi (26, 27). Commercially available extraction reagents (e.g., TRIzol® or TRI Reagent®) yield the same results at a significantly higher cost. GITC-phenol extraction performs well for the isolation of peptidyl-tRNAs carrying nascent chains up to 80 residues in length (Fig. 1), but is not suitable for the recovery of peptidyl-tRNAs with longer nascent chains. Presumably, larger peptidyl-tRNAs partition to the organic phenol phase during extraction due to their higher protein content (see Note 2 for analysis of large peptidyl-tRNAs). GITC-phenol reagents are typically buffered with acetate to pH ~ 5, so the labile ester bonds within aminoacyl- and peptidyl-tRNAs are stabilized during extraction. Once isolated, RNA samples must be stored at acidic pH to preserve aminoacyl-tRNA. We use dilute sodium acetate (10 mM, pH 5.0) supplemented with 1 mM EDTA for long-term storage of RNAs. In contrast to the original report by Ho & Kan (7), we have observed no appreciable tRNA deacylation in samples that have been subjected to multiple freeze–thaw cycles.

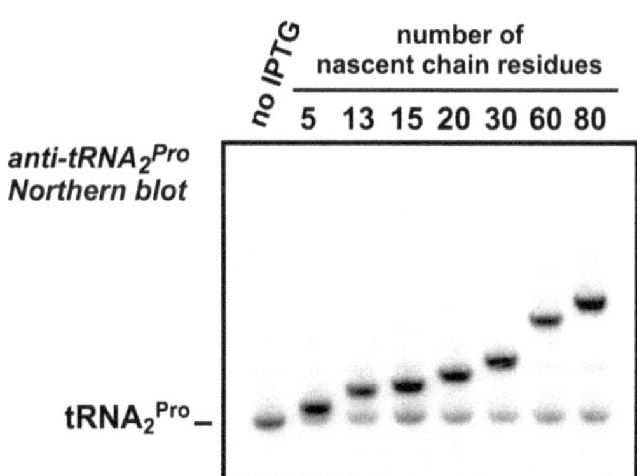

Fig. 1. Analysis of peptidyl-tRNAs by Northern blot hybridization. Total RNA was extracted from *E. coli* cells expressing peptides containing a C-terminal Pro-Pro motif. The Pro-Pro nascent peptide interferes with translation termination, causing the accumulation of peptidyl prolyl-tRNA$_2^{Pro}$ on paused ribosomes (20, 34). RNA samples were resolved on an acid-urea 6%-polyacrylamide gel followed by Northern blot hybridization with a radiolabeled oligonucleotide probe specific for *E. coli* tRNA$_2^{Pro}$. The number of nascent chain residues encoded by each expression construct is indicated above the lanes. Peptidyl prolyl-tRNA$_2^{Pro}$ is not detected in the uninduced sample (*no IPTG*).

3.1. Alkaline pH Treatment of RNA

1. Pipette 50 μg of total *E. coli* RNA into a microfuge tube (e.g., 25 μL of a 2 mg/mL RNA solution). Add water to 40 μL, then 5 μL of 10 mM EDTA (pH 8.0) and 5 μL of 1 M Tris–HCl (pH 8.9). Spot 2 μL of the sample onto pH indicator paper to confirm pH > 8.0.
2. Incubate at 37°C for 60 min.
3. Add 3 M sodium acetate (pH 5.0) to a final concentration of 0.3 M.
4. Add 2.5-volumes of ice-cold 95% ethanol and incubate on ice for 10 min.
5. Precipitate RNA by centrifugation in a microcentrifuge at 13,000 × *g* for 10 min at 4°C.
6. Remove the supernatant and wash once with 70% ethanol followed by centrifugation at 13,000 × *g* at 4°C.
7. Remove supernatant and dissolve the RNA pellet in 10–20 μL of AcE storage buffer for gel electrophoresis. We typically load 2.5–10 μg of total RNA per lane on a 1.0 mm thick mini-gel. Therefore, the concentration of the re-dissolved RNA should be at least 0.5 μg/mL to minimize the volume of the loaded sample.
8. Perform gel electrophoresis and Northern blot analysis as described in Subheadings 3.2 and 3.3. Representative results of this procedure are presented in Fig. 2.

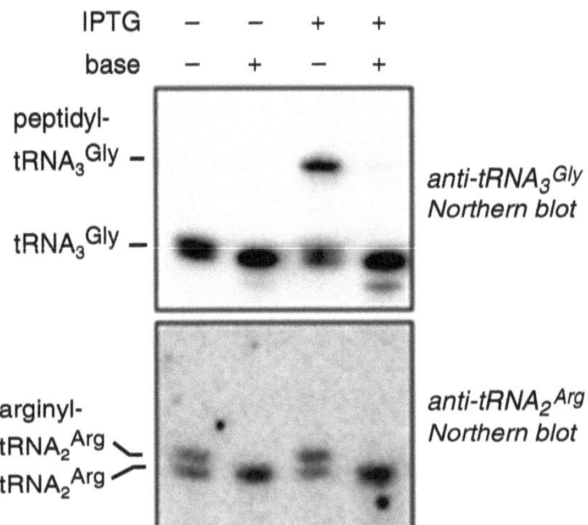

Fig. 2. Northern blot analysis of aminoacyl- and peptidyl-tRNAs. Total RNA from *E. coli* cells carrying an inducible *secM'* mini-gene construct was extracted and analyzed by acid-urea gel electrophoresis. Subsequent Northern blot hybridization was conducted with probes for tRNA$_3^{Gly}$ and tRNA$_2^{Arg}$. The SecM nascent peptide induces a site-specific translational arrest with the codon corresponding to Gly^{165} in the ribosome P site (21, 35). The resulting peptidyl glycyl-tRNA$_3^{Gly}$ is detected by Northern blot upon induction with IPTG. Treatment at alkaline pH (+*base*) promotes the deacylation of both peptidyl glycyl-tRNA$_3^{Gly}$ and arginyl-tRNA$_2^{Arg}$.

3.2. Copper Sulfate Treatment of RNA

1. Pipette 30 μg of total RNA into a fresh 1.5 mL microfuge tube (e.g., 15 μL of a 2 mg/mL RNA solution), then add AcE storage buffer to 27 μL.
2. Add 3 μL of 100 mM $CuSO_4$ to the RNA solution (10 mM $CuSO_4$ final concentration).
3. Incubate at 37°C for 1 h.
4. The deacylated samples may be analyzed directly. Add 5 μL of AcE gel-loading buffer to 5 μL of $CuSO_4$-treated RNA and perform acid-urea gel electrophoresis and Northern blot hybridization (Subheadings 3.3 and 3.4). Figure 3 shows an application of this procedure to detect peptidyl-tRNA drop-off products induced by erythromycin. Alternatively, deacylated samples may be precipitated for long-term storage following steps 5 through 10 below.
5. Add 1.0 mL 95% ethanol, invert to mix and incubate at –80°C for 1 h.
6. Centrifuge at 13,000 × *g* for 15 min at 4°C.
7. Remove supernatant and add 1.0 mL 75% ethanol.
8. Centrifuge at 13,000 × *g* for 10 min at 4°C.
9. Carefully remove supernatant with a pipette and allow the RNA pellet to air dry.
10. Dissolve RNA pellet in AcE storage buffer.

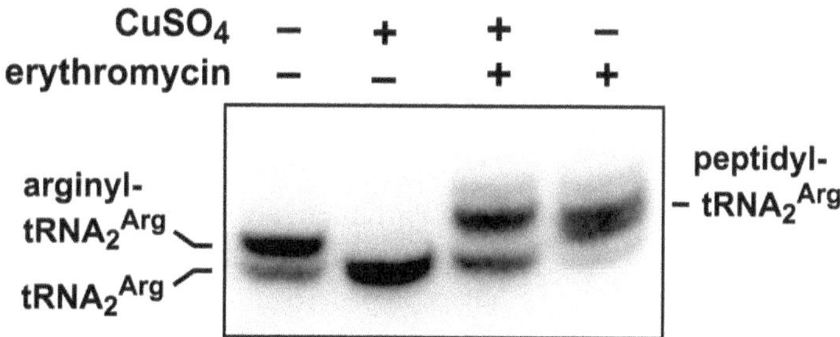

Fig. 3. Deacylation of aminoacyl-tRNA with copper sulfate. Total RNA was isolated from *E. coli pth*(ts) cells grown at the permissive temperature (30°C) with and without 100 μg/mL of erythromycin. Erythromycin induces drop-off of small peptidyl-tRNAs from the ribosome (36). These drop-off products are rapidly hydrolyzed in *pth*+ cells, but can be detected in *pth*(ts) cells even at the permissive temperature. RNA samples were treated with copper sulfate ($CuSO_4$) as described in Subheading 3.2 and run directly on an acid-urea 12% polyacrylamide gel. Northern blot hybridization was performed with a probe for $tRNA_2^{Arg}$. The migration positions of deacylated $tRNA_2^{Arg}$, arginyl-$tRNA_2^{Arg}$ and peptidyl arginyl-$tRNA_2^{Arg}$ are indicated.

3.3. Acid-Urea Polyacrylamide Gel Electrophoresis

The following procedures exploit gel mobility shifts to separate deacylated tRNA from aminoacyl- and peptidyl-tRNAs. We routinely use 12% polyacrylamide mini-gels (70 mm × 90 mm) to analyze aminoacyl-tRNAs. This system can resolve tRNAs carrying large (e.g., Trp) and basic (Arg, Lys and His) amino acid residues from the corresponding deacylated tRNAs (8, 28). However, larger gel systems are required to separate most other aminoacyl-tRNAs from their deacylated counterparts. A portion of each RNA sample should be incubated at alkaline pH to hydrolyze aminoacyl- and peptidyl-tRNA esters. These base-treated samples should be included in each experiment to determine the gel migration position of deacylated tRNA. In general, peptidyl-tRNAs are easily resolved from deacylated and aminoacyl-tRNAs, and therefore 6% polyacrylamide gels and shorter periods of electrophoresis may be used for their analysis. Small peptidyl-tRNAs (two to four amino acid residues) can be differentiated from aminoacyl-tRNA by treating samples with 10 mM $CuSO_4$ (29). Cu^{2+} ions promote the deacylation of aminoacyl-tRNA, but have little effect on peptidyl-tRNA. Therefore, direct comparison of untreated and $CuSO_4$-treated samples can reveal peptidyl-tRNAs that migrate near aminoacyl-tRNA.

1. Prepare the acid-urea polyacrylamide gel solution in a graduated 15 mL conical polypropylene tube.
2. Assemble the Mini-Protean 3 gel-casting stand according to manufacturer's instructions.
3. Add 70–100 μL of 10% APS to the gel solution and invert gently to mix.
4. Add 7–10 μL of TEMED to initiate gel polymerization. Invert gently to mix and pour (or pipette) the solution into the gel cast. Insert a 10-well comb and ensure that no bubbles are trapped beneath the comb. Polymerization is typically complete within 20–30 min at room temperature (see Note 3).
5. After polymerization is complete, carefully remove the comb from the gel and assemble into the Bio-Rad Mini-Protean 3 apparatus.
6. Add *cold* sodium acetate-EDTA running buffer to the upper and lower electrophoresis chambers. Use a needle and syringe to carefully wash the un-polymerized acrylamide and urea from the wells. Because urea continually diffuses from the gel into the sample wells, it is important to remove the urea with a needle and syringe *immediately* prior to sample loading. Failure to do so results in diffuse samples and significantly reduces resolution during electrophoresis.
7. Load the gel with 2.5–10 μg of total RNA. Samples may be denatured at 95°C for 3 min if desired (see Note 4).

8. Perform electrophoresis at 100 V (constant voltage) at 4°C in a cold room (see Note 5). Run 6% polyacrylamide gels for 105 min to analyze peptidyl-tRNAs, and 12% gels for up to 6 h to analyze aminoacyl-tRNAs.

3.4. Northern Blot Analysis

1. After electrophoresis, remove the gel sandwich from the apparatus. Carefully remove the gel and soak in 0.5× TBE in a small plastic container.
2. Soak the nylon membrane in 0.5× TBE buffer in a separate container.
3. While the gel and membrane are soaking, assemble the transfer apparatus: wet a piece of Whatman filter paper in 0.5× TBE and place on the transfer apparatus; roll a 15 × 100 mm glass test tube (or equivalent) over the wetted filter paper to remove air bubbles; pour a small amount (1–2 mL) of 0.5× TBE buffer onto the filter paper; carefully hold the gel in the Tupperware container and pour out all excess buffer.
4. Carefully place a dry piece of filter paper onto the gel and lift. Place the gel (filter-paper side down) onto the first wetted filter paper. Roll carefully to remove air bubbles. Place the wetted nylon membrane directly onto the gel, followed by the remaining two wetted filter papers. Roll carefully to remove air bubbles. Use a large Kim-wipe to remove excess transfer buffer.
5. Place the top electrode (anode) onto the sandwich and transfer at 650 mA (constant current) for 30 min.
6. Disassemble the transfer apparatus. Remove the nylon membrane and incubate in 5× SSC briefly (1–2 min).
7. Cross-link the RNA to the nylon membrane using a Stratalinker or other UV light source for 2 min.
8. Wash the blotted membrane with 10–15 mL of 0.1× SSC, 0.1% SDS at 65°C in a rotisserie hybridization oven for 30 min. We perform hybridizations in disposable 50 mL polypropylene conical tubes.
9. Remove the wash solution and add 10–15 mL of prehybridization solution.
10. Incubate for 1 h at the hybridization temperature. The hybridization temperature is 15°C below the calculated melting temperature (T_m) of the oligonucleotide probe.
11. Label the oligonucleotide probe with [γ-^{32}P]-ATP and T4 PNK. In a microfuge tube add: 33 μL water, 4 μL 10× PNK buffer, 2 μL oligonucleotide (10 μM) and 1 μL [γ-^{32}P]-ATP. Add 1 μL of PNK and mix gently by pipette. Incubate at 37°C

for 40 min. Stop the reaction by incubating at 70°C in a water bath for 10 min, and collect reaction by centrifugation. Pipette the reaction onto a microspin G-25 gel filtration column to remove free radiolabeled ATP.

12. Add the radiolabeled oligonucleotide probe to the hybridization solution and incubate in the rotisserie oven overnight.
13. Decant the hybridization solution into a 50 mL polypropylene conical tube. Hybridization solutions can be frozen and reused several times.
14. Wash the membrane with 10–15 mL of 6× SSC, 0.1% SDS for 5 min at room temperature. Repeat this wash step two more times. Decant the wash solutions into the appropriate [^{32}P]-waste container.
15. Wash the membrane with 10–15 mL of 6× SSC, 0.1% SDS for 20 min at 10°C below the calculated T_m for the oligonucleotide probe.
16. Remove the membrane with tweezers and blot onto a KimWipe to remove excess fluid. Wrap in plastic food wrap and expose to a phosphorimager screen (2–48 h) or X-ray film (1–3 days).
17. Visualize by phosphorimager or film developer.

3.5. Pulse-Chase Protocol

The following pulse-chase approach is used to monitor peptidyl-tRNA turnover in vivo. *E. coli* cells are grown in defined media and pulsed with [^{35}S]-L-methionine/L-cysteine to radiolabel nascent peptide chains. The decay of [^{35}S]-labeled peptidyl-tRNA is monitored following the addition of excess unlabeled L-methionine/L-cysteine. For this analysis, peptidyl-tRNAs are purified from bulk cellular proteins using CTABr precipitation. Once isolated, the peptidyl-tRNAs can be hydrolyzed and specific nascent chains immunoprecipitated for subsequent SDS-PAGE analysis as described (30, 31). Alternatively, plasmid-borne over-expression systems can be used to increase the level of peptidyl-tRNAs of interest. We have used this approach to study specific peptidyl-tRNAs in unfractionated CTABr precipitates (20). If multiple peptidyl-tRNAs are observed in [^{35}S]-labeled samples, then the species of interest can be identified by Northern blot hybridization (see Note 6).

1. Grow *E. coli* cells overnight at 37°C in MOPS-glucose defined media supplemented with the appropriate antibiotics (see Note 7).
2. Resuspend the overnight culture in 10 mL of pre-warmed MOPS media to an optical density at 600 nm (OD_{600}) of 0.05.

3. Place flask in a rotary shaker and incubate at 37°C with shaking (250–300 rpm) until the culture grows to OD_{600} ~ 0.5.
4. Prepare prechilled microfuge tubes containing 20 μL of 50% TCA. Prepare one tube for each time point in your experiment. Label the tubes and place on ice for use in step 7 below.
5. Add 15 μL of 1 M IPTG to induce expression of the plasmid construct. Incubate culture for 15–30 min.
6. Add 20 μL of [^{35}S]-L-methionine/L-cysteine solution (final concentration of 20 μCi/mL) to the culture.
7. Incubate for 2 min with shaking.
8. Add 80 μL of chase solution (final concentration 0.2 mg/mL, see Note 8) to the flask and rapidly mix by swirling.
9. Remove 1.0 mL of the culture and rapidly transfer into a microfuge tube containing 20 μL of 50% TCA on ice. Mix briefly, cap tube, and place into ice.
10. Continue to swirl culture by hand or on shaker. Remove 1.0 mL aliquots every 15 s for the first minute, then at 30–60 s intervals for the next 2 min.
11. After all culture aliquots have been removed, centrifuge all the sample tubes at 13,000 × *g* for 2 min in a microcentrifuge to collect cells (see Note 9).
12. Carefully remove the supernatant by pipette and dispense into a [^{35}S]-waste container.
13. Add 1.0 mL of cell wash buffer to each tube—*do not resuspend the cell pellet.*
14. Centrifuge at 13,000 × *g* for 5 min and remove the supernatant.
15. Resuspend the cell pellet in 100 μL of cell lysis buffer, and allow cells to lyse at room temperature for at least 20 min. Cell lysates can be frozen at this point and the protocol can be resumed at step 16 at a later time.
16. Remove cell debris by centrifugation at 13,000 × *g* for 10 min. Carefully remove supernatant and transfer to a fresh microfuge tube.
17. Remove 45 μL of each cell lysate and place in a fresh microfuge tube; freeze the remaining 55 μL of cell lysate as a backup sample. Add 0.5 mL of 2% CTABr solution and 0.5 mL of 0.5 M sodium acetate (pH 5.0) solution. Cap and invert tube to mix.
18. Place in an ice water bath for 20 min.
19. Thaw precipitates at 30°C for 10 min.

20. Centrifuge at 13,000 × *g* for 30 min.
21. Carefully remove supernatant and add 1.0 mL of 100% acetone.
22. Centrifuge 13,000 × *g* for 10 min. Carefully remove supernatants and allow pellet to air-dry until the acetone evaporates.
23. Resuspend pellets in 25 μL of AcE gel-loading buffer (see Note 10).
24. Determine the absorbance at 260 nm (A_{260}) of each reaction. Load equivalent A_{260} units onto the acid-urea polyacrylamide gel.
25. Load 10 μL of each sample onto an acid-urea polyacrylamide gel and perform electrophoresis (100 V constant voltage) at 4°C for 105 min.
26. Disassemble the Mini-Protean 3 apparatus and carefully separate the glass plates.
27. Submerge the glass plate to which the gel is attached in 30% methanol.
28. Gently shake the gel for 10 min at room temperature to remove urea.
29. Decant the methanol into an [^{35}S]-waste container.
30. Place a piece of dry filter paper onto the gel and carefully lift the gel from the glass plate.
31. Cover the gel with plastic food wrap and place in a gel dryer at 75°C under vacuum until the gel is completely dehydrated.
32. Expose the dried gel to a phosphorimaging screen for 8–16 h.
33. Image the screen and quantify the appropriate peptidyl-tRNA species using phosphorimager software. Fit exponential decay equations to the experimental data using DeltaGraph (RedRock Software) or equivalent software package:

$$\text{single-exponential decay}: N(t) = N_0 e^{-\lambda t}$$

$$\text{single-exponential half-life}: t_{1/2} = \ln 2 / \lambda$$

$$\text{double-exponential decay}: N(t) = N_0 e^{-(\lambda_a + \lambda_b)t}$$

$$\text{double-exponential half-life}: T_{1/2} = \ln 2 / (\lambda_a + \lambda_b)$$

where N_0 = peptidyl-tRNA at t = 0 s, λ is the decay constant, t is time, λ_a and λ_b are constants for two independent decay processes. Multiple exponential decay equations generally yield better fits, but should only be used if the decay process is known to occur through multiple independent pathways. Typical results from these procedures are presented in Fig. 4.

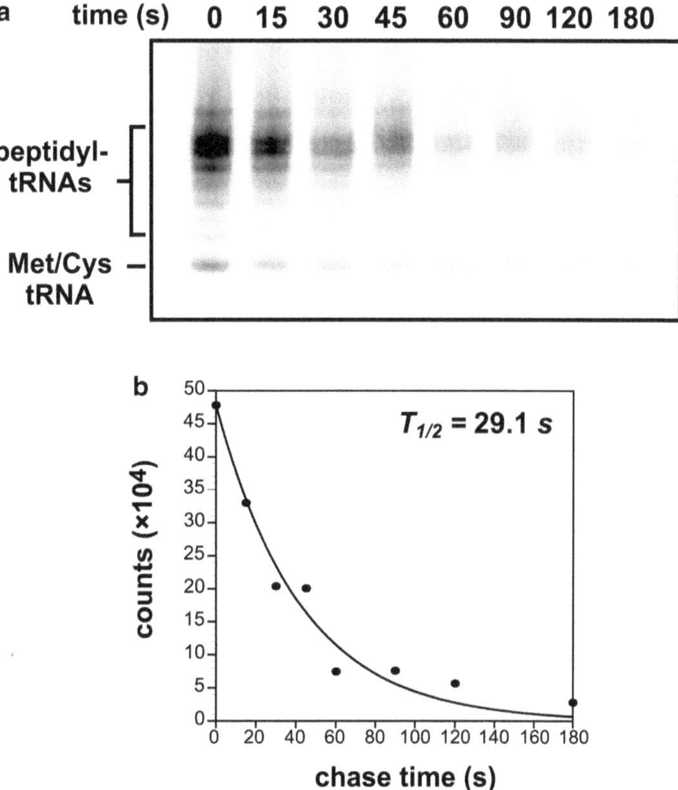

Fig. 4. Pulse-chase analysis of peptidyl-tRNA turnover. (**a**) Autoradiogram illustrating peptidyl-tRNA turnover using [^{35}S] pulse-labeling and acid-urea gel electrophoresis. Samples were taken at the indicated time points for tRNA isolation as described in Section 3.5. Radiolabeled species corresponding to methionyl-/cysteinyl-tRNAs and peptidyl-tRNAs are indicated. (**b**) Fitting of exponential decay equations. Peptidyl-tRNA signal intensities in (**a**) were quantified using Quantity One software (BioRad) and the data fitted to a double-exponential decay equation to calculate composite half-life ($T_{1/2}$) as described in Subheading 3.5.

4. Notes

1. Pre-packed, disposable Sephadex G-25 gel filtration spin columns are available from GE Healthcare and a number of other commercial vendors. We prepare our own spin columns using recycled silica-based mini-prep spin columns (e.g., Qiagen or Promega). Used mini-prep columns are loaded with up to 750 μL of Sephadex G-25 slurry equilibrated in 10 mM Tris–HCl (pH 7.5), 1 mM EDTA. Centrifuge the column at 3,000 × *g* for 2 min to remove the void volume. Pipette the labeling reaction (40–50 μL) onto the resin and spin at 3,000 × *g* for 2 min to collect the [^{32}P]-labeled oligonucleotide

probe. Single-stranded oligonucleotides bind poorly to the silica resin and are therefore recovered efficiently.

2. Although larger peptidyl-tRNAs are difficult to isolate with GITC-phenol, they can be precipitated from whole-cell SDS lysates using CTABr. Follow the procedures in Subheading 3.5, steps 11 through 16, and analyze by Northern blot hybridization. This procedure has been used successfully to isolate tRNAs carrying up to 166 amino acid residues (20, 30, 31).

3. Deprotonated TEMED is required to initiate acrylamide/bisacrylamide polymerization. Therefore, acid-urea polyacrylamide gels polymerize more slowly than Tris-buffered gels. Riboflavin/visible light and other initiator strategies are sometimes used to initiate polymerization at acidic pH (32, 33). However, we find that higher concentrations of both APS (100 μL of 10% solution per 6 mL gel) and TEMED (10–15 μL per 6 mL gel) are sufficient to promote polymerization at acidic pH. If polymerization requires more than 20 min, prepare a fresh 10% APS solution.

4. Nucleic acid samples are usually heat denatured prior to polyacrylamide gel electrophoresis. In principle, high temperatures could promote deacylation, but we have seen no difference between unheated samples and those incubated at 95°C for 3 min. Peptidyl-tRNAs are typically more stable than aminoacyl-tRNAs and can be heated without degradation. The acid-urea gels described here possess no stacking mechanism and therefore it is critical to minimize the loaded sample volume. For optimal resolution of aminoacyl-tRNA from deacylated tRNA, we recommend a maximum volume of 10 μL (5 μL sample + 5 μL AcE loading buffer). If using larger gels, then sample volumes may be increased accordingly.

5. Polyacrylamide gels buffered with 100 mM sodium acetate draw high currents during electrophoresis. This results in substantial Joule heating, which can readily fracture glass plates. We conduct electrophoresis in a 4°C environmental room using currents ranging from 17 to 22 mA (corresponds to 100 V for the Bio-Rad Mini-Protean 3 apparatus). Although the Mini-Protean 3 and other electrophoresis apparatuses can run multiple gels simultaneously, we recommend running individual gels to reduce Joule heating.

6. Multiple peptidyl-tRNA species may be observed in [^{35}S]-labeled CTABr precipitates (see Fig. 4). Northern blot analysis can be used to unambiguously identify these radiolabeled peptidyl-tRNAs through the correlation of gel migration positions (20). Briefly, peptidyl-tRNA is isolated from [^{35}S]-labeled and unlabeled cells. The labeled and unlabeled samples are run in adjacent lanes on an acid-urea polyacrylamide gel, followed by

electrotransfer to nylon membrane. The membrane is then cut between the two lanes, and the unlabeled section is subjected to Northern blot hybridization with a [^{32}P]-labeled oligonucleotide probe. After hybridization and washing, the two membranes are carefully realigned with one another and exposed to a phosphorimager screen.

7. *E. coli* cells undergo growth arrest for several hours after transfer from rich broth into defined MOPS-glucose medium. Therefore, we recommend overnight culture in MOPS-glucose medium prior to running the experiment. Most prototrophic *E. coli* strains will reach mid-log phase within 4–5 h if seeded from an overnight MOPS-glucose culture.
8. In our experience, L-methionine/L-cysteine chase solutions lose their effectiveness within hours after preparation. Treatment of the chase solution with reducing agents (e.g., dithiothreitol or β-mercaptoethanol) exacerbates the problem. Therefore, we recommend preparing fresh L-methionine/L-cysteine chase solution immediately prior to each pulse-chase experiment.
9. Radioactive contamination of the rotor and interior surfaces of the microcentrifuge commonly occurs during this procedure. To minimize contamination, stretch a small piece of Parafilm® M around the microcentrifuge tube caps prior to centrifugation.
10. CTABr precipitates can be distributed diffusely over the interior surfaces of microfuge tubes. To dissolve the entire sample, take care to pipette gel-loading buffer onto all surfaces that may have collected the precipitate. Adherence to this procedure will significantly increase the tRNA concentration in the samples.

Acknowledgments

This work was supported by grant R01 GM078634 from the National Institutes of Health.

References

1. Banerjee R, Chen S, Dare K, Gilreath M, Praetorius-Ibba M, Raina M et al (2010) tRNAs: cellular barcodes for amino acids. FEBS Lett 584:387–395
2. Cashel M, Gentry DR, Hernandez VJ, Vinella D (1996) In: Neidhardt FC (ed) *Escherichia coli* and *Salmonella*: cellular and molecular biology, vol 1. American Society for Microbiology, Washington, DC, The stringent response. pp 1458–1524.
3. Potrykus K, Cashel M (2008) (p)ppGpp still magical? Annu Rev Microbiol 62:35–51
4. Dittmar KA, Sorensen MA, Elf J, Ehrenberg M, Pan T (2005) Selective charging of tRNA isoacceptors induced by amino-acid starvation. EMBO Rep 6:151–157

5. Elf J, Nilsson D, Tenson T, Ehrenberg M (2003) Selective charging of tRNA isoacceptors explains patterns of codon usage. Science 300:1718–1722
6. Sorensen MA, Elf J, Bouakaz E, Tenson T, Sanyal S, Bjork GR et al (2005) Over expression of a tRNA(Leu) isoacceptor changes charging pattern of leucine tRNAs and reveals new codon reading. J Mol Biol 354:16–24
7. Ho YS, Kan YW (1987) In vivo aminoacylation of human and *Xenopus* suppressor tRNAs constructed by site-specific mutagenesis. Proc Natl Acad Sci U S A 84:2185–2188
8. Ataide SF, Jester BC, Devine KM, Ibba M (2005) Stationary-phase expression and aminoacylation of a transfer-RNA-like small RNA. EMBO Rep 6:742–747
9. Kohrer C, Rajbhandary UL (2008) The many applications of acid urea polyacrylamide gel electrophoresis to studies of tRNAs and aminoacyl-tRNA synthetases. Methods 44:129–138
10. Varshney U, Lee CP, RajBhandary UL (1991) Direct analysis of aminoacylation levels of tRNAs in vivo. Application to studying recognition of *Escherichia coli* initiator tRNA mutants by glutaminyl-tRNA synthetase. J Biol Chem 266:24712–24718
11. Cruz-Vera LR, Magos-Castro MA, Zamora-Romo E, Guarneros G (2004) Ribosome stalling and peptidyl-tRNA drop-off during translational delay at AGA codons. Nucleic Acids Res 32:4462–4468
12. Garcia-Villegas MR, De La Vega FM, Galindo JM, Segura M, Buckingham RH, Guarneros G (1991) Peptidyl-tRNA hydrolase is involved in lambda inhibition of host protein synthesis. EMBO J 10:3549–3555
13. Heurgue-Hamard V, Karimi R, Mora L, MacDougall J, Leboeuf C, Grentzmann G et al (1998) Ribosome release factor RF4 and termination factor RF3 are involved in dissociation of peptidyl-tRNA from the ribosome. EMBO J 17:808–816
14. Heurgue-Hamard V, Mora L, Guarneros G, Buckingham RH (1996) The growth defect in *Escherichia coli* deficient in peptidyl-tRNA hydrolase is due to starvation for Lys-tRNA(Lys). EMBO J 15:2826–2833
15. Menninger JR (1976) Peptidyl transfer RNA dissociates during protein synthesis from ribosomes of *Escherichia coli*. J Biol Chem 251: 3392–3398
16. Menninger JR (1979) Accumulation of peptidyl tRNA is lethal to *Escherichia coli*. J Bacteriol 137:694–696
17. Menninger JR, Walker C, Tan PF (1973) Studies on the metabolic role of peptidyl-tRNA hydrolase. I. Properties of a mutant *E. coli* with temperature-sensitive peptidyl-tRNA hydrolase. Mol Gen Genet 121:307–324
18. Shiloach J, Lapidot Y, de Groot N (1975) The specificity of peptidyl-tRNA hydrolase from *E. coli*. FEBS Lett 57:130–133
19. Vogel Z, Vogel T, Zamir A, Elson D (1971) The protection by 70 S ribosomes of N-acyl-aminoacyl-tRNA against cleavage by peptidyl-tRNA hydrolase and its use to assay ribosomal association. Eur J Biochem 21:582–592
20. Janssen BD, Hayes CS (2009) Kinetics of paused ribosome recycling in *Escherichia coli*. J Mol Biol 394:251–267
21. Garza-Sánchez F, Janssen BD, Hayes CS (2006) Prolyl-tRNA(Pro) in the A-site of SecM-arrested ribosomes inhibits the recruitment of transfer-messenger RNA. J Biol Chem 281:34258–34268
22. Gong F, Ito K, Nakamura Y, Yanofsky C (2001) The mechanism of tryptophan induction of tryptophanase operon expression: tryptophan inhibits release factor-mediated cleavage of TnaC-peptidyl-tRNA(Pro). Proc Natl Acad Sci U S A 98:8997–9001
23. Gong F, Yanofsky C (2001) Reproducing tna operon regulation in vitro in an S-30 system. Tryptophan induction inhibits cleavage of TnaC peptidyl-tRNA. J Biol Chem 276: 1974–1983
24. Yao S, Blaustein JB, Bechhofer DH (2008) Erythromycin-induced ribosome stalling and RNase J1-mediated mRNA processing in *Bacillus subtilis*. Mol Microbiol 69:1439–1449
25. Neidhardt FC, Bloch PL, Smith DF (1974) Culture medium for enterobacteria. J Bacteriol 119:736–747
26. Chomczynski P, Sacchi N (1987) Single-step method of RNA isolation by acid guanidinium thiocyanate-phenol-chloroform extraction. Anal Biochem 162:156–159
27. Chomczynski P, Sacchi N (2006) The single-step method of RNA isolation by acid guanidinium thiocyanate-phenol-chloroform extraction: twenty-something years on. Nat Protoc 1:581–585
28. Garza-Sánchez F, Gin JG, Hayes CS (2008) Amino acid starvation and colicin D treatment induce A-site mRNA cleavage in *Escherichia coli*. J Mol Biol 378:505–519
29. Schofield P, Zamecnik PC (1968) Cupric ion catalysis in hydrolysis of aminoacyl-tRNA. Biochim Biophys Acta 155:410–416

30. Nakatogawa H, Ito K (2001) Secretion monitor, SecM, undergoes self-translation arrest in the cytosol. Mol Cell 7:185–192
31. Nakatogawa H, Ito K (2002) The ribosomal exit tunnel functions as a discriminating gate. Cell 108:629–636
32. Andrews AT (1990) Electrophoresis theory, techniques, and biochemical and clinical applications. Oxford Science, Oxford, pp 141–143
33. Shi Q, Jackowski G (1998) In: Hames B, Rickwood D (eds) Gel electrophoresis of proteins: a practical approach. One-dimensional polyacrylamide gel electrophoresis. Oxford University Press, Oxford, pp 1–52
34. Hayes CS, Bose B, Sauer RT (2002) Proline residues at the C terminus of nascent chains induce SsrA tagging during translation termination. J Biol Chem 277:33825–33832
35. Muto H, Nakatogawa H, Ito K (2006) Genetically encoded but nonpolypeptide prolyl-tRNA functions in the A site for SecM-mediated ribosomal stall. Mol Cell 22:545–552
36. Menninger JR, Coleman RA, Tsai LN (1994) Erythromycin, lincosamides, peptidyl-tRNA dissociation, and ribosome editing. Mol Gen Genet 243:225–233

Chapter 20

In Vitro *Trans*-Translation Assays

Daisuke Kurita, Akira Muto, and Hyouta Himeno

Abstract

Trans-translation is a bacterial quality control system in protein synthesis facilitated by transfer-messenger RNA (tmRNA). Here, we describe the in vitro system using purified factors to evaluate the two steps of *trans*-translation: peptidyl-transfer from peptidyl-tRNA to alanyl-tmRNA and decoding of the resume codon on tmRNA.

Key words: tmRNA, SmpB, Ribosome, Translation, *Trans*-translation, Cell-free protein synthesis

1. Introduction

During protein synthesis, ribosome may stall on an incomplete mRNA for which the stop codon is missing. Bacteria employ a rescue system called *trans*-translation mediated by tmRNA to resolve the stalled ribosome (1–4). tmRNA is a unique molecule acting as both tRNA and mRNA. tmRNA enters the A-site of the stalled ribosome in an alanine-charged form to receive the nascent polypeptides from the P-site peptidyl-tRNA. Subsequently, tmRNA serves as mRNA using a short open reading frame with a termination codon within tmRNA. By switching the template from the stop codon-less mRNA to the mRNA domain on tmRNA, protein synthesis is completed and the nascent polypeptide is tagged with a specific sequence as the degradation signal, allowing the stalled ribosomes to be recycled for a new round of translation (5, 6).

Several factors, such as SmpB (7–9) and EF-Tu (10–12), have been identified to be involved in *trans*-translation. SmpB is essential for tmRNA association with the ribosome. EF-Tu delivers alanyl-tmRNA to the ribosomal A-site in a GTP-dependent manner like aminoacyl-tRNA in the canonical translation. S1, a component of the small subunit of the ribosome, has also been identified as a

Kenneth C. Keiler (ed.), *Bacterial Regulatory RNA: Methods and Protocols*, Methods in Molecular Biology, vol. 905,
DOI 10.1007/978-1-61779-949-5_20, © Springer Science+Business Media, LLC 2012

tmRNA binding factor (13–15), although its involvement in *trans*-translation remains controversial (16–18).

Several kinds of cell-free *trans*-translation systems have been developed in *Escherichia coli* and *Thermus thermophilus* (2, 16, 19–22). In an earlier study from our group, an in vitro *trans*-translation system has been developed to evaluate tag-peptide synthesis using cell extracts from *E. coli* (2, 19, 21). Later, we have developed the assay system composed of purified factors from *E. coli*, which we describe in this chapter, allowing us to evaluate the activity of two steps of *trans*-translation (23–25) (Fig. 1).

Initially, [^{14}C]polyphenylalanine is synthesized from [^{14}C]phenylalanyl-tRNA^{Phe} and a synthetic mRNA $(UUC)_{10}$, which has ten consecutive UUC codons, to produce a ribosome stalled at the 3′ end of mRNA. To measure the activity of *trans*-transfer (peptidyl-transfer from polyphenylalanyl-tRNA to alanyl-tmRNA), the stalled ribosome is incubated with [^{3}H]alanyl-tmRNA. Reaction is quenched with hot 5% trichloroacetic acid (TCA). The precipitated polypeptide ([^{14}C]polyphenylalanyl-[^{3}H]alanine) is recovered by filtration with a mixed cellulose membrane and the radioactivity on the membrane is measured by a liquid scintillation counter. Incorporation of [^{3}H]alanine into the polypeptide fraction is observed in the presence of SmpB and reaches a plateau within 10 min (Fig. 2).

After *trans*-transfer reaction, translocation of tmRNA from the A-site to the P-site should occur, allowing the resume codon on tmRNA to be set at the A-site. To measure the activity of decoding of the resume codon, the stalled ribosome is incubated with unlabeled alanyl-tmRNA and [^{3}H]alanyl-tRNA^{Ala} together. The polypeptide (polyphenylalanyl-alanyl-[^{3}H]alanine) is recovered by the same procedure as that in the *trans*-transfer reaction. This system can be applied to the initiation shift assay of resume codon by using corresponding aminoacyl-tRNA instead of [^{3}H]alanyl-tRNA^{Ala} (see Note 1).

2. Materials

1. *E. coli* strain: W3110Δ*ssrA*Δ*smpB* (in which the *ssrA* and *smpB* genes are disrupted by a chloramphenicol-resistant gene), BL21(DE3)Δ*ssrA* (in which the *ssrA* gene is disrupted by a kanamycin-resistant gene).
2. pGEMEX-II expression vector.
3. LB broth (10 g tryptone, 5 g yeast extract and 5 g NaCl/1 L).
4. Alumina.

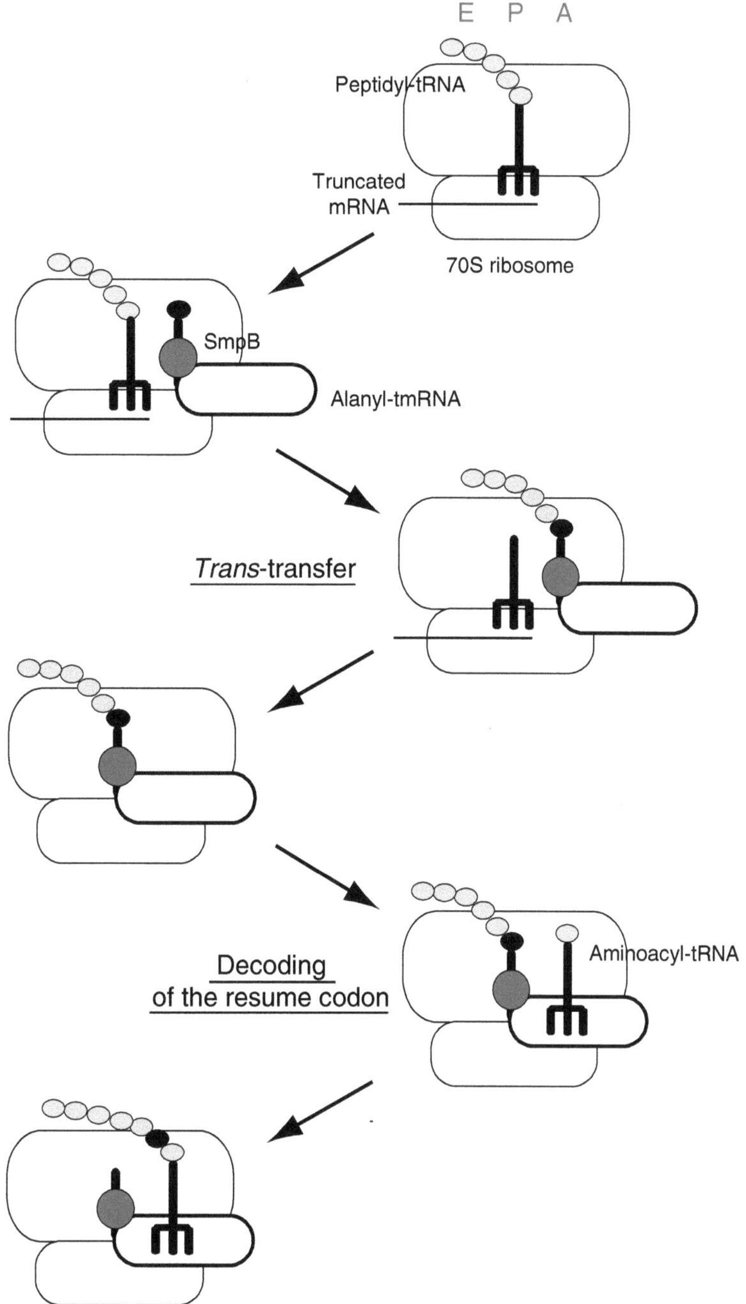

Fig. 1. Schematic model of *trans*-translation. Ribosome stalled at the 3′ end of mRNA contains peptidyl-tRNA in the P-site. Alanyl-tmRNA/SmpB enters the empty A-site to receive the nascent polypeptide from the peptidyl-tRNA in the P-site (*trans*-transfer). The resulting peptidyl-alanyl-tmRNA/SmpB moves from the A-site to the P-site to set the resume codon on tmRNA in the A-site. And then aminoacyl-tRNA corresponding to the resume codon enters the A-site to receive peptidyl-alanine from P-site peptidyl-alanyl-tmRNA/SmpB.

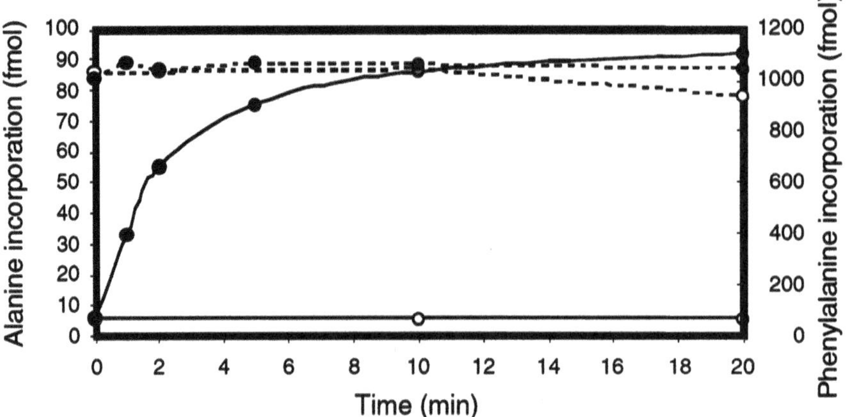

Fig. 2. Time course of *trans*-transfer. [^{14}C]phenylalanine (*dotted line*) and [^{3}H]alanine (*solid line*) incorporated into the polypeptide fraction are quantified in the presence (*closed circle*) and absence (*open circle*) of SmpB. The experimental details are described in Subheading 3.8.

5. Mortar.
6. Centrifuge Ware (Hitachi).
7. 70S A-buffer: 10 mM Tris–HCl (pH 7.5), 10 mM $MgCl_2$, 60 mM NH_4Cl and 7 mM 2-mercaptoethanol.
8. 70S B-buffer: 10 mM Tris–HCl (pH 7.5), 10 mM $MgCl_2$, 1,000 mM NH_4Cl, and 7 mM 2-mercaptoethanol.
9. Sucrose cushion: 10 mM Tris–HCl (pH 7.5), 10 mM $MgCl_2$, 1,000 mM NH_4Cl, 20% sucrose, and 7 mM 2-mercaptoethanol.
10. Gradient Mate (Towa Kagaku).
11. 70S stock buffer: 10 mM Tris–HCl (pH 7.5), 10 mM $MgCl_2$, 60 mM NH_4Cl, 10% glycerol, and 7 mM 2-mercaptoethanol.
12. ARS-A buffer: 20 mM Tris–HCl (pH 7.5), 10 mM $MgCl_2$, 50 mM KCl, 10% glycerol, and 1 mM dithiothreitol.
13. ARS-B buffer: 20 mM Tris–HCl (pH 7.5), 10 mM $MgCl_2$, 300 mM KCl, 10% glycerol, and 1 mM dithiothreitol.
14. ARS stock buffer: 60 mM Tris–HCl (pH 7.5), 10 mM $MgCl_2$, 50% glycerol, and 10 mM dithiothreitol.
15. tmRNA buffer: 10 mM Tris–HCl (pH 7.5), 10 mM $MgCl_2$, and 60 mM NH_4Cl.
16. tmRNA folding buffer: 10 mM HEPES-KOH (pH 7.5), 5 mM $MgCl_2$, and 20 mM NH_4Cl.
17. TM buffer: 10 mM Tris–HCl (pH 7.5) and 10 mM $MgCl_2$.
18. Isopropyl-β-thiogalactopyranoside (IPTG).
19. RNase-free DNase.

20. Urea dye: 0.1% xylene cyanol, 0.1% bromophenol blue, and 7 M urea.
21. 10× TBE buffer: 890 mM Tris-borate, 890 mM boric acid, and 20 mM EDTA.
22. Handheld UV lamp (254 nm).
23. Fluorescent TLC plate.
24. Seamless cellulose tubing (molecular cut off of 14,000 Da).
25. Vacuum filter system.
26. Radiolabeled amino acids: [^{14}C]L-phenylalanine, [^{3}H]L-alanine, [^{3}H]L-arginine, and [^{3}H]L-glutamine.
27. Unlabeled L-alanine.
28. SP-sepharose.
29. Ni-NTA agarose.
30. Amicon Ultra-15 (Millipore).
31. SmpB-A buffer: 50 mM HEPES-KOH (pH 7.5), 100 mM KCl, and 7 mM 2-mercaptoethanol.
32. SmpB-B buffer: 50 mM HEPES-KOH (pH 7.5), 1,000 mM KCl, and 7 mM 2-mercaptoethanol.
33. SmpB-C buffer: 50 mM HEPES-KOH (pH 7.5), 200 mM KCl, 20 mM imidazole, and 7 mM 2-mercaptoethanol.
34. SmpB-D buffer: 50 mM HEPES-KOH (pH 7.5), 200 mM KCl, 500 mM imidazole, and 7 mM 2-mercaptoethanol.
35. SmpB wash buffer: 50 mM HEPES-KOH (pH 7.5), 200 mM KCl, 1,000 mM NH_4Cl, and 7 mM 2-mercaptoethanol.
36. SmpB stock buffer: 50 mM HEPES-KOH (pH 7.5), 100 mM KCl, 10% glycerol, and 7 mM 2-mercaptoethanol.
37. Sodium dodecyl sulfate-polyacrylamide gel electrophoresis (SDS-PAGE) equipment.
38. 10% Ammonium persulfate (APS) solution.
39. TEMED.
40. Silver staining kit.
41. 10× Aminoacylation buffer: 600 mM Tris–HCl (pH 7.5), 100 mM $MgCl_2$, 300 mM KCl, and 100 mM dithiothreitol.
42. Aminoacylation mixture: 1× aminoacylation buffer, 3 mM ATP, 10 μM radiolabeled or unlabeled amino acid, 25 A_{260} of tRNA or 4 A_{260} tmRNA with 1 μM SmpB (see Note 2), and corresponding aminoacyl-tRNA synthetase.
43. DEAE sepharose.
44. EF-A buffer: 50 mM Tris–HCl (pH 7.8), 20 mM NaCl, 7 mM $MgCl_2$, 7 mM 2-mercaptoethanol, and 5 μM GDP.

45. EF-B buffer: 50 mM Tris–HCl (pH 7.8), 500 mM NaCl, 7 mM $MgCl_2$, 7 mM 2-mercaptoethanol, and 5 μM GDP.
46. EF-C buffer: 50 mM Tris–HCl (pH 7.8), 500 mM NaCl, 7 mM $MgCl_2$, 300 mM imidazole, 7 mM 2-mercaptoethanol, and 5 μM GDP.
47. EF stock buffer: 50 mM Tris–HCl (pH 7.8), 20 mM NaCl, 7 mM $MgCl_2$, 10% glycerol, 7 mM 2-mercaptoethanol, and 40 μM GDP.
48. 10× TMND buffer: 800 mM Tris–HCl (pH 7.8), 70 mM $MgCl_2$, 1,500 mM NH_4Cl, and 25 mM dithiothreitol.
49. *E. coli* $tRNA^{Phe}$ (Sigma).
50. Synthetic mRNA: $(UUC)_{10}$.
51. Stalled ribosome mixture: 1× TMND buffer, 200 μM GTP, 2 mM spermidine, 2 μM $(UUC)_{10}$ mRNA, 200 nM 70S ribosome (see Note 3), 200 nM [^{14}C]phenylalanyl-tRNA, 2 μM EF-Tu, and 60 nM EF-G.
52. tmRNA mixture: 1× TMND buffer, 200 μM GTP, 2 mM spermidine, 100 nM [^{3}H]alanyl-tmRNA, and 1 μM SmpB.
53. tmRNA + tRNA mixture: 1× TMND buffer, 200 μM GTP, 2 mM spermidine, 100 nM unlabeled alanyl-tmRNA, 1 μM SmpB, and 100 nM [^{3}H]alanyl-tRNA.
54. Trichloroacetic acid.
55. Mixed cellulose ester membrane (Advantec).
56. Sampling Manifold (Millipore).
57. Liquid scintillation counter.

3. Methods

3.1. Preparation of 70S Ribosomes

2-Mercaptoethanol should be added to buffers immediately before use. All procedures should be carried out at 4°C.

1. Grow cells (*E. coli* W3110Δ*ssrA*Δ*smpB*) in 15 L of LB broth at 37°C in the absence of chloramphenicol (see Note 4) until mid-log phase (A_{600} = 0.5).
2. Harvest the cells by centrifugation at 3,000 × *g* (Hitachi R9AF rotor) for 5 min at 4°C and wash the cell pellet with 20 mL of 70S-A buffer.
3. Store the cells at –80°C overnight.
4. Crush the cells in a cold mortar with double volume of alumina in a cold room until cells get wet (approximately 15 min).

5. Dissolve in 25 mL of 70S-A buffer.
6. To remove alumina and cell debris, centrifuge at 9,000×*g* (Hitachi R10A2 rotor) for 30 min at 4°C. Keep the supernatant and discard the pellet.
7. Centrifuge twice at 25,000×*g* (Hitachi P65A rotor) for 30 min at 4°C. Keep the supernatant and discard the pellet each time.
8. Centrifuge at 300,000×*g* (Hitachi RP80AT rotor) for 30 min at 4°C.
9. Discard the supernatant and rinse the pellet with 8 mL of 70S-A buffer rapidly.
10. Dissolve in 16 mL of 70S-B buffer with the help of a spatula and gently stir for 60 min at 4°C.
11. Lay 2 mL of this solution on 1 mL of sucrose cushion solution in a Centrifuge Ware.
12. Centrifuge at 300,000×*g* (Hitachi RP80AT rotor) for 90 min at 4°C.
13. Discard the supernatant and rinse the pellets with 8 mL of 70S-A buffer rapidly.
14. Dissolve in 8 mL of 70S-A buffer.
15. Gently lay an appropriate amount of solution (about 100 A_{260} units per bucket) on a 5–20% sucrose density gradient preformed using Gradient Mate (Towa Kagaku) in 70S-A buffer.
16. Centrifuge at 80,000×*g* (Hitachi P28S rotor) for 4 h at 4°C.
17. Measure A_{260} to identify the fractions containing 70S ribosomes.
18. Collect the 70S peak fraction and dilute it with an equal volume of 70S-A buffer.
19. Centrifuge at 300,000×*g* (Hitachi RP80AT) for 2 h at 4°C.
20. Discard the supernatant and dissolve the precipitate in 1 mL of 70S stock buffer.
21. Measure A_{260} to determine the concentration (1 A_{260} unit = 24 pmol).
22. Freeze small aliquots (100 μL) with liquid nitrogen and store at −80°C (see Note 5).

3.2. Preparation of Alanyl-, Arginyl-, or Glutamyl-tRNA Synthetase

Aminoacyl-tRNA synthetases (ARS) were roughly purified from cell extract.

1. Grow cells (W3110Δ*ssrA*Δ*smpB*) in 12 L of LB broth with 34 μg/mL chloramphenicol until late-log phase (A_{600} = 0.8–1.0).
2. Harvest the cells by centrifugation at 3,000×*g* (Hitachi R9AF rotor) for 5 min at 4°C and wash thcm with 20 mL of ARS buffer.

3. Store at −80°C overnight.
4. Crush the frozen cells in a cold mortar with double volume of alumina in cold room for 15 min.
5. Dissolve in 25 mL of ARS buffer.
6. To remove alumina and cell debris, centrifuge at 9,000 × *g* (Hitachi R10A2 rotor) for 30 min at 4°C. Keep the supernatant and discard the pellet.
7. Centrifuge at 25,000 × *g* (Hitachi P65A rotor) for 30 min at 4°C. Keep the supernatant and discard the pellet.
8. Centrifuge at 300,000 × *g* (Hitachi RP80AT rotor) for 30 min at 4°C.
9. Recover the supernatant and apply it to 30 mL of DEAE-Sepharose column that has been equilibrated with ARS-A buffer. Elute with a linear gradient of KCl from 50 to 300 mM in ARS buffer (AlaRS is typically eluted at 80–140 mM KCl).
10. Determine the fractions containing ARS by aminoacylation assay as described below (see Subheading 3.6).
11. Concentrate the solution and exchange it with ARS stock buffer using an Amicon Ultra-15 once.
12. Store at −30°C (see Note 5).

3.3. Preparations of tmRNA

All procedure must be done with gloves to avoid contamination of RNases.

1. Grow overproducing cells (*E. coli* BL21(DE3)Δ*ssrA* harboring pGEMEX-II in which the gene for tmRNA is placed under the control of a T7 promoter) in 12 L of LB broth with 50 μg/mL ampicillin and 25 μg/mL kanamycin until early log phase ($A_{600} = 0.3$).
2. Add 12 mL of 0.5 M IPTG and grow at 37°C for another 2 h.
3. Harvest the cells by centrifugation at 3,000 × *g* (Hitachi R9AF rotor) for 5 min at 4°C and wash them with 20 mL of tmRNA buffer.
4. Store at −80°C overnight.
5. Crush the frozen cells (expected mass: 15 g) in a cold mortar with double volume of alumina (30 g) in cold room until cells become wet (typically 15 min).
6. Add 60 mL of tmRNA buffer.
7. Add 60 mL of phenol and 6 mL of 10% SDS.
8. Shake at 37°C for 30 min.
9. Centrifuge at 11,000 × *g* (Hitachi R10A2 rotor) for 10 min at room temperature and collect the upper layer.
10. Add another 60 mL of phenol and shake at room temperature for 10 min.

11. Centrifuge at 11,000 × *g* (Hitachi R10A2 rotor) for 10 min at room temperature and collect the upper layer.
12. Add 6 mL of 3 M sodium acetate (pH 5.2) and 180 mL of ethanol.
13. Vortex and cool at −30°C for 30 min.
14. Centrifuge at 11,000 × *g* (Hitachi R10A2 rotor) for 20 min at 4°C and discard the supernatant.
15. Dissolve the pellet in 120 mL of water and add 12 mL of 3 M sodium acetate (pH 5.2).
16. Cool the resulting nucleic acid fraction on ice.
17. Add 310 μL of 2-propanol per 1 mL of nucleic acid fraction in 1.5 mL microcentrifuge tubes (approximately 150 tubes).
18. Mix the contents by inversion.
19. Centrifuge at 9,000 × *g* (Hitachi T15A39 rotor) for 5 min at 4°C and transfer the supernatant to a new tube.
20. Add 520 μL of 2-propanol per tube.
21. Mix the contents by inversion.
22. Centrifuge at 15,000 × *g* (Hitachi T15A39 rotor) for 10 min at 4°C and discard the supernatant.
23. Rinse the pellets with 70% ethanol.
24. Dissolve in 100 μL of TM buffer per tube.
25. Add 3 μL of RNase-free DNase (5 U/μL) per tube and incubate at 37°C for 1 h.
26. Add 50 μL of Urea dye per tube and electrophorese on a 5% polyacrylamide gel containing 7 M urea until xylene cyanol reaches the bottom of the gel.
27. Detect tmRNA band by UV shadowing on a fluorescent TLC plate: Cover the gel with plastic wrap. Place the plastic-wrapped gel on the fluorescent TLC plate. Visualize nucleic acid bands by a handheld UV light source (254 nm). Cut the tmRNA band.
28. Recover the RNA by electroelution: Put the gel into Seamless Cellulose Tubing (molecular cut-off of 14,000 Da). Fill the tube with 0.1× TBE (approximately 40 mL). Run at 250 V for 30 min. Filter the resulting solution using Vacuum Filter System.
29. Add 4 mL of 3 M sodium acetate (pH 5.2) and 100 mL of ethanol.
30. Cool the solution at −80°C for 15 min.
31. Centrifuge at 15,000 × *g* (Hitachi T15A39 rotor) for 20 min at 4°C.
32. Rinse the pellets with 20 mL of 70% ethanol.

33. Dissolve in 10 mL of tmRNA folding buffer.
34. Heat at 75°C for 5 min and cool slowly back to room temperature for 1 h.
35. Add 1 mL of 3 M sodium acetate (pH 5.2) and 25 mL of ethanol.
36. Centrifuge at 8,000 × *g* (Hitachi R10A2 rotor) for 20 min at 4°C.
37. Rinse the pellets with 5 mL of 70% ethanol.
38. Dissolve in 2 mL of H_2O.
39. Measure A_{260} to determine the concentration (1 A_{260} unit = 330 pmol).
40. Store at −30°C.

3.4. Preparation of SmpB

1. Grow overproducing cells of *E. coli* BL21(DE3)Δ*ssrA* harboring pGEMEX-II in which the gene for His-tagged SmpB is placed under the control of a T7 promoter in 6 L of LB broth with 50 μg/mL ampicillin and 25 μg/mL kanamycin until mid-log phase (A_{600} = 0.5–0.7).
2. Add 6 mL of 0.5 M IPTG and grow at 37°C for another 2 h.
3. Harvest the cells by centrifugation at 3,000 × *g* (Hitachi R9AF rotor) for 5 min at 4°C and wash them with 20 mL of SmpB-A buffer.
4. Store at −80°C overnight.
5. Resuspend the cells in 25 mL of SmpB-A buffer and lyse the cells by sonication.
6. To remove cell debris, centrifuge at 9,000 × *g* (Hitachi R10A2 rotor) for 30 min at 4°C. Keep the supernatant and discard the pellet.
7. Centrifuge at 25,000 × *g* (Hitachi P65A rotor) for 30 min at 4°C. Keep the supernatant and discard the pellet.
8. Centrifuge at 300,000 × *g* (Hitachi RP80AT) for 30 min at 4°C.
9. Recover the supernatant.
10. Apply the supernatant to 30 mL of SP-Sepharose column that has been equilibrated with SmpB-A buffer. Elute SmpB with a linear gradient of KCl from 100 to 1,000 mM in SmpB-A buffer (SmpB is typically eluted at 500–700 mM KCl).
11. Carry out SDS-PAGE to detect the fractions containing SmpB by means of band mobility.
12. Apply the fractions containing SmpB to 10 mL of Ni-NTA agarose column that has been equilibrated with SmpB-C buffer (no buffer exchange is required). Wash the column with 30 mL of SmpB-wash buffer. Elute SmpB with a linear gradient of imidazole from 20 to 500 mM (SmpB is typically eluted at 130–240 mM imidazole).

13. Carry out SDS-PAGE to detect the fractions containing SmpB by means of band mobility.
14. Concentrate the solution and exchange the buffer with SmpB stock buffer using an Amicon Ultra-15. The final concentration is typically 20–200 μM (in 500 μL).
15. Freeze the small aliquots (100 μL) of SmpB in liquid nitrogen and store at –80°C (see Note 5).

3.5. Preparation of tRNA

E. coli $tRNA^{Phe}$ is available from Sigma. Preparation of other tRNAs is described below. All procedures must be done with gloves to avoid contamination of RNases.

1. Grow cells (*E. coli* W3110Δ*ssrA*Δ*smpB*) in 12 L of LB broth with 34 μg/mL chloramphenicol until mid-log phase (A_{600} = 0.5).
2. Harvest the cells by centrifugation at 3,000 × *g* (Hitachi R9AF rotor) for 5 min at 4°C and wash them with 20 mL of tmRNA buffer.
3. Store at –80°C overnight.
4. Follow the steps 5 through 16 of Subheading 3.3.
5. Add 450 μL of 2-propanol per 1 mL of nucleic acid fraction in 1.5 mL microcentrifuge tubes.
6. Mix the contents by inversion.
7. Centrifuge at 9,000 × *g* (Hitachi T15A39 rotor) for 5 min at 4°C and transfer the supernatant to a new tube.
8. Add 380 μL of 2-propanol per tube.
9. Mix the contents by inversion.
10. Follow the steps 22 through 38 of Subheading 3.3.
11. Store at –30°C.

3.6. Aminoacylation

$tRNA^{Phe}$ and tmRNA are aminoacylated with radiolabeled-phenylalanine and -alanine, respectively, using their corresponding aminoacyl-tRNA synthetases. For *trans*-transfer assay, different kinds of radioisotopes should be used to distinguish between the two amino acids. Typically, [^{14}C]phenylalanine and [^{3}H]alanine are used. In the case of resuming assay, tmRNA and $tRNA^{Ala}$ are aminoacylated with unlabeled- and ^{3}H-labeled alanines, respectively.

1. Incubate 400 μL of reaction mixture at 37°C for 15 min.
2. Add 8 μL of acetic acid (see Note 6), 80 μL of 7.5 M ammonium acetate, and 400 μL of phenol.
3. Vortex and centrifuge at 11,000 × *g* (Hitachi T15A39 rotor) for 3 min at room temperature.
4. Recover the upper (aqueous) layer.
5. Add 1 mL of ethanol and vortex.

6. Cool the mixture at –80°C for 5 min.
7. Centrifuge at 11,000 × *g* (Hitachi T15A39 rotor) for 15 min at 4°C.
8. Discard the supernatant and dissolve the pellet in 300 μL of 0.3 M sodium acetate (pH 5.2).
9. Add 900 μL of ethanol and cool at –80°C for 5 min.
10. Centrifuge at 11,000 × *g* (Hitachi T15A39 rotor) for 15 min at 4°C.
11. Discard the supernatant.
12. Dissolve the pellet in 50 μL of 2 mM sodium acetate (pH 5.2).
13. Withdraw 1 μL and spot on a cellulose ester membrane. Dry the membrane and measure the radioactivity to assess aminoacylation efficiency.
14. Store at –30°C (see Note 5).

3.7. Preparation of Elongation Factors

Protocols for purifying EF-Tu and EF-G are the same.

1. Grow overproducing cells of *E. coli* BL21(DE3)Δ*ssrA* harboring pGEMEX-II in which the gene for EF-Tu or EF-G is placed under the control of a T7 promoter in 6 L of LB broth with 50 μg/mL ampicillin and 25 μg/mL kanamycin until mid-log phase (A_{600} = 0.5–0.7).
2. Add 6 mL of 0.5 M IPTG and grow at 37°C for another 2 h.
3. Harvest the cells by centrifugation at 3,000 × *g* (Hitachi R9AF rotor) for 5 min at 4°C and wash them with 20 mL of EF-A buffer.
4. Store at –80°C overnight.
5. Resuspend the cells in 25 mL of EF-A buffer and lyse the cells by sonication.
6. To remove cell debris, centrifuge at 9,000 × *g* (Hitachi R10A2 rotor) for 30 min at 4°C. Keep the supernatant and discard the pellet.
7. Centrifuge at 25,000 × *g* (Hitachi P65A rotor) for 30 min at 4°C. Keep the supernatant and discard the pellet.
8. Centrifuge at 300,000 × *g* (Hitachi RP80AT) for 30 min at 4°C.
9. Recover the supernatant.
10. Apply it to 30 mL of DEAE Sepharose column that has been equilibrated with EF-A buffer. Elute EF-Tu or EF-G with a linear gradient of NaCl from 20 to 500 mM (EF-Tu is typically eluted at 130–180 mM NaCl).
11. Carry out SDS-PAGE to detect the fractions containing EF-Tu or EF-G by means of band mobility.

12. Apply fractions to 10 mL of Ni-NTA agarose column that has been equilibrated with EF-A buffer. Elute EF-Tu or EF-G with a linear gradient of imidazole from 0 to 300 mM (EF-Tu is typically eluted at 50–90 mM imidazole).
13. Carry out SDS-PAGE to detect the fractions containing EF-Tu or EF-G by means of band mobility.
14. Concentrate the solution and exchange the buffer with EF stock buffer using Amicon Ultra-15. The final concentration is typically 100–1,000 μM (in 500 μL).
15. Freeze the small aliquots (100 μL) of EF-Tu or EF-G in liquid nitrogen and store at –80°C (see Note 5).

3.8. Trans-Transfer Assay

1. Incubate 40 μL of "stalled ribosome mixture" for 10 min at 37°C.
2. Add 10 μL of "tmRNA mixture" to "stalled ribosome mixture."
3. After different periods of incubation at 37°C, withdraw a 10 μL aliquot of the mixture.
4. Put the aliquot into a test tube containing 5 mL of hot 5% TCA, which has been preheated at 90°C for 5 min.
5. Incubate for 10 min at 90°C to hydrolyze the ester bond between polypeptide and tmRNA.
6. After incubation, put the test tube on ice for 10 min.
7. Recover the precipitated peptide by filtration with a mixed cellulose membrane using a Millipore Sampling Manifold.
8. Wash the membrane with 5 mL of cold 5% TCA twice.
9. Dry the membrane and measure the radioactivity by a liquid scintillation counter.
10. Plot the level of radioactivity (fmol that has been calculated from dpm) vs. incubation time.

3.9. Resuming Assay

1. Incubate "stalled ribosome mixture" (see step 1 of Subheading 3.8) for 10 min at 37°C.
2. Add 10 μL of "tmRNA + tRNA mixture" to "stalled ribosome mixture."
3. Follow the steps 3 through 10 of Subheading 3.6.

4. Notes

1. It has been reported that mutation(s) at the upstream region of the tag-encoding sequence or the addition of aminoglycoside such as paromomycin or neomycin causes the initiation shift of the resume codon (19, 21, 24). In this system, such an

initiation shift can be detected by using [^{3}H]arginyl-tRNA or [^{3}H]glutaminyl-tRNA (corresponding to the −1 or +1 shifted resume codon, respectively, in *E. coli* tmRNA) instead of [^{3}H] alanyl-tRNA (20).

2. Aminoacylation of tmRNA is enhanced by the presence of SmpB at the appropriate concentration (8, 9).
3. It is critical that ribosomes are added just prior to incubation to form the stalled ribosome.
4. To exclude the effect of chloramphenicol on the 50S subunit, cells should be grown in the absence of chloramphenicol.
5. Refrozen factors (aminoacyl-tRNA, -tmRNA, ribosome, and purified proteins) retain sufficient activities.
6. Aminoacylated-tmRNA or -tRNA is stable under an acidic condition.

Acknowledgements

This work was supported by a Research Fellowship of Hirosaki University (D.K.), a Grant-in-Aid for Young Scientists from the Japan Society for the Promotion of Science (D.K., no. 23780099), a Grant-in-Aid for Scientific Research from the Japan Society for the Promotion of Science (A.M. and H.H., no. 23380045), a Grant-in-Aid for Scientific Research from the Japan Society for the Promotion of Science (H.H., no. 22020001), and a Grant for Hirosaki University Institutional Research (H.H.).

References

1. Keiler KC, Waller PR, Sauer RT (1996) Role of a peptide tagging system in degradation of proteins synthesized from damaged messenger RNA. Science 271:990–993
2. Himeno H, Sato M, Tadaki T, Fukushima M, Ushida C, Muto A (1997) In vitro *trans* translation mediated by alanine-charged 10Sa RNA. J Mol Biol 268:803–808
3. Muto A, Sato M, Tadaki T, Fukushima M, Ushida C, Himeno H (1996) Structure and function of bacterial 10Sa RNA: *trans*-translation system. Biochimie 78:985–991
4. Muto A, Ushida C, Himeno H (1998) A bacterial RNA that functions as both a tRNA and an mRNA. Trends Biochem Sci 23:25–29
5. Hayes CS, Keiler KC (2010) Beyond ribosome rescue: tmRNA and co-translational processes. FEBS Lett 584:413–419
6. Kurita D, Muto A, Himeno H (2011) tRNA/mRNA mimicry by tmRNA and SmpB in *trans*-translation. J Nucleic Acids 2011: 130581
7. Karzai AW, Susskind MM, Sauer RT (1999) SmpB, a unique RNA-binding protein essential for the peptide-tagging activity of SsrA (tmRNA). EMBO J 18:3793–3799
8. Barends S, Karzai AW, Sauer RT, Wower J, Kraal B (2001) Simultaneous and functional binding of SmpB and EF-Tu-GTP to the alanyl acceptor arm of tmRNA. J Mol Biol 314:9–21

9. Hanawa-Suetsugu K, Takagi M, Inokuchi H, Himeno H, Muto A (2002) SmpB functions in various steps of *trans*-translation. Nucleic Acids Res 30:1620–1629
10. Rudinger-Thirion J, Giegé R, Felden B (1999) Aminoacylated tmRNA from *Escherichia coli* interacts with procaryotic elongation factor Tu. RNA 5:989–992
11. Barends S, Wower J, Kraal B (2000) Kinetic parameters for tmRNA binding to alanyl-tRNA synthetase and elongation factor Tu from *Escherichia coli*. Biochemistry 39:2652–2658
12. Stepanov VG, Nyborg J (2003) tmRNA from *Thermus thermophilus*. Interaction with alanyl-tRNA synthetase and elongation factor Tu. Eur J Biochem 270:463–475
13. Wower IK, Zwieb CW, Guven SA, Wower J (2000) Binding and cross-linking of tmRNA to ribosomal protein S1, on and off the *Escherichia coli* ribosome. EMBO J 19: 6612–6621
14. Karzai AW, Sauer RT (2001) Protein factors associated with the SsrA.SmpB tagging and ribosome rescue complex. Proc Natl Acad Sci U S A 98:3040–3044
15. McGinness KE, Sauer RT (2004) Ribosomal protein S1 binds mRNA and tmRNA similarly but plays distinct roles in translation of these molecules. Proc Natl Acad Sci U S A 101: 13454–13459
16. Takada K, Takemoto C, Kawazoe M, Konno T, Hanawa-Suetsugu K, Lee S, Shirouzu M, Yokoyama S, Muto A, Himeno H (2007) In vitro *trans*-translation of *Thermus thermophilus*: ribosomal protein S1 is not required for the early stage of *trans*-translation. RNA 13: 503–510
17. Qi H, Shimizu Y, Ueda T (2007) Ribosomal protein S1 is not essential for the *trans*-translation machinery. J Mol Biol 368:845–852
18. Saguy M, Gillet R, Skorski P, Hermann-Le Denmat S, Felden B (2007) Ribosomal protein S1 influences *trans*-translation in vitro and in vivo. Nucleic Acids Res 35:2368–2376
19. Lee S, Ishii M, Tadaki T, Muto A, Himeno H (2001) Determinants on tmRNA for initiating efficient and precise *trans*-translation: Some mutations upstream of the tag-encoding sequence of *Escherichia coli* tmRNA shift the initiation point of *trans*-translation in vitro. RNA 7:999–1012
20. Shimizu Y, Ueda T (2002) The role of SmpB protein in *trans*-translation. FEBS Lett 514:74–77
21. Takahashi T, Konno T, Muto A, Himeno H (2003) Various effects of paromomycin on tmRNA-directed *trans*-translation. J Biol Chem 278:27672–27680
22. Ivanova N, Pavlov MY, Felden B, Ehrenberg M (2004) Ribosome rescue by tmRNA requires truncated mRNAs. J Mol Biol 338:33–41
23. Asano K, Kurita D, Takada K, Konno T, Muto A, Himeno H (2005) Competition between *trans*-translation and termination or elongation of translation. Nucleic Acids Res 33: 5544–5552
24. Konno T, Kurita D, Takada K, Muto A, Himeno H (2007) A functional interaction of SmpB with tmRNA for determination of the resuming point of *trans*-translation. RNA 13: 1723–1731
25. Kurita D, Muto A, Himeno H (2010) Role of the C-terminal tail of SmpB in the early stage of *trans*-translation. RNA 16:980–990

Index

A

B

Kenneth C. Keiler (ed.), *Bacterial Regulatory RNA: Methods and Protocols*, Methods in Molecular Biology, vol. 905,
DOI 10.1007/978-1-61779-949-5, © Springer Science+Business Media, LLC 2012

T

V

W

Y

MIX
Papier aus verantwortungsvollen Quellen
Paper from responsible sources
FSC® C105338

If you have any concerns about our products,
you can contact us on
ProductSafety@springernature.com

In case Publisher is established outside the EU,
the EU authorized representative is:
Springer Nature Customer Service Center GmbH
Europaplatz 3, 69115 Heidelberg, Germany

Printed by Libri Plureos GmbH
in Hamburg, Germany